Communications in Computer and Information Science 2499

Series Editors

Rationale
The CCIS series is devoted to the publication of proceedings of computer science conferences. Its aim is to efficiently disseminate original research results in informatics in printed and electronic form. While the focus is on publication of peer-reviewed full papers presenting mature work, inclusion of reviewed short papers reporting on work in progress is welcome, too. Besides globally relevant meetings with internationally representative program committees guaranteeing a strict peer-reviewing and paper selection process, conferences run by societies or of high regional or national relevance are also considered for publication.

Topics
The topical scope of CCIS spans the entire spectrum of informatics ranging from foundational topics in the theory of computing to information and communications science and technology and a broad variety of interdisciplinary application fields.

Information for Volume Editors and Authors
Publication in CCIS is free of charge. No royalties are paid, however, we offer registered conference participants temporary free access to the online version of the conference proceedings on SpringerLink (http://link.springer.com) by means of an http referrer from the conference website and/or a number of complimentary printed copies, as specified in the official acceptance email of the event.

CCIS proceedings can be published in time for distribution at conferences or as post-proceedings, and delivered in the form of printed books and/or electronically as USBs and/or e-content licenses for accessing proceedings at SpringerLink. Furthermore, CCIS proceedings are included in the CCIS electronic book series hosted in the SpringerLink digital library at http://link.springer.com/bookseries/7899. Conferences publishing in CCIS are allowed to use Online Conference Service (OCS) for managing the whole proceedings lifecycle (from submission and reviewing to preparing for publication) free of charge.

Publication process
The language of publication is exclusively English. Authors publishing in CCIS have to sign the Springer CCIS copyright transfer form, however, they are free to use their material published in CCIS for substantially changed, more elaborate subsequent publications elsewhere. For the preparation of the camera-ready papers/files, authors have to strictly adhere to the Springer CCIS Authors' Instructions and are strongly encouraged to use the CCIS LaTeX style files or templates.

Abstracting/Indexing
CCIS is abstracted/indexed in DBLP, Google Scholar, EI-Compendex, Mathematical Reviews, SCImago, Scopus. CCIS volumes are also submitted for the inclusion in ISI Proceedings.

How to start
To start the evaluation of your proposal for inclusion in the CCIS series, please send an e-mail to ccis@springer.com

Supreeti Kamilya · Sukanta Das ·
Enrico Formenti
Editors

Cellular Automata Technology

4th Asian Symposium, ASCAT 2025
Ranchi, India, March 6–8, 2025
Revised Selected Papers

Springer

Editors
Supreeti Kamilya
Birla Institute of Technology
Ranchi, Jharkhand, India

Sukanta Das
Indian Institute of Engineering Science
and Technology
Howrah, West Bengal, India

Enrico Formenti
Université Côte d'Azur
Nice, France

ISSN 1865-0929 ISSN 1865-0937 (electronic)
Communications in Computer and Information Science
ISBN 978-3-031-94120-7 ISBN 978-3-031-94121-4 (eBook)
https://doi.org/10.1007/978-3-031-94121-4

This Springer imprint is published by the registered company Springer Nature Switzerland AG
The registered company address is: Gewerbestrasse 11, 6330 Cham, Switzerland

Preface

This volume contains 16 papers that were accepted for presentation at the Fourth Asian Symposium on Cellular Automata Technology (ASCAT 2025), hosted by the Department of Computer Science and Engineering, Birla Institute of Technology, Mesra, Ranchi, India on March 6–8, 2025. The ASCAT series serves as a platform for exchanging ideas, insights and innovations in cellular automata research, keeping an eye on the development of cellular automata technology. So, the series encompasses all theoretical aspects of cellular automata and their applications across various domains. ASCAT 2025 was the fourth edition of the series that fostered the same goal.

ASCAT 2025 received 35 high-quality submissions, each of which went through a rigorous double-blind peer review process. Each of the submissions received at least 3 reviews, and then submissions and the reviews were made open among Program Committee (PC) members for discussion. Through an online PC meeting held on December 23, 2024, 8 papers were accepted and 10 papers were sent for revision. After the revision, another 8 papers were accepted for presentation at ASCAT 2025. This volume contains all the 16 papers. We are indebted to our reviewers for their timely effort and commitment to rigorously reviewing the submitted papers. We extend our sincere appreciation to the authors for their contributions that made the event possible.

ASCAT 2025 was happy to feature six distinguished invited speakers: Supratik Chakraborty (IIT Bombay, India), Rolf Hoffmann (Technical University of Darmstadt, Germany), Pedro Paulo Balbi (Mackenzie Presbyterian University, Brazil), Pabitra Pal Choudhury (ISI Kolkata, India), Raja Natarajan (TIFR Mumbai, India) and Kamalika Bhattacharjee (NIT Trichy, India). On behalf of the organizing committee of ASCAT 2025, we express our sincere gratitude to all the speakers. ASCAT 2025 also included a PhD Forum, where two PhD synopses were accepted for oral presentation in a special session, offering scholars the opportunity to discuss their research with leading experts and receive valuable feedback. We congratulate the young researchers of the PhD forum for their interesting research.

The ASCAT series is marked as an annual event of Cellular Automata India (CAI) (https://www.cellularautomata.in/), a virtual open research community which was founded during the COVID 19 pandemic and dedicated to encouraging and advancing work in cellular automata. The group took several initiatives to successfully organize the event. As part of the regular activities of CAI, the group conducts an annual summer school named the Indian Summer School on Cellular Automata, a rigorous research based-internship program for encouraging students and faculty members to explore research in cellular automata. This time 10 papers were accepted at ASCAT 2025 which originated from the research initiated in the 2024 edition of the summer school. As a result, ASCAT 2025 recorded a number of excellent young student presenters who originated as participants of the summer school. We congratulate the organizers, mentors and participants of the summer school for their dedication, meticulousness and tenacity in successfully converting internship projects into research publications.

Finally, we extend our appreciation to the Department of Computer Science and Engineering, BIT Mesra for their continuous support in organizing this event. Our heartfelt thanks to Springer for publishing the ASCAT 2025 proceedings in the Springer CCIS Series and to the Springer Nature EquinOCS system for streamlining the submission and review process. The event was supported by BIT Mesra, DST-ANRF, DST-PURSE, SCUBE Scientific Software Solutions and Micropoint Computers. We are indebted to each of the sponsors for their contributions. Last but not the least, we express our deep appreciation to the entire team of volunteers, supporting staff and participants who remained as the silent architects of the success of the event.

March 2025

Supratim Biswas
Biplab K. Sikdar
Pedro Paulo Balbi
Supreeti Kamilya
Sukanta Das
Enrico Formenti

Organization

General Chairs

Supratim Biswas	Birla Institute of Technology, Mesra, India
Biplab K. Sikdar	Indian Institute of Engineering Science and Technology Shibpur, India
Pedro Paulo Balbi de Oliveira	Universidade Presbiteriana Mackenzie, Brazil

Program Chairs

Supreeti Kamilya	Birla Institute of Technology, Mesra, India
Sukanta Das	Indian Institute of Engineering Science and Technology Shibpur, India
Enrico Formenti	Université Côte d'Azur, France

Program Committee Members

Kenichi Morita	Hiroshima University, Japan
Supratim Biswas	Birla Institute of Technology, Mesra, India
Biplab K. Sikdar	IIEST, Shibpur, India
Sukanta Das	IIEST, Shibpur, India
Pabitra Pal Chaudhuri	ISI, Kolkata, India
Jarkko Kari	University of Turku, Finland
Enrico Formenti	Université Côte d'Azur, France
Stefano Nichele	Østfold University College, Norway
Teijiro Isokawa	University of Hyogo, Japan
Hiroshi Umeo	Osaka Electro-Communication University, Japan
Rezki Chemlal	University of Bejaia, Algeria
Pedro Paulo Balbi de Oliveira	Universidade Presbiteriana Mackenzie, Brazil
Sudhakar Sahoo	Institute of Mathematics and Applications, India
Nazma Naskar	Kalinga Institute of Industrial Technology, India
Jimmy Jose	NIT, Calicut, India
Raju Hazari	NIT, Calicut, India
Kamalika Bhattacharjee	NIT, Tiruchirappalli, India
Souvik Roy	Ahmedabad University, India
Supreeti Kamilya	BIT, Mesra, India

Sukanya Mukherjee	Institute of Engineering and Management, India
Sumit Adak	Technical University of Denmark, Denmark
Bibhash Sen	NIT, Durgapur, India
Mamata Dalui	NIT, Durgapur, India
Sutapa Chatterjee Sarkar	Brainware University, India
Krishna Kumari R.	SRM Institute of Science and Technology, India
Ummity Srinivasa Rao	Vellore Institute of Technology, Chennai, India
Rinkaj Goyal	GGS Indraprastha University, India
Sumita Basu	Bethune College, India
Guillaume Theyssier	CNRS & Université Aix-Marseille, France
Bernard De Baets	Ghent University, Belgium
Nazim Fatès	Inria-Loria, France
Franco Bagnoli	University of Florence, Italy
Stefania Bandini	University of Milano-Bicocca, Italy
R. Ramanujam	Azim Premji University, India
Mihir K. Chakraborty	Jadavpur University, India
K. Ramachandra Rao	Indian Institute of Technology, Delhi, India
Alonso Castillo-Ramirez	Universidad de Guadalajara, Mexico
Hector Zenil	Kings College London, UK
Katsunobu Imai	Fukuyama University, Japan
N. Raja	TIFR, Mumbai, India
Andreas Deutsch	TU Dresden, Germany
Luca Mariot	Radboud University, Netherlands
Prince Gideon Kubendran Amos	NIT, Trichy, India
Sylvain Sené	Aix-Marseille University, France
Hasan Akin	Harran University, Turkey
Jan M. Baetens	Ghent University, Belgium
Giuliamaria Menara	University of Trieste, Italy
Anna Lawniczak	University of Guelph, Canada
Genaro Juarez Martinez	National Polytechnic Institute, ESCOM, Mexico

Organizing Committee Members

Supreeti Kamilya	Birla Institute of Technology, Mesra, India
Richa	Birla Institute of Technology, Mesra, India
Jamimamul Bakas	Birla Institute of Technology, Mesra, India
Jayasree Sengupta	Birla Institute of Technology, Mesra, India
Sandip Ghosal	Birla Institute of Technology, Mesra, India
Saikat Chakraborty	Birla Institute of Technology, Mesra, India
Shruti Garg	Birla Institute of Technology, Mesra, India

Publicity Chairs

Kamalika Bhattacharjee	National Institute of Technology, Trichy, India
Mamata Dalui	National Institute of Technology, Durgapur, India
Sumit Adak	Technical University of Denmark, Denmark

Web Team

Sandip Ghosal	Birla Institute of Technology, Mesra, India
Subrata Paul	IIEST Shibpur, India

Registration Chairs

Richa	Birla Institute of Technology, Mesra, India
Sukanya Mukherjee	Institute of Engineering and Management, India

Contents

Self-synchronization of Temporally Stochastic Cellular Automata

Subrata Paul(✉)

Department of Information Technology, Indian Institute of Engineering Science and Technology, Shibpur, Howrah, West Bengal 711103, India
psubrata.it@gmail.com

Abstract. This work explores the dynamics of a newly proposed cellular automata model, called *Temporally Stochastic Cellular Automata* (TSCAs), where two local rules, say f and g are applied to all cells with a probability. The first rule, say f, acts as a default rule of the system, whereas the second rule, say g, acts as a perturbation rule that perturbed the system with a probability, say τ. This work aims to get an approach solution of a relaxed version of the Global Synchronization Problem, named the Self-synchronization Problem on TSCA. An n-cell self-synchronized TSCA reaches a homogeneous configuration (0^n or 1^n) from any initial configuration and oscillates between two homogeneous configurations (0^n and 1^n) with some probability. The main focus of this work is to classify the whole set of TSCAs based on their self-synchronization capabilities.

Keywords: Temporally Stochastic Cellular Automata (TSCAs) · Global synchronization problem · Self-synchronization · Markov chain

1 Introduction

A Cellular Automaton (CA) is a discrete dynamical system consisting of a grid of cells called a lattice. Each cell of the lattice is a finite state machine. Classically, each cell uses a local rule to update its state during the evolution [2]. One of the prior works shows the self-replication cellular automata [9], which is an example of the basic simple behavior of a discrete dynamical system. Despite this simplicity, cellular automata are able to exhibit a wide range of complex behaviors [11,13].

This work explores a special class of non-uniform cellular automata, named Temporally Stochastic Cellular Automata (TSCAs) [11]. This class of cellular automata uses two local rules, say f and g. At a time, all the cells update uniformly using any of one rule (f or g). Here, one rule, say f, acts as a default rule of the system, and another rule, say g, acts as a perturbation rule that is applied with a perturbation rate τ. That is, the rule g is applied with probability τ, and rule f is applied with probability $1-\tau$. The notation of a TSCA is $(f, g)[\tau]$.

Previously, many researchers targeted to construct exact or approached solutions to the density classification problem [3,5,7,8]. Previously, a generalization

S. Kamilya et al. (Eds.): ASCAT 2025, CCIS 2499, pp. 1–18, 2025.
https://doi.org/10.1007/978-3-031-94121-4_1

of density classification problem, called *Affinity Classification Problem*, was targeted to solve using 2-D Temporally Stochastic Cellular Automata in [1]. This work targets to construct an approached solution to a relaxed version of *Global Synchronization Problem* called *Self-synchronization Problem* using 1-D Temporally Stochastic Cellular Automata. Formally, the problem can be stated as, *Given any initial configuration, find an n-cell Temporally Stochastic Cellular Automaton (TSCA) that reaches a recurrent set consisting of two homogeneous configurations* (0^n and 1^n) *which oscillates.*

This problem can easily be generalized in such a way that the system reaches one configuration (some parts of convergent CAs that contain only one fixed point attractor) or more than two configurations. However, this paper sticks to a binary case, where the system reaches a configuration $c \in S$, where $S = \{0^n, 1^n\}$, n is the CA size.

The model is described in the next section. Section 3.1 describes the self-synchronized property of TSCA, where the TSCA is compared with absorbing Markov chain and obtain the similarities between them. Theoretical analysis of self-synchronization of TSCA is discussed in Sect. 3.2 and Sect. 4. Section 5 concludes the work.

2 Model Description

This work considers Elementary Cellular Automaton (ECA) rules, where the CA is finite and follows one-dimensional, two-state and three-neighborhood settings with periodic boundary conditions. The array of cells is called a lattice that is denoted as $\mathscr{L} = \mathbb{Z}/n\mathbb{Z}$, where n is the number of cells present in the lattice. At a time step, a state $s \in \mathscr{S}$ is assigned to each cell, where $\mathscr{S} = \{0, 1\}$. The collection of the state of all cells at a time is called configuration. Let c be a configuration, $c = (c_i)_{i \in \mathscr{L}}$ where c_i is the state of cell $i \in \mathscr{L}$. The set of all configurations is denoted as $\mathscr{S}^{\mathscr{L}}$.

According to an ECA, each cell updates its state based on its left, self and right neighbors, where the updates are made uniformly according to a local update rule $f : \{0, 1\}^3 \to \{0, 1\}$. Hence for the given local rule f and the lattice $\mathscr{L}$, one can define the global function $G : \mathscr{S}^{\mathscr{L}} \to \mathscr{S}^{\mathscr{L}}$ such that, a configuration $x = (x_i)_{i \in \mathscr{L}} = G(c)$ where each $c_i \in \mathscr{S}^{\mathscr{L}}$, $\forall i \in \mathscr{L}$, $x_i = f(c_{i-1}, c_i, c_{i+1})$.

Table 1. Look-up table of ECAs

Current State (CS)	111	110	101	100	011	010	001	000	ECAs
RMTs	(7)	(6)	(5)	(4)	(3)	(2)	(1)	(0)	
Next State (NS)	0	0	0	1	1	1	1	0	30
Next State (NS)	0	1	0	1	1	0	1	0	90
Next State (NS)	1	0	1	0	1	1	1	1	175

Table 1 shows the look-up table of ECAs, where the first row indicates the possible combination of the current state of the left, self and right neighbors of a cell. Such a combination can be described using Rule Min Term (RMT). The Next State (NS) rows depict the next states of ECAs 30, 90, and 175.

A Temporally Stochastic Cellular Automaton (TSCA) utilizes two ECAs, say f and g, to update each cell. All the cells can be updated using one of the rules (f or g) at a time step. That is, one rule (say f) is applied over all the cells uniformly as a default rule, and the other rule (say g) is applied as a perturbation rule with a probability τ. Alternatively, let the probability of applying g over all the cells be $\tau \in (0, 1)$, whereas the application probability of applying f is $1 - \tau$. Then,

$$x = \begin{cases} G_g(c) & \text{with probability } \tau \\ G_f(c) & \text{with probability } 1 - \tau \end{cases}$$

where, $G_f(c) = G_f((c_i)_{i \in \mathscr{L}}) = (f(c_{i-1}, c_i, c_{i+1}))_{i \in \mathscr{L}}$ and $G_g(c) = G_g((c_i)_{i \in \mathscr{L}})$ $= (g(c_{i-1}, c_i, c_{i+1}))_{i \in \mathscr{L}}$. The proposed TSCA model can be identified using a notation $(f, g)[\tau]$. For example, let a TSCA $(20, 45)[0.5]$, where ECA 20 is con-

a) ECA 20 **(b) ECA 45** **(c) TSCA $(20, 45)[0.5]$**

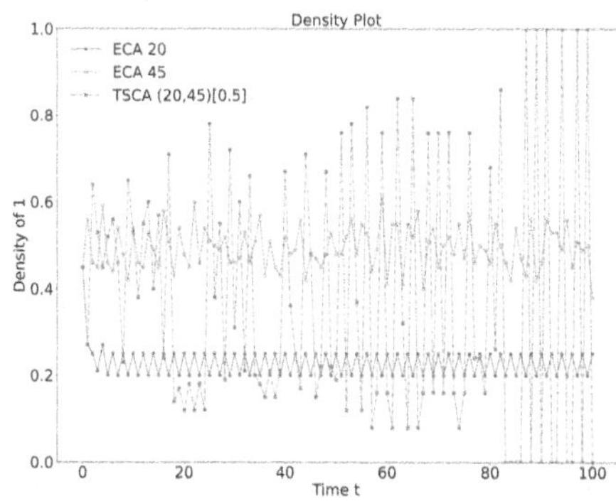

(d) Density plots of ECA 20, ECA 45 and TSCA $(20, 45)[0.5]$

Fig. 1. Dynamics of ECA 20, ECA 45 and TSCA $(20, 45)[0.5]$ with random initial configuration, where density of 1 (ρ_1) = 0.44. Here, a configuration consists of 100 cells and the model runs up to 100 time steps; (a) Space-time diagram of ECA 20; (b) Space-time diagram of ECA 45; (c) Space-time diagram of TSCA $(20, 45)[0.5]$; (d) Density plots of ECA 20, ECA 45 and TSCA $(20, 45)[0.5]$ (Color figure online) sidered as a default rule f of the CA and ECA 45 is considered as g, the perturbation rule which is applied with probability 0.5. The space-time diagram of the TSCA is shown in Fig. 1(c). A configuration of 100 cells with density of 1 (ρ_1[1])= 0.45, is considered as initial configuration to all the CAs in Fig. 1(a)-1(c). Here, the cells with state 1 (resp. state 0) are indicated by blue (resp. light blue), and the TSCAs evolve up to 100 time steps. Figure 1(a), 1(b) and 1(c) show the space-time diagrams of ECA 20, ECA 45 and TSCA $(20, 45)[0.5]$ respectively. Figure 1(d) shows the density plots of respective space-time diagrams, where the blue line, red line, and greed line indicate the changes in ρ_1 (density plots) with respect to time of ECA 20, ECA 45 and TSCA $(20, 45)[0.5]$, respectively. Here, the X axis and Y axis indicate the time t and density ρ_1, respectively.

In the density plot of the TSCA $(20, 45)[0.5]$ (Fig. 1(d)), the maximum (resp. minimum) ρ_1 reaches 1.0 (resp. 0.0) after the 82th time steps. Now, the density plot of ECA 20 shows that there are fixed changes in ρ_1 after 5th time step (blue line), whereas in case of ECA 45, the changes in the density plot are not fixed (red line). Interestingly, the TSCA $(20, 45)[0.5]$ reaches lower (resp. higher) than the existing minimum (resp. maximum) ρ_1 achieved by ECA 20 (resp. ECA 45), which is indicated by the green line. In the first few time steps, the trajectory of the TSCA density plots is similar to the trajectory of ECA 20. However, The trajectory of the TSCA changes when another rule 45 is applied (green line).

Observation 1. *If there exists a configuration for which a TSCA reaches the maximum ρ_1 (which is 1.0) and the minimum ρ_1 (which is 0.0) and its alternates, then one can consider the TSCA is synchronized for this configuration. If the TSCA is synchronized for all configurations, then it can be called self-synchronized TSCA.*

Hence, the density plot in Fig. 1(d) indicates that the TSCA can be an ideal choice for solving self-synchronization problems, where a TSCA settles in a two-length cycle of two configurations $0^{\mathscr{L}}$ and $1^{\mathscr{L}}$.

3 Self-synchronization of TSCA

During the evolution, a CA reaches a recurrent set of configurations called attractors. If an attractor set consists of only two configurations and it alternates during evolution, then it is called *absorbing configurations* (or blinking attractors). If all the configuration reaches the *absorbing configurations*, then the CA is called self-synchronized. Classically, an ECA cannot solve global synchronization problem [4]. This section discusses the self-synchronization of TSCA, where a rule is applied to all cells with some probability.

The theoretical aspects of TSCA are discussed using a notation called Rule Min Term (RMT). A Rule Min Term (RMT) is the combination of left, self and right neighbors –that is, 'xyz' with respect to $f(x, y, z)$, where f is the local update rule. An RMT is generally expressed as its decimal equivalent value

[1] $\rho_1 = \frac{\text{Number of cells with state 1}}{\text{Total Number of cells}}$.

$r = 4x + 2y + z$, using which a rule can be expressed its subsequent state values (Table 1). An RMT is referred to as active when the RMT changes the state in the subsequent steps during the evolution of a cell. The RMT is passive otherwise. For example, in ECA 30, the RMTs $1, 4, 6$ and 7 are active, whereas the RMTs $0, 2, 3$ and 5 are passive (Table 1). A configuration can be represented using an RMT sequence $(r_i)_{i \in \mathscr{L}}$ and a configuration $c = (c_i)_{i \in \mathscr{L}}$. For example, a configuration '001001' can be represented as an RMT sequence '412412' (it follows periodic boundary conditions).

Definition 1. *Two RMTs r_1 and r_2 of an ECA rule are said to be equivalent to each other if $2r_1 \equiv 2r_2 \pmod{2^3}$.*

Definition 2. *Two RMTs r_1' and r_2' of an ECA rule are said to be siblings to each other if $\lfloor \frac{r_1'}{2} \rfloor = \lfloor \frac{r_2'}{2} \rfloor$.*

An ECA consists of 4 sets of equivalent and sibling RMTs – $Equi_i = \{i, i+2^2\}$ and $Sibl_j = \{2j, 2j+1\}$. There exists a relation between RMT sequence $(r_i)_{i \in \mathscr{L}}$ –that is, $r_{i+1} \in Sibl_j$ if $r_i \in Equiv_j$.

Definition 3. *A set of RMTs $P = \{r_1, r_2, \cdots, r_p\}$ is said to be a primary RMT set if the following conditions are satisfied*

1. *If $r \in P$, then $r' \in P$, where $r' \in Sibl_k$, $r \equiv k (mod\ 8)$*
2. *No proper subset of P can form an RMT sequence.*

There exist 6 primary RMT sets in an ECA – $\{0\}$, $\{7\}$, $\{2, 5\}$, $\{1, 2, 4\}$, $\{3, 6, 5\}$ and $\{1, 3, 6, 4\}$ corresponds to the six elementary cycles in the De Bruijn graph. The random arrangement of RMTs cannot be the representation of a configuration [12]. A configuration can form with the RMTs of one or more primary RMT sets. A configuration is called a homogeneous configuration if its corresponding RMT sequence consists of the RMTs of one primary RMT set only. For example, for an n cell CA, the configurations 0^n and 1^n use the RMT sequence $(0)^n$ and $(7)^n$ (i.e., the primary RMT set $\{0\}$ and $\{7\}$), respectively. Similarly, the configurations $(01)^{\frac{n}{2}}$ uses the RMTs 2 and 5 only (i.e., the primary RMT set $\{2, 5\}$ and the sequence is $(52)^{\frac{n}{2}}$); $(001)^{\frac{n}{3}}$ uses the RMTs $1, 2$ and 4 only (i.e., $(412)^{\frac{n}{3}}$); $(011)^{\frac{n}{3}}$ uses the RMTs $3, 5$ and 6 only (i.e., $(536)^{\frac{n}{3}}$); $(0011)^{\frac{n}{4}}$ uses the RMTs $1, 3, 4$ and 6 only (i.e., $(4136)^{\frac{n}{4}}$).

Definition 4. *A primary RMT set Γ is called a Passive Primary RMT Set (PPRS) if all RMTs $r \in \Gamma$ are passive.*

For example, the PPRSs of ECA 104 are $\{0\}$ and $\{1, 3, 6, 4\}$, where the RMTs $0, 1, 3, 4$ and 6 are passive in rule 104.

3.1 Relationship Between TSCA and Markov Chain

Since, in a TSCA, the sequence of applying two rules f and g are random, TSCA maps to the Markov chain model, with the following situations–

- Markov chain model and TSCA both systems are discrete-time systems, and the proposed work considers a finite number of states.
- Both the systems move to the next configuration (the configuration at time $t+1$) only depending on the present configuration (the configuration at time t).
- Both the systems are stochastic in nature.

Hence, the transition of a TSCA can be analyzed with the help of the Markov chain model. This work considers the Markov chain model to explore the behavior of a TSCA.

Given that an absorbing state is a state that cannot be left, a Markov chain is absorbing if it admits at least one absorbing state and if any state can reach at least one absorbing state in finite time [6].

Let a absorbing set of configurations be S where, it is impossible to move to any configuration $s_1 \notin S$ from a configuration $s_2 \in S$. Here, s_1 is a transient configuration. This work considers $S = \{0^{\mathscr{L}}, 1^{\mathscr{L}}\}$, and a self-synchronized TSCA oscillates between $0^{\mathscr{L}}$ and $1^{\mathscr{L}}$ with a probability τ, after it reaches $0^{\mathscr{L}}$ or $1^{\mathscr{L}}$.

Theorem 1. *In a self-synchronized TSCA, the probability that the TSCA reaches $0^{\mathscr{L}}$ or $1^{\mathscr{L}}$ is 1.*

Proof. It is possible to reach $0^{\mathscr{L}}$ or $1^{\mathscr{L}}$ from each configuration $y \notin \{0^{\mathscr{L}}, 1^{\mathscr{L}}\}$. Let m be the minimum number of steps required to reach $0^{\mathscr{L}}$ or $1^{\mathscr{L}}$, starting from y. Let p be the probability that, starting from y, the TSCA does not reach $0^{\mathscr{L}}$ or $1^{\mathscr{L}}$ in m steps then, $p < 1$. Let m_l be the maximum of m and p_l be the maximum of p. This implies, the probability of not reach $0^{\mathscr{L}}$ or $1^{\mathscr{L}}$ in m_l steps is less than or equal to p_l, in $2m_l$ steps is less than or equal to p_l^2, etc. Since $p_l < 1$, the probability of not reach $0^{\mathscr{L}}$ or $1^{\mathscr{L}}$ in the given steps is monotonically decreasing. Hence, these probabilities of not reach $0^{\mathscr{L}}$ or $1^{\mathscr{L}}$ tend to 0.

Hence, there exists similarities between a self-synchronized TSCA and an absorbing Markov chain model.

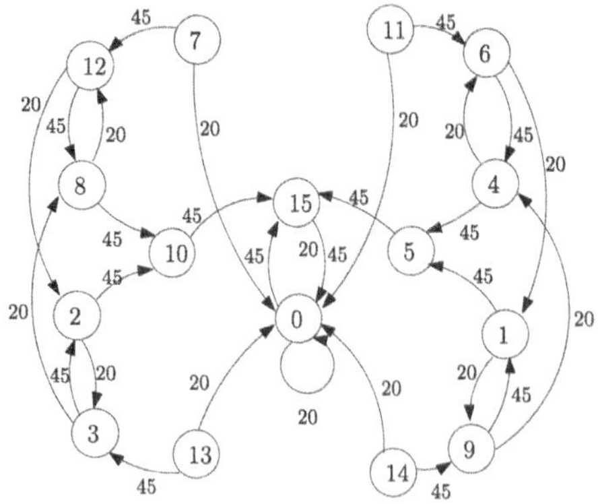

Fig. 2. Transition diagram of a 4-cell TSCA $(20, 45)[\tau]$

For example, Fig. 2 shows the transition diagram of a 4-cell TSCA $(20,45)[\tau]$, where every node represents the decimal equivalent of a configuration, and the label of each edge indicates the transition by a rule. Here, from all configurations, there exists a path to reach a homogeneous configuration 0 (0^4) or 15 (1^4). Hence, the TSCA is a self-synchronized TSCA.

3.2 Properties of Self-synchronized TSCA

This section investigates the properties of self-synchronized TSCAs.

Proposition 1. *A TSCA $(f,g)[\tau]$, where $\tau \in (0,1)$ is not a self-synchronized TSCA if any of the RMT 0 or 7 is passive in both f and g.*

Proof. This proposition can be proved by the definition of self-synchronization in TSCA. Let's assume that a TSCA, where the RMT 0 (or 7) is passive in both f and g. This implies, after reaching $0^{\mathscr{L}}$ (or $1^{\mathscr{L}}$) the TSCA remains there. Hence, the TSCA is not a self-synchronized TSCA.

Corollary 1. *A TSCA $(f,g)[\tau]$, where $\tau \in (0,1)$ is not a self-synchronized TSCA if it satisfies any of the following conditions –*

- *both f and g are even;*
- *both f and g are greater than 127;*

The ECAs, where RMT 0 is passive ('000'→ 0), depict the even rules, and the ECAs, where RMT 7 is passive ('111'→ 1), depict the rule is greater than 127. Hence, Corollary 1 can be derived from the Proposition 1.

Fig. 3 shows the space-time diagrams and the density plots of TSCAs $(10,60)[0.4]$, $(131,168)[0.4]$ and $(132,150)[0.4]$, where the density of 1 (ρ_1) in the initial configuration is 0.46, and the 1s are placed randomly. The dimensions of each image are 100×100, i.e., a configuration consists of 100 cells, and the model evolves up to 100 time steps. The TSCAs in Fig. 3(a) and 3(b) satisfy the first and second condition of the Corollary 1, respectively. The TSCA in Fig. 3(c) satisfies both the conditions of the Corollary 1.

Proposition 2. *A TSCA $(f,g)[\tau]$, where $\tau \in (0,1)$ is not self-synchronized if, for any $n \in 2\mathbb{N}$, the RMTs 2 and 5 both are passive or active in both f and g.*

Proof. Let the CA size of a TSCA $(f,g)[\tau]$ be $n \in 2\mathbb{N}$;

There exists a homogeneous configuration $(01)^{\frac{n}{2}}$ that corresponds to an RMT sequence $(52)^{\frac{n}{2}}$. Let both the RMTs 2 and 5 be passive in both f and g. Then, the configuration becomes a fixed point for the TSCA. On the other hand, if both the RMTs 2 and 5 are active in both f and g, then the configuration $(01)^{\frac{n}{2}}$ reaches another homogeneous configuration $(10)^{\frac{n}{2}}$ which consists of two RMTs 2 and 5. In this case, the configuration constructs a two-length cycle. Hence, in both cases, the TSCA cannot be self-synchronized.

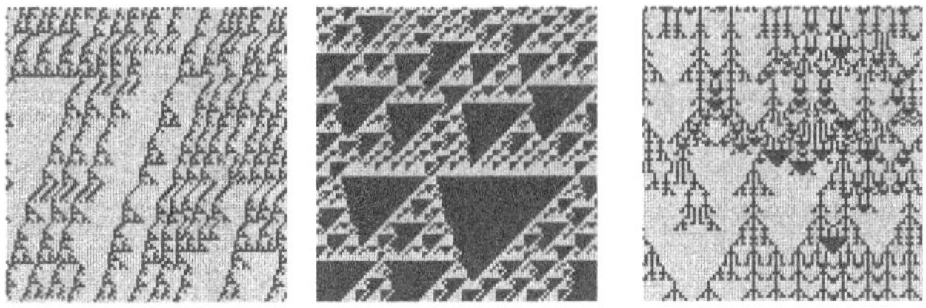

(a) TSCA $(10, 60)[0.4]$ (c) TSCA $(131, 168)[0.4]$ (b) TSCA $(132, 150)[0.4]$

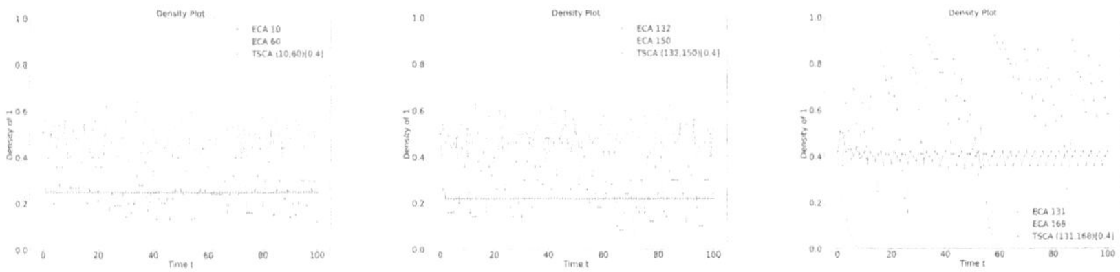

(d) TSCA $(10, 60)[0.4]$ (f) TSCA $(132, 150)[0.4]$ (e) TSCA $(131, 168)[0.4]$

Fig. 3. Dynamics of TSCAs $(10, 60)[0.4]$,$(131, 168)[0.4]$ and $(132, 150)[0.4]$, with random initial configuration where $\rho_1 = 0.46$; (a) Space-time diagram of TSCA $(10, 60)[0.4]$; (b) Space-time diagram of TSCA $(131, 168)[0.4]$; (c) Space-time diagram of TSCA $(132, 150)[0.4]$; (d) Density plots of TSCA $(10, 60)[0.4]$; (e) Density plots of TSCA $(131, 168)[0.4]$; (f) Density plots of TSCA $(132, 150)[0.4]$

Proposition 3. *A TSCA $(f, g)[\tau]$ is not self-synchronized for any $n \in 3\mathbb{N}$ if any of the following properties hold.*

- *The RMTs* 1, 2 *and* 4 *(resp. RMTs* 3, 5 *and* 6*) are passive in both f and g.*
- *The RMTs* 1, 2 *and* 4 *(resp. RMTs* 3, 5 *and* 6*) are active in f or g and the RMT* 3 *or* 6 *(resp.* 1 *or* 4*) are passive in both f and g.*

Proof. Let the CA size of a TSCA $(f, g)[\tau]$ be $n \in 3\mathbb{N}$;
Case 1: Let a homogeneous configuration $(001)^{\frac{n}{3}}$ that corresponds to an RMT sequence $(412)^{\frac{n}{3}}$. Let the RMTs 1, 2 and 4 be passive in f. If the RMTs 1, 2 and 4 are passive in g, then the configuration is a fixed point for the TSCA (Condition 1). However, if the RMTs 1, 2 and 4 are active in g then, the configuration reaches another homogeneous configuration $(110)^{\frac{n}{3}}$ that corresponds to an RMT sequence $(365)^{\frac{n}{3}}$, where each RMT $r \in \{3, 6, 5\}$ (Case 2), when g is applied.

On the other hand, let the RMTs 1, 2 and 4 be active in f. When f is applied, then the configuration $(001)^{\frac{n}{3}}$ reaches another homogeneous configuration $(110)^{\frac{n}{3}}$ that corresponds to the RMT sequence $(365)^{\frac{n}{3}}$ (Case 2).
Case 2: Let a homogeneous configuration $(110)^{\frac{n}{3}}$ that corresponds to an RMT sequence $(365)^{\frac{n}{3}}$. Let the RMTs 3, 5 and 6 be passive in f, then the configuration that corresponds to an RMT sequence where each RMT $\in \{3, 5, 6\}$ is a fixed point for f. Now, if RMTs 3, 5 and 6 are passive in g, then the configuration is a fixed point for the TSCA. Let the RMTs 3, 5 and 6 be active in g

then, the configuration reaches another homogeneous configuration $(001)^{\frac{n}{3}}$ that corresponds to the RMT sequence $(412)^{\frac{n}{3}}$ (Case 1), when g is applied.

On the other hand, if RMTs 3, 5 and 6 are active in f. When f is applied, then the configuration reaches another configuration that corresponds to the RMT sequence $(412)^{\frac{n}{3}}$ (Case 1).

Hence, in both cases, the TSCA is unable to synchronize. For the respective situations, these cases can be proved with a similar logic because RMTs 1, 2 and 4 are 0/1 exchanges of the RMTs 3, 5 and 6.

Proposition 4. *A TSCA $(f, g)[\tau]$, where $\tau \in (0, 1)$ is not self-synchronized for any $n \in 4\mathbb{N}$ if any of the following properties hold.*

- *if the RMTs 1, 3, 4 and 6 are passive (or active) in both f and g.*
- *if the RMTs 1 and 6 (resp. 4 and 3) are active and RMTs 3 and 4 (resp. 6 and 1) are passive in both f and g.*

Proof. Let the CA size of a TSCA $(f, g)[\tau]$ be $n \in 4\mathbb{N}$;
Case 1: Let the RMTs 1, 3, 4 and 6 be passive in both f and g, then a homogeneous configuration $(0011)^{\frac{n}{4}}$ that corresponds to an RMT sequence $(4136)^{\frac{n}{4}}$, where each RMT $r \in \{1, 3, 4, 6\}$ is a fixed point for the TSCA. This implies the TSCA cannot be self-synchronized.

Let the RMTs 1, 3, 4 and 6 be active in both f and g, then two consecutive 0s (resp. 1s) switch to two consecutive 1s (resp. 0s) during evolution. Hence, the configuration $(0011)^{\frac{n}{4}}$ reaches another configuration $(1100)^{\frac{n}{4}}$ that corresponds to an RMT sequence $(3641)^{\frac{n}{4}}$, contains same RMTs.. This implies the TSCA cannot be reached $0^{\mathscr{L}}$ or $1^{\mathscr{L}}$ and the TSCA cannot be self-synchronized.
Case 2: Let the RMTs 1 and 6 be active (resp. passive) and RMTs 3 and 4 are passive (resp. active) in both f and g, then a configuration with the RMTs 1, 3, 4 and 6 that consists of two consecutive 1s followed by two consecutive 0s shifted left (resp. right) in one place when f or g is applied. Hence, the TSCA cannot be reached to $0^{\mathscr{L}}$ or $1^{\mathscr{L}}$ and the TSCA cannot be self-synchronized.

For the respective situations, these cases can be proved with a similar logic because RMTs 1 and 4 are 0/1 exchanges of the RMTs 6 and 3.

Corollary 2. *A TSCA $(f, g)[\tau]$, where $\tau \in (0, 1)$ is not self-synchronized for any $n \in 3\mathbb{N}$, if both f and g satisfy any of the following condition,*

- *RMTs 1 and 2 (resp. 5 and 6) are active and RMT 4 (resp. 3) is passive.*
- *RMTs 4 and 2 (resp. 5 and 3) are active and RMT 1 (resp. 6) is passive.*
- *if RMTs 2, 3, 6 and 1 (resp. 4) are active and RMT 4 (resp. 1) is passive.*
- *if RMTs 1, 4, 5 and 3 (resp. 6) are active and RMT 6 (resp. 3) is passive.*

The TSCAs that are not self-synchronized, are ensured by the aforementioned propositions. Corollary 2 can be proven using a similar fashion where there exists at least one configuration that cannot reach $0^{\mathscr{L}}$ or $1^{\mathscr{L}}$.

Theorem 2. *An n-cell TSCA $(f, g)[\tau]$, where $\tau \in (0, 1)$ cannot be self-synchronized if there exists a Passive Primary RMT set (PPRS), say $\mathscr{J}$ in both the rules, and $n \in |\mathscr{J}| \times \mathbb{N}$.*

Proof. Let there exists a common passive primary RMT set (PPRS), say $\mathscr{J}$ in both f and g of an n-cell TSCA $(f, g)[\tau]$. Let $n \in x\mathbb{N}$, where $x = |\mathscr{J}|$, then there exists a homogeneous configuration that corresponds to an RMT sequence where each RMT $r \in \mathscr{J}$. The homogeneous configuration is a fixed point for the TSCA. Hence, the TSCA cannot be self-synchronized.

The ECAs that satisfy the above propositions and theory are listed in Appendix B and Appendix C.

4 The Self-synchronized TSCAs

There exist no such ECAs that are self-synchronized under periodic boundary conditions. However, the temporal stochastic update scheme allows to change the applied ECA rule during evolution, which may change the dynamics of an ECA.

Lemma 1. *A TSCA $(f, g)[\tau]$, where $\tau \in (0, 1)$ with one rule is* 204 *can not be self-synchronized with any τ value.*

Proof. Let a TSCA with a default rule f and the perturbation rule be 204. The ECA 240 is a self-replicating rule where all the RMTs are passive. Hence, the application of ECA 204 has no influence on the dynamics of f. Since there exists no ECA rules that are self-synchronized, the TSCA cannot be self-synchronized

The pairs of ECA rules that may be candidates for self-synchronized TSCA are investigated in this section. According to Theorem 1, a TSCA is self-synchronized if there exists at least one rule sequence for which the TSCA reaches the *absorbing configurations* $(\{0^{\mathscr{L}}, 1^{\mathscr{L}}\})$.

4.1 Self-synchronized TSCA Where One Rule is Convergent

An ECA is convergent if, from any configuration, the ECA converges a fixed point attractor. In a convergent ECA, there exists at least one passive primary RMT set (PPRS). This section considers one rule f of a TSCA $(f, g)[\tau]$ is convergent.

Proposition 5. *Let a TSCA $(f, g)[\tau]$, where $\tau \in (0, 1)$ and f be a convergent rule. The TSCA is self-synchronized if it satisfies any of the following conditions:*

- *The PPRS of f is $\{0\}$ (resp. $\{7\}$) and the RMT* 0 *(resp.* 7*) is active in g.*
- *The PPRSs of f are $\{0\}$ and $\{1, 2, 4\}$ (resp. $\{7\}$ and $\{3, 5, 6\}$), and the RMTs* 0,1 *and* 4 *(resp.* 7, 3 *and* 6*) are active and RMT* 2 *(resp.* 5*) is passive in g.*

- *The PPRSs of f are $\{0\}$ and $\{1,3,4,6\}$(resp. $\{7\}$ and $\{1,3,4,6\}$) and the RMTs* 0, 1 *and* 4 *(resp.* 7, 3 *and* 6*) are active and the RMTs* 3 *and* 6 *(resp.* 1 *and* 4*) are passive in g.*

Proof. Let a TSCA $(f, g)[\tau]$, where $\tau \in (0, 1)$ and f be a convergent rule. A TSCA can be considered as self-synchronized TSCA if there exists at least one rule sequence for which the TSCA reaches $0^{\mathscr{L}}$ or $1^{\mathscr{L}}$ (Theorem 1).

Case 1: Let the passive primary RMT set (PPRS) of f be $\{0\}$. There exists a rule sequence for which the TSCA reaches configuration $0^{\mathscr{L}}$ with probability 1. According to the first condition, the RMT 0 is active in g, hence the configuration $0^{\mathscr{L}}$ reaches $1^{\mathscr{L}}$ when g is applied, and these repeat. Hence, the TSCA is self-synchronized.

Case 2: Let the passive primary RMT sets (PPRSs) of f be $\{0\}$ and $\{1,2,4\}$. This implies the converging configurations of f correspond to the RMT sequences where each RMT $r \in \{0,1,2,4\}$. There exists a rule sequence for which the TSCA reaches any of the converging configurations of f.

Let the RMTs 0, 1, and 4 be active, and RMT 2 is passive in g. Hence, all the 0s in the converging configurations switch to 1s. However, the 1s in the converging configurations are not changed when g is applied, and the configurations reach $1^{\mathscr{L}}$. The configuration $1^{\mathscr{L}}$ reaches $0^{\mathscr{L}}$ if f is applied and these repeat. Hence, the TSCA is self-synchronized.

Case 3: Similar to the Case 2, the passive primary RMT sets (PPRSs) of f are $\{0\}$ and $\{1,3,4,6\}$ implies that the converging configurations of f correspond to the RMT sequences where each RMT $r \in \{0,1,3,4,6\}$. There exists a rule sequence for which the TSCA reaches any of the converging configurations of f.

Let the RMTs 0, 1, and 4 be active, and RMTs 3 and 6 are passive in g. Hence, all the 0s in the converging configurations switch to 1s. However, the 1s in the configuration are not changed when g is applied, and the configurations reach $1^{\mathscr{L}}$. The configuration $1^{\mathscr{L}}$ reaches $0^{\mathscr{L}}$ if f is applied and these repeat. Hence, the TSCA is self-synchronized.
For the respective situations, these cases can be proved with a similar logic because RMTs 0, 1, 2 and 4 are 0/1 exchanges of the RMTs 7, 3, 5 and 6.

Proposition 6. *Let a TSCA $(f, g)[\tau]$, where $\tau \in (0, 1)$ and f be the convergent rule. The TSCA is self-synchronized if it satisfies any of the following conditions:*

- *The PPRSs of f are $\{0\}$, $\{1,2,4\}$ and $\{2,5\}$(resp. $\{7\}$, $\{2,5\}$ and $\{3,5,6\}$) and the RMTs* 0, 1, 4 *and* 5 *(resp.* 2, 3, 6 *and* 7*) are active and RMT* 2 *(resp.* 5*) is passive in g.*
- *The PPRSs of f are $\{0\}$, $\{2,5\}$ and $\{1,3,4,6\}$(resp. $\{7\}$, $\{2,5\}$ and $\{1,3,4,6\}$) and the RMTs* 0, 1, 4 *and* 5 *(resp.* 2, 3, 6 *and* 7*) are active and RMTs* 2, 3 *and* 6 *(resp.* 1, 4 *and* 5*) are passive in g.*

Proof. Let a TSCA $(f, g)[\tau]$, where $\tau \in (0, 1)$ and f be convergent. A TSCA can be considered as self-synchronized TSCA if there exists at least one rule sequence for which the TSCA reaches $0^{\mathscr{L}}$ or $1^{\mathscr{L}}$ (Theorem 1).

Case 1: Let the passive primary RMT sets (PPRSs) of f be $\{0\}$, $\{1,2,4\}$ and $\{2,5\}$. Here, the converging configurations of f correspond to the RMT sequences where each RMT $r \in \{0,1,2,4,5\}$. There exists a rule sequence for which the TSCA reaches any of the converging configurations of f.

According to the first condition, the RMTs $0,1,4$ and 5 are active, and RMT 2 is passive in g. Hence, all the 0s (resp. 1s) in the converging configurations switch to 1s (resp. 1s) when g is applied, and the configurations reach $1^{\mathscr{L}}$. The configuration $1^{\mathscr{L}}$ reaches $0^{\mathscr{L}}$ if f is applied and these repeat. Hence, the TSCA is self-synchronized.

Case 2: Similar to Case 1, let the passive primary RMT sets (PPRSs) of f be $\{0\}$, $\{2,5\}$ and $\{1,3,4,6\}$. The converging configurations of f correspond to the RMT sequences where each RMT $r \in \{0,1,2,4,5,6\}$. There exists a rule sequence for which the TSCA reaches any of the converging configurations of f.

Let the RMTs $0,1,4$ and 5 be active, and RMTs $2,3$ and 6 are passive in g. Hence, all the 0s (resp. 1s) in the converging configurations switch to 1s (resp. 1s) when g is applied, and the configurations reach $1^{\mathscr{L}}$. The configuration $1^{\mathscr{L}}$ reaches $0^{\mathscr{L}}$ when f is applied and this repeats. Hence, the TSCA is self-synchronized. For the respective situations, these cases can be proved with a similar logic because RMTs $0,1,2$ and 4 are 0/1 exchanges of the RMTs $7,3,5$ and 6.

Corollary 3. *Let a TSCA $(f,g)[\tau]$, where $\tau \in (0,1)$ and f be the convergent rule. The TSCA is not self-synchronized if the PPRSs of f are $\{0\}$ $\{2,5\}$, $\{1,3,4,6\}$ and $\{7\}$.*

Proof. Let a TSCA with one rule f, where the PPRSs of f be $\{0\}$, $\{2,5\}$, $\{1,3,4,6\}$ and $\{7\}$. This implies all the RMTs in f are passive and the f is a self-replicating rule, i.e., ECA 204. The application of f has no influence on the behavior of the other rule in the TSCA. Hence, the TSCA can not be self-synchronized (Lemma 1).

Appendix D shows a list of ECAs based on the above propositions. From the lists of ECAs in Appendix D, one can construct self-synchronized TSCAs.

4.2 Self-synchronized TSCA Where the Rules Are Non-convergent

This work considers two important properties of cellular automata – non-increasing and non-decreasing properties. An ECA holds a non-increasing (resp. non-decreasing) property if the number of 1s in a configuration does not increase (resp. decrease) during the evolution.

Let a rule, say f be a non-increasing (or non-decreasing) rule. Each cycle in f consists of cyclic configurations where the number of 1*s* in each configuration is same (since non-increasing or non-decreasing). Then, all the cyclic configurations in a cycle utilize the same primary RMT sets in their corresponding RMT sequences.

Observation 2. *Let a TSCA $(f,g)[\tau]$, where $\tau \in (0,1)$ and f be non-increasing or non-decreasing. Suppose there exists a cycle in f that consists of k cyclic configurations, and each configuration corresponds to an RMT sequence where each*

RMT $r \in \gamma$, $\gamma \subseteq \{0,1,2,3,4,5,6,7\}$ (ECA). Let each cyclic configuration c_1 in a cycle $\mathscr{C}$ represents an RMT sequence where each RMT $r \in \gamma_1$. Similarly, for a k length cycle of configurations $c_1, c_2, \cdots, c_k$ represent the RMT sets $\gamma_1, \gamma_2, \cdots, \gamma_k$. Now, $\gamma_1 = \gamma_2 = \cdots = \gamma_k$ if the ECA is non-increasing or non-decreasing.

Appendix A reports the lists of non-increasing and non-decreasing ECAs.

Proposition 7. *In a TSCA $(f, g)[\tau]$, where $\tau \in (0, 1)$, let f be a non-increasing or non-decreasing rule, then the TSCA is self-synchronized –*

- *If the RMTs that are included in the cyclic configurations in f are $0, 1, 2$ and 4 (resp. $7, 3, 5$ and 6), and the RMTs $0, 1$ and 4 (resp. $7, 3$ and 6) are active in g, and RMT 2 (resp. 5) is passive in g.*
- *If the RMTs that are included in the cyclic configurations in f are $0, 2$ and 5 (resp. $7, 2$ and 5), and the RMTs 0 and 5 (resp. 7 and 2) are active in g, and RMT 2 (resp. 5) is passive in g.*
- *If the RMTs that are included in the cyclic configurations in f are $0, 1, 2, 4$ and 5 (resp. $7, 3, 5, 6$ and 2), and the RMTs $0, 1, 4$ and 5 (resp. $7, 3, 6$ and 2) are active in g, and RMT 2 (resp. 5) is passive in g.*
- *If the RMTs that are included in the cyclic configurations in f are $0, 1, 2, 3, 4$ and 6 (resp. $7, 1, 3, 4, 5$ and 6), and the RMTs $0, 1$ and 4 (resp. $7, 3, 6$ and 2) are active in g and RMTs $2, 3$ and 6 (resp. $1, 4$ and 5) are passive in g.*
- *If the RMTs that are included in the cyclic configurations in f are $0, 1, 2, 3, 4, 5, 6$ and 7, and the RMTs $0, 1, 4, 5$ and 7 (resp. $0, 2, 3, 6$ and 7) are active in g and RMTs $2, 3$ and 6 (resp. $1, 4$ and 5) are passive in g.*

Proof. Let a notation of 1-region (resp. 0-region) be introduced, which corresponds to a maximal set of contiguous cells with state 1 (resp. 0) in a given configuration. Now, in any configuration c (except $0^{\mathscr{L}}$), there exist three possibilities: 1-region of sizes one, two, and three or more. The first three conditions in this proposition deal with the 1-region of size one.

Case 1: Let the RMTs $0, 1, 2$ and 4 (resp. $7, 3, 5$ and 6) be included in a cyclic configuration in f. Let the RMT 2 be passive and the RMTs $0, 1$ and 4 be active in g then, the 0s are switched to 1s when g is applied. However, the isolated 1s are not changed. This implies the configuration reaches $1^{\mathscr{L}}$. Next, the $1^{\mathscr{L}}$ reaches $0^{\mathscr{L}}$ (given) when f is applied, and these repeat.

Case 2: Let the RMTs $0, 2$ and 5 (resp. $7, 2$ and 5) be included in a cyclic configuration in f for the CA size $n \in 2\mathbb{N}$. Let RMT 2 be passive and the RMTs 0 and 5 are active in g then, the 0s are switched to 1s when g is applied. However, the isolated 1s are not changed. The configuration reaches $1^{\mathscr{L}}$. Next, the $1^{\mathscr{L}}$ reaches $0^{\mathscr{L}}$ (given) when f is applied, and these repeat.

Case 3: This case is the combination of Case 1 and Case 2. Hence, a cyclic configuration may contain consecutive 0*s* (for RMTs $0, 1$ and 4) and isolated 0 (for RMT 5). In both cases, the 0s are switched to 1s when g is applied (since the RMTs $0, 1, 4$ and 5 are active g) –that is, the configuration reaches $1^{\mathscr{L}}$. Next, the $1^{\mathscr{L}}$ reaches $0^{\mathscr{L}}$ when f is applied, and these repeat.

Case 4: This condition considers that the RMTs $0, 1, 2, 3, 4$ and 6 (resp. $7, 1, 3, 4, 5$ and 6) are included in a cyclic configuration in f –that is the configuration contains 1-region of sizes one and two. Let the RMTs $2, 3$ and 6 be passive in g –that is, 1-region of sizes one and two remain unchanged during the application of g. However, the RMTs $0, 1$ and 4 are active in g then, the 0s are switched to 1s when g is applied. This implies the configuration reaches $1^{\mathscr{L}}$. Next, the $1^{\mathscr{L}}$ reaches $0^{\mathscr{L}}$ when f is applied, and these repeat.

Case 5: This condition is the critical one, where a cyclic configuration in f may include 1-region of sizes one, two, and three or more. Let the RMTs $2, 3$ and 6 be passive in g –that is, 1-region of sizes one and two remain unchanged during the application of g. However, the RMT 7 is active in g implies that the 1-region of sizes three or more is switched to 1-region of sizes one or two (the middle 1s in the 1-region of three or more that corresponds to the RMT 7 switch to 0s). On the other hand, let the RMTs $0, 1, 4$ and 5 be active in g then, the 0s are switched to 1s during the application of g. This implies the configuration reaches $1^{\mathscr{L}}$. Next, the $1^{\mathscr{L}}$ reaches $0^{\mathscr{L}}$ (given) when f is applied, and these repeat.
For the respective situations, these cases can be proved with a similar logic because RMTs $0, 1, 2$ and 4 are 0/1 exchanges of the RMTs $7, 3, 5$ and 6.

The above cases deal with all possible combinations of 1-regions in a configuration. For all such cases, the TSCA reaches $1^{\mathscr{L}}$ or $0^{\mathscr{L}}$.

5 Conclusion

A recently proposed class of cellular automata called Temporally Stochastic Cellular Automata (TSCAs) allows to combine two separate rules to achieve some interesting characteristics of cellular automata. This work investigated an interesting characteristic of TSCA, called *self-synchronization*, which is a relaxed version of the global synchronization problem. A list of TSCAs has been explored that exhibit self-synchronization even when the ECAs used in it are not self-synchronized.

This work looked at TSCAs with a probability τ and two rules that are temporally mixed. However, one may adjust the rule sequence to mix two different rules. The study of *self-synchronization* of a Temporally Non-uniform Cellular Automata (t-NUCAs), where the rule sequences depend not only on a probability τ but also on whether they are periodic or aperiodic (a given rule sequence) [10], is therefore a natural extension of this work.

Acknowledgement. The author is grateful to Dr. Sukanta Das for his insightful comments and discussions which have been useful for this work.

Appendix A:

A list of ECAs with ca size $n > 4$. The second row (resp. third row) reports a set of non-increasing (resp. non-decreasing) ECAs.

Type	ECAs (size $n > 4$)
Non-increasing	0, 2, 4, 8, 10, 12, 16, 24, 32, 34, 40, 42, 48, 56, 64, 66, 68, 72, 76, 80, 96, 98, 112, 128, 130, 132, 136, 138, 140, 144, 152, 160, 162, 168, 170, 176, 184, 192, 194, 196, 200, 204, 208, 224, 226, 240
Non-decreasing	170, 171, 174, 175, 184, 185, 186, 187, 188, 189, 190, 191, 204, 205, 206, 207, 220, 221, 222, 223, 226, 227, 230, 231, 234, 235, 236, 237, 238, 239, 240, 241, 242, 243, 244, 245, 246, 247, 248, 249, 250, 251, 252, 253, 254, 255

Appendix B:

The first and second columns indicate the propositions and the conditions of the corresponding propositions, respectively. The third column indicates the ECAs that satisfy the corresponding conditions of the Propositions. Here, a TSCA $(f, g)[\tau]$ is not self-synchronized if it considers f and g both are taken from the same row (same set of ECAs) of the third column. For example, a TSCA $(4, 165)[\tau]$ is not a self-synchronized TSCA, where both 4 and 165 are placed in the first row of Condition 1 of Proposition 3. Each condition represents the ECA sets that satisfy the conditions. However, the ECAs are distributed in more than one row for each condition. The first row of each condition represents the set of ECAs that satisfy the condition. The other rows of the condition represent the respective situations.

		Set of ECAs
Proposition 2		4, 5, 6, 7, 12, 13, 14, 15, 20, 21, 22, 23, 28, 29, 30, 31, 68, 69, 70, 71, 76, 77, 78, 79, 84, 85, 86, 87, 92, 93, 94, 95, 132, 133, 134, 135, 140, 141, 142, 143, 148, 149, 150, 151, 156, 157, 158, 159, 196, 197, 198, 199, 204, 205, 206, 207, 212, 213, 214, 215, 220, 221, 222, 223
		32, 33, 34, 35, 40, 41, 42, 43, 48, 49, 50, 51, 56, 57, 58, 59, 96, 97, 98, 99, 104, 105, 106, 107, 112, 113, 114, 115, 120, 121, 122, 123, 160, 161, 162, 163, 168, 169, 170, 171, 176, 177, 178, 179, 184, 185, 186, 187, 224, 225, 226, 227, 232, 233, 234, 235, 240, 241, 242, 243, 248, 249, 250, 251
Proposition 3	Condition 1	4, 5, 12, 13, 36, 37, 44, 45, 68, 69, 76, 77, 100, 101, 108, 109, 132, 133, 140, 141, 164, 165, 172, 173, 196, 197, 204, 205, 228, 229, 236, 237
		72, 73, 74, 75, 76, 77, 78, 79, 88, 89, 90, 91, 92, 93, 94, 95, 200, 201, 202, 203, 204, 205, 206, 207, 216, 217, 218, 219, 220, 221, 222, 223
	Condition 2	26, 27, 58, 59, 90, 91, 122, 123, 154, 155, 186, 187, 218, 219, 250, 251
		82, 83, 90, 91, 114, 115, 122, 123, 210, 211, 218, 219, 242, 243, 250, 251
		32, 33, 36, 37, 48, 49, 52, 53, 160, 161, 164, 165, 176, 177, 180, 181
		32, 33, 34, 35, 36, 37, 38, 39, 160, 161, 162, 163, 164, 165, 166, 167
Proposition 4	Condition 1	72, 73, 76, 77, 104, 105, 108, 109, 200, 201, 204, 205, 232, 233, 236, 237
		18, 19, 22, 23, 50, 51, 54, 55, 146, 147, 150, 151, 178, 179, 182, 183
	Condition 2	10, 11, 14, 15, 42, 43, 46, 47, 138, 139, 142, 143, 170, 171, 174, 175
		80, 81, 84, 85, 112, 113, 116, 117, 208, 209, 212, 213, 240, 241, 244, 245

Appendix C:

The first column indicates the theorem. The second column indicates the Passive Primary RMT Set (PPRS). In the third column, each row contains a set of ECAs that has the corresponding PPRS (mentioned in Column 2). The ECAs in the first row have a PPRS $\{0\}$; The ECAs in the second row has a PPRS $\{2,5\}$; The ECAs in third row has a PPRS $\{1,2,4\}$; The ECAs in fourth row has a PPRS $\{3,5,6\}$; The ECAs in fifth row has a PPRS $\{1,3,4,6\}$; The ECAs in sixth row has a PPRS $\{7\}$. There exists an ECA that has more than one PPRS, such ECAs are located in different rows (different sets of ECAs). For example, ECA 4 is in the first, second and third rows, hence, ECA 4 has PPRSs $\{0\}$, $\{2,5\}$ and $\{1,2,4\}$. On the other hand, to find a set of ECAs that contains same a set of PPRSs, then one need to intersect the sets of ECAs of the corresponding PPRSs. Similar to the previous Appendix B, a TSCA $(f,g)[\tau]$ is not self-synchronized if f and g both are taken from the same set of ECAs, where each ECA has a given set of PPRSs. For example, a TSCA $(12,76)[\tau]$ is not a self-synchronized TSCA, where both 12 and 76 have a set of PPRSs $\{\{0\},\{2,5\},\{1,2,4\}\}$.

	PPRS	Set of ECAs
Theorem 2	$\{0\}$	0, 2, 4, 6, 8, 10, 12, 14, 16, 18, 20, 22, 24, 26, 28, 30, 32, 34, 36, 38, 40, 42, 44, 46, 48, 50, 52, 54, 56, 58, 60, 62, 64, 66, 68, 70, 72, 74, 76, 78, 80, 82, 84, 86, 88, 90, 92, 94, 96, 98, 100, 102, 104, 106, 108, 110, 112, 114, 116, 118, 120, 122, 124, 126, 128, 130, 132, 134, 136, 138, 140, 142, 144, 146, 148, 150, 152, 154, 156, 158, 160, 162, 164, 166, 168, 170, 172, 174, 176, 178, 180, 182, 184, 186, 188, 190, 192, 194, 196, 198, 200, 202, 204, 206, 208, 210, 212, 214, 216, 218, 220, 222, 224, 226, 228, 230, 232, 234, 236, 238, 240, 242, 244, 246, 248, 250, 252, 254
	$\{2,5\}$	4, 5, 6, 7, 12, 13, 14, 15, 20, 21, 22, 23, 28, 29, 30, 31, 68, 69, 70, 71, 76, 77, 78, 79, 84, 85, 86, 87, 92, 93, 94, 95, 132, 133, 134, 135, 140, 141, 142, 143, 148, 149, 150, 151, 156, 157, 158, 159, 196, 197, 198, 199, 204, 205, 206, 207, 212, 213, 214, 215, 220, 221, 222, 223
	$\{1,2,4\}$	4, 5, 12, 13, 36, 37, 44, 45, 68, 69, 76, 77, 100, 101, 108, 109, 132, 133, 140, 141, 164, 165, 172, 173, 196, 197, 204, 205, 228, 229, 236, 237
	$\{3,5,6\}$	72, 73, 74, 75, 76, 77, 78, 79, 88, 89, 90, 91, 92, 93, 94, 95, 200, 201, 202, 203, 204, 205, 206, 207, 216, 217, 218, 219, 220, 221, 222, 223
	$\{1,3,6,4\}$	72, 73, 76, 77, 104, 105, 108, 109, 200, 201, 204, 205, 232, 233, 236, 237
	$\{7\}$	128, 129, 130, 131, 132, 133, 134, 135, 136, 137, 138, 139, 140, 141, 142, 143, 144, 145, 146, 147, 148, 149, 150, 151, 152, 153, 154, 155, 156, 157, 158, 159, 160, 161, 162, 163, 164, 165, 166, 167, 168, 169, 170, 171, 172, 173, 174, 175, 176, 177, 178, 179, 180, 181, 182, 183, 184, 185, 186, 187, 188, 189, 190, 191, 192, 193, 194, 195, 196, 197, 198, 199, 200, 201, 202, 203, 204, 205, 206, 207, 208, 209, 210, 211, 212, 213, 214, 215, 216, 217, 218, 219, 220, 221, 222, 223, 224, 225, 226, 227, 228, 229, 230, 231, 232, 233, 234, 235, 236, 237, 238, 239, 240, 241, 242, 243, 244, 245, 246, 247, 248, 249, 250, 251, 252, 253, 254, 255

Appendix D:

The list of ECAs that are based on the conditions of Proposition 5 and 6 are reported here. This analysis considers the CA size is 7. The first and second columns indicate the Propositions and the conditions of the corresponding propositions, respectively. Each condition shows the results in its first row, along with its respective case in its second row. Here, the first row contains the headings of the table. The second row satisfies the first condition of proposition 5, whereas the third row indicates the resp. case of the first condition of the proposition 5. A TSCA $(f, g)[\tau]$ is self-synchronized if f belongs to a set of ECAs from the third column of row r_1 and g is taken from the fourth column of the same row r_1. For example, a TSCA $(36, 87)[\tau]$ is a self-synchronized TSCA, where ECA 36 is located in the third column of the fourth row, and 87 is located in the fourth column of the fourth row. Interestingly, there exists no f for TSCA size 7 that satisfy the second condition of proposition 6.

Proposition	Condition	ECA f	ECA g
5	1	0, 8, 18, 32, 64, 122, 126	1, 3, 5, 7, 9, 11, 13, 15, 17, 19, 21, 23, 25, 27, 29, 31, 33, 35, 37, 39, 41, 43, 45, 47, 49, 51, 53, 55, 57, 59, 61, 63, 65, 67, 69, 71, 73, 75, 77, 79, 81, 83, 85, 87, 89, 91, 93, 95, 97, 99, 101, 103, 105, 107, 109, 111, 113, 115, 117, 119, 121, 123, 125, 127, 129, 131, 133, 135, 137, 139, 141, 143, 145, 147, 149, 151, 153, 155, 157, 159, 161, 163, 165, 167, 169, 171, 173, 175, 177, 179, 181, 183, 185, 187, 189, 191, 193, 195, 197, 199, 201, 203, 205, 207, 209, 211, 213, 215, 217, 219, 221, 223, 225, 227, 229, 231, 233, 235, 237, 239, 241, 243, 245, 247, 249, 251, 253, 255
		129, 161, 183, 239, 251, 253, 255	0, 1, 2, 3, 4, 5, 6, 7, 8, 9, 10, 11, 12, 13, 14, 15, 16, 17, 18, 19, 20, 21, 22, 23, 24, 25, 26, 27, 28, 29, 30, 31, 32, 33, 34, 35, 36, 37, 38, 39, 40, 41, 42, 43, 44, 45, 46, 47, 48, 49, 50, 51, 52, 53, 54, 55, 56, 57, 58, 59, 60, 61, 62, 63, 64, 65, 66, 67, 68, 69, 70, 71, 72, 73, 74, 75, 76, 77, 78, 79, 80, 81, 82, 83, 84, 85, 86, 87, 88, 89, 90, 91, 92, 93, 94, 95, 96, 97, 98, 99, 100, 101, 102, 103, 104, 105, 106, 107, 108, 109, 110, 111, 112, 113, 114, 115, 116, 117, 118, 119, 120, 121, 122, 123, 124, 125, 126, 127
	2	36, 44, 100	23, 31, 55, 63, 87, 95, 119, 127, 151, 159, 183, 191, 215, 223, 247, 255
		203, 217, 219	0, 1, 2, 3, 4, 5, 6, 7, 16, 17, 18, 19, 20, 21, 22, 23
	3	104	91, 95, 123, 127, 219, 223, 251, 255
		233	0, 1, 4, 5, 32, 33, 36, 37
6	1	4, 12, 68	55, 63, 119, 127, 183, 191, 247, 255
		207, 221, 223	0, 1, 2, 3, 16, 17, 18, 19
	2		127, 255
			0, 1

References

1. Bhattacharjee, K., Paul, S., Das, S.: Affinity classification problem by stochastic cellular automata. Complex Syst. **32**(3), 271–288 (2023)
2. Bhattacharjee, K., Naskar, N., Roy, S., Das, S.: A survey of cellular automata: types, dynamics, non-uniformity and applications. Nat. Comput. **19**, 433–461 (2020)
3. Fatès, N.: Stochastic cellular automata solutions to the density classification problem - when randomness helps computing. Theory Comput. Syst. **53**(2), 223–242 (2013)
4. Fatès, N.: Remarks on the cellular automaton global synchronisation problem: deterministic versus stochastic models. Nat. Comput. **18**(3), 429–444 (2019)
5. Fukś, H.: Solving two-dimensional density classification problem with two probabilistic cellular automata. J. Cell. Automata **10** (2015)
6. Grinstead, C.M., Snell, J.L.: Introduction to probability, the chance project version dated 4 July 2006 of the second revised edition, 1997. American Mathematical Society, Rhode Island, USA (2003)

7. Kari, J., Le Gloannec, B.: Modified traffic cellular automaton for the density classification task. Fund. Inform. **116**(1–4), 141–156 (2012)
8. Land, M., Belew, R.K.: No perfect two-state cellular automata for density classification exists. Phys. Rev. Lett. **74**(25), 5148 (1995)
9. von Neumann, J.: Theory of Self-Reproducing Automata. University of Illinois Press, USA (1966)
10. Paul, S., Das, S.: Temporally non-uniform cellular automata (t-nuca): reversibility and cyclic behavior (2024). https://arxiv.org/abs/2411.17421
11. Roy, S., Paul, S., Das, S.: Temporally stochastic cellular automata: classes and dynamics. Int. J. Bifurcat. Chaos **32**, 2230029 (2022)
12. Sethi, B., Roy, S., Das, S.: Convergence of asynchronous cellular automata: does size matter? J. Cell. Automata **13**(5–6), 527–542 (2018)
13. Wolfram, S.: A New kind of Science. Wolfram-Media (2002)

Modeling of Crack Propagation in Composite Materials Using Cellular Automata

V. R. Priyadarshini(✉) and P. G. Kubendran Amos

Theoretical Metallurgy Group, Department of Metallurgical and Materials Engineering, National Institute of Technology Tiruchirappalli, Tiruchirappalli, India
priyadarshinivr19@gmail.com

Abstract. In recent years, composite materials have garnered significant attention due to their exceptional mechanical properties and versatile applications in various industries. However, the intricate behavior of crack propagation within these materials remains a complex challenge. This research presents a novel approach to model crack propagation in composite materials using a two-dimensional cellular automata framework. The proposed model leverages hexagonal discretization to accurately represent the composite's concentric structure, consisting of a core and a surrounding matrix. By introducing an initial crack within the material, the CA model simulates the evolution of crack growth under varying stress conditions. Two primary modes of crack propagation are observed: penetration through the core when the stress concentration is greater in the core than in the matrix, and deviation along the interface between the core and the matrix when the stress concentration is higher in the matrix. The model integrates computational mechanics and stress analysis to evaluate the stress concentration at each cell, utilizing stress data to establish rules that govern the transition of cell states from intact to cracked.

Keywords: Crack Propagation · Composite Materials · Stress Concentration · Material Failure · Dynamic Crack Growth · Elastic-Plastic Transition · Cellular Automata · Fracture Mechanics

1 Introduction

The study of crack propagation in materials is fundamental for predicting material failure, particularly in structural applications. Early work by Griffith [4] laid the foundation for understanding crack initiation, which was later extended by Irwin [6] to account for plastic deformation at the crack tip. In composite materials, crack behavior is more complex due to interactions between the matrix

This work is carried out as a project in the Indian Summer School on Cellular Automata 2024.

S. Kamilya et al. (Eds.): ASCAT 2025, CCIS 2499, pp. 19–33, 2025.
https://doi.org/10.1007/978-3-031-94121-4_2

and inclusions, as discussed by Greco et al. [3], who analyzed crack propagation using multiscale techniques and moving mesh methods.

[?] Cellular automata (CA) have emerged as a promising method for modeling crack propagation, offering advantages such as reduced computational effort and avoidance of remeshing [10]. Yan et al. [11] introduced a continuous-discontinuous cellular automaton for dynamic crack propagation, demonstrating its efficiency in handling crack growth without the need for remeshing. Helm et al. [5] provided a comprehensive overview of CA and their applications in material science, highlighting the ability of CA to model complex behaviors through simple local rules.

Several studies have applied CA specifically to crack propagation. Gobron and Chiba [2] used CA to simulate crack patterns on 3D surfaces, while Li et al. [8] successfully applied CA to simulate rock failure, incorporating local equilibrium and constitutive equations to model the failure process. The model developed by Sadek et al. [9] for composite patches in marine structures further demonstrates the applicability of CA in different domains.

Gao [1] presented an analytical solution for the stress field around a hard spherical particle embedded in a finite elastic-plastic metal matrix composite. The study incorporates size and finite volume effects through the use of strain gradient plasticity theory. This work is particularly significant because it captures the particle size effect and accounts for both dilute and non-dilute particle distributions, unlike previous models. The results show that the stress concentration factor at the particle/matrix interface increases with the particle volume fraction and decreases with particle size at the micron scale, making this solution especially relevant for understanding failure mechanisms in metal matrix composites.

These contributions illustrate the versatility of CA in modeling crack propagation across various materials, making it a suitable framework for further exploration in this field.

Building on these foundational studies, this paper introduces a refined CA-based model specifically tailored for crack propagation in composite materials. The proposed model addresses the unique challenges posed by composite structures, such as matrix-inclusion interactions and varying fracture mechanics at different scales. By leveraging cellular automata, this model offers an efficient and scalable approach for predicting crack initiation and propagation in composite materials, aiming to enhance accuracy and reduce computational requirements. Additionally, this work introduces new parameters for capturing the heterogeneity of composites, thus expanding the applicability of CA to more complex material systems.

2 Background

Consider a unit cell consisting of a spherical matrix with an outer radius b and an inner radius a, surrounding a spherical particle (inclusion) of radius a, both concentric. The unit cell is subjected to hydrostatic tension σ_0 (force per unit area) applied on the outer surface of the matrix at $r = b$, as depicted in Fig. 1. Both the matrix and particle materials are assumed to be homogeneous and isotropic. Under the applied load σ_0, the hard particle behaves elastically and remains perfectly bonded to the surrounding elastic-plastic matrix, with an interface normal tension p_i. In this model, the inclusion (particle) is stiffer than the matrix material.

When σ_0 is relatively small, the entire matrix remains elastic. However, as σ_0 increases, yielding occurs in the matrix near the inner surface at $r = a$, due to the stress concentration caused by the hard inclusion. As the load continues to increase, the plastic zone within the matrix expands outward. Due to the spherical symmetry, the elastic-plastic interface in the matrix remains a spherical surface for any load σ_0 that produces plastic deformation. Let r_c represent the radius of this elastic-plastic boundary, and p_c the corresponding normal tension at the interface under a generic load σ_0. The region between $a \leq r \leq r_c$ undergoes plastic deformation, while the matrix material in the region $r_c \leq r \leq b$ remains elastic.

The material properties for both the inclusion and the matrix are critical in determining the stress and strain distribution within the composite material. The Young's modulus of the inclusion, denoted as E^I, and the Poisson's ratio of the inclusion, denoted as ν^I, are key parameters for the inclusion. Similarly, for the matrix, the Young's modulus is represented as E, and the Poisson's ratio is ν. These parameters influence the interaction between the inclusion and the matrix, affecting the overall mechanical behavior under load. The radial stress is denoted as σ_{rr} and the normal stress is denoted as $\sigma_{\theta\theta}$.

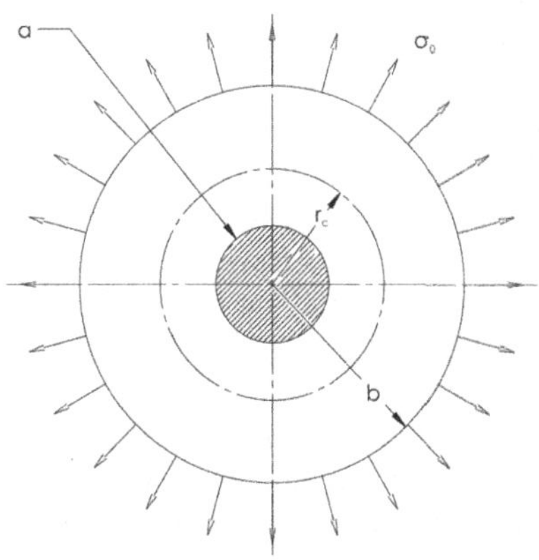

Fig. 1. Domain Configuration

Elasticity and Plasticity Solutions

The elasticity and plasticity solutions provide critical insights into the stress and strain distributions within the material under different loading conditions. These solutions form the basis for understanding the behavior of composite materials in the CA model. The detailed equations and explanations for the classical elasticity and plasticity solutions are provided in the Appendix.

The elasticity solution is based on the assumption that the material behaves elastically, meaning it deforms under stress but returns to its original shape once the stress is removed. It is used when the load is small and the material remains within its elastic limit. The classical elasticity equations describe the stress (σ) and strain (ε) in the material, considering factors like Young's modulus (E), Poisson's ratio (ν), and the geometry of the material.

The plasticity solution applies when the load exceeds the elastic limit, causing permanent deformation in the material. This solution incorporates yield stress (σ_y) and accounts for the material's hardening behavior. The equations describe how the plastic zone evolves within the material, where stresses redistribute beyond the elastic region. This approach is critical for understanding the onset and progression of cracks in composite materials.

3 Cellular Automaton Model

Cellular Automata (CA) are discrete computational systems with local interactions that can emulate complex physical and biological phenomena due to their parallel and homogeneous nature. Each cell within a CA system updates its state based on the states of neighboring cells, following a uniform rule set. This parallelism and simplicity make CA especially suited for simulating crack propagation in composite materials, where local material behavior has significant influence on global properties. Kari [7] provides a comprehensive overview of CA's theoretical aspects, highlighting their suitability in representing local interactions and conservation laws critical for simulating physical systems, such as crack propagation in materials.

In the present model, the material domain is represented as a hexagonal grid where each hexagonal cell acts as a discrete element of the composite material. This grid format allows the CA to effectively model stress distribution and crack growth, as the local rules in the CA framework naturally capture the interactions between cells, simulating how cracks propagate through the material under stress concentration.

In the cellular automata model representing composite materials, the material domain is discretized into a hexagonal grid, where each hexagonal cell represents a small region of the material. This hexagonal representation is particularly effective because it closely approximates the natural arrangement of composite materials at the microscopic level. The algorithm begins by creating a grid with hexagonal cells, where each cell has an associated status. The board is initialized based on three predefined states, where cells near the center of the grid are marked as "core," simulating the elastic inclusion, while cells near the edges are set as "matrix" to form a boundary and "cracked" state when the crack as propagated to the cell. The third state of each cell can evolve over time based on local rules governing interaction with neighboring cells, which mimic physical processes like crack propagation. The hexagonal layout ensures that each cell has six neighbors, reflecting a realistic interaction model in the composite structure. This approach provides a computational framework for studying the material's response to crack propagation (Fig. 2).

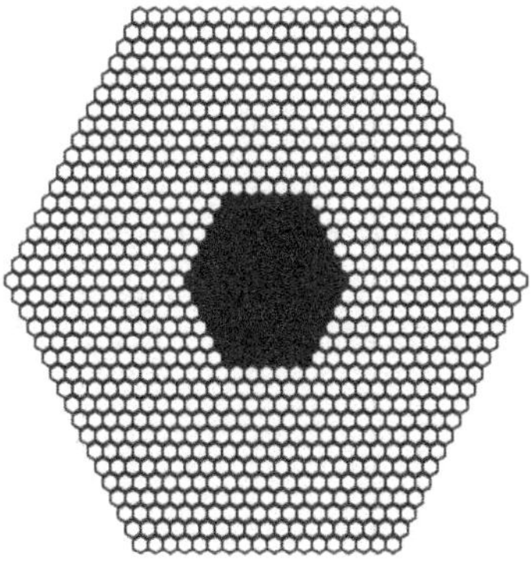

Fig. 2. Hexagonal grid representation of a cellular automata model with a central inclusion. The dark hexagonal region represents the inclusion, surrounded by the matrix modeled as individual cells.

In the hexagonal grid-based cellular automata, two types of neighborhood definitions were employed to simulate different interaction dynamics between cells: the Moore neighborhood and the Von Neumann neighborhood.

3.1 Simulation Setup

The cellular automaton (CA) model for simulating crack propagation in composite materials relies on an open boundary condition. This configuration allows cracks to propagate outward without artificial constraints at the model's edges, closely mimicking real-world conditions in unbounded materials.

The simulation is initialized with a hexagonal grid where each cell represents a small, distinct region within the composite material. Cells are assigned one of three states: "core" for regions simulating the elastic inclusion, "matrix" for surrounding boundary cells, and "cracked" for areas where the material has undergone damage. This initialization provides the model with a structured yet flexible framework to capture stress distribution and crack progression accurately.

Key parameters used in the CA include:

- **Material properties** such as the Young's modulus and Poisson's ratio for both matrix and inclusion.
- **Stress thresholds** for transitioning from elastic to plastic deformation, governing crack initiation.

- **Neighborhood dynamics** through the Moore and Von Neumann neighborhood schemes as two separate cases, which define the interaction radius and thus influence how stress propagates between adjacent cells.

The state vector for each cell can be expressed as:

$$S = \{c, m, \sigma_r, E_a, E_s\}$$

where:

- **Cracked Status** (c):
  - $c = 1$: Cracked (the cell has failed or fractured).
  - $c = 0$: Not cracked.
- **Material Type** (m): Represents whether the cell belongs to the core or the matrix:
  - $m = 0$: Core
  - $m = -1$: Matrix.
- **Radial Stress** (σ_r): Stress acting radially within the material.
- **Area Energy** (E_a): Represents energy related to the interface, such as the energy required for crack propagation.
- **Strain Energy** (E_s): Represents the deformation energy stored in the cell due to external stress.

3.2 Neighborhood Definitions

Moore Neighborhood. For each hexagonal cell, the Moore neighborhood consists of all six surrounding cells that share a common edge with the central cell. These six neighboring cells represent the full interaction domain in the immediate vicinity, capturing how stress or strain propagates across the material through adjacent regions.

In the code, this is represented by checking all six neighboring offsets for each cell, depending on whether the row is even or odd, ensuring proper alignment within the grid.

The Moore neighborhood includes all six cells that are directly adjacent to the central cell. If the central cell is at position (i, j) in the hexagonal grid, the neighbors are defined as:

$$N_{\text{Moore}}(i, j) = \{(i', j') : (i', j') \in \text{Adjacency}(i, j)\}$$

where Adjacency(i, j) is defined as:

$$\text{If } i \text{ is even:} \quad \{(i-1, j), (i-1, j+1), (i, j-1), (i, j+1), (i+1, j), (i+1, j+1)\}$$

$$\text{If } i \text{ is odd:} \quad \{(i-1, j-1), (i-1, j), (i, j-1), (i, j+1), (i+1, j-1), (i+1, j)\}$$

This ensures proper alignment of neighbors in the hexagonal grid depending on whether the row index i is even or odd.

Von Neumann Neighborhood. The Von Neumann neighborhood, a simpler form of neighborhood, considers only three of the six neighboring cells. This reduces the complexity of interactions, simulating more localized stress transfer dynamics. In this setup, the cell checks three possible adjacent cells, which could be thought of as a simplified model for certain types of anisotropic material behavior or directional stress propagation.

The Von Neumann neighborhood considers only three of the six neighbors. For the central cell at position (i, j), the neighbors are defined as:

$$N_{\text{Von Neumann}}(i, j) = \{(i', j') : (i', j') \in \text{Subset}(i, j)\}$$

where $\text{Subset}(i, j)$ is defined as:

$$\text{If } i \text{ is even:} \quad \{(i-1, j), (i, j+1), (i+1, j)\}$$

$$\text{If } i \text{ is odd:} \quad \{(i-1, j), (i, j+1), (i+1, j-1)\}$$

This subset simplifies the neighborhood to three cells, simulating directional or localized stress propagation.

3.3 Radial Stress Computation

For the computation of radial stresses, the algorithm assigns a radial distance (r) from the center of the material to each cell, which is based on the geometric layout of the hexagonal grid. The radial stress is calculated using a combination of material properties, including Young's modulus (E), Poisson's ratio (ν), and the internal/external radii (a and b).The radial stress σ_{rr}, the circumferential stresses $\sigma_{\theta\theta}$ and $\sigma_{\phi\phi}$, as well as the corresponding strain components, are computed using the classical elasticity solutions derived in the previous section (Equations (1)-(7) can be found in the appendix).
Additionally, the properties of the inclusion were assumed to be that of boron, while the matrix properties were considered to be those of aluminium during the initial phase of the computation.

3.4 Rule and Motivation Behind

For each cell with an existing crack, identify its Moore's neighborhood (which consists of the surrounding cells) and von Neumann neighborhood (which consists of alternate surrounding cells). Calculate the stress in each neighboring cell. The cell with the highest stress is then selected, and a crack is introduced at that cell in the subsequent time step.

Cracks in materials tend to propagate in regions of high stress concentration. When a material is under load, the stress is not evenly distributed. Instead, it concentrates at certain points, especially near existing defects or cracks. The physical principle behind this is that materials will fail where they are weakest, and these weak points are where the stress is highest. By introducing a crack in the cell with the maximum stress, you are simulating this natural tendency for cracks to propagate through the material.

Materials and physical systems tend to move towards configurations that minimize their potential energy. In the context of crack propagation, introducing a crack in the cell with the highest stress helps to relieve that localized stress. By redistributing the stress after introducing a crack, the system's overall potential energy is reduced. This reduction in energy is a natural process that the material undergoes to reach a more stable state.

3.5 Development of the Rule by Comparing the Area Energy with the Strain Energy

In this cellular automata framework, crack propagation is governed by a rule that incorporates both the strain energy and stress of neighboring cells. The goal of this rule is to simulate the natural tendency of materials to crack in regions with the highest stress concentrations while ensuring the overall stability of the system. For each cell, its neighboring cells, which can be from either Moore's or von Neumann's neighborhood, are first identified. The strain energy of each neighboring cell is compared with its corresponding area energy. If the strain energy is less than the area energy, the neighboring cell is considered to be in a relatively elastic state and becomes a candidate for further evaluation. However, if the strain energy exceeds the area energy, the neighboring cell is excluded from further consideration, as it may already be in a critical or plastic state. Among the eligible neighboring cells, the one with the maximum stress is identified, since cracks tend to propagate through areas with the highest stress concentration, where the material is most likely to fail. Once the cell with the maximum stress is identified, a crack is introduced at that location. This crack relieves the localized stress in the system, redistributing the forces across the material, and as a result, the overall potential energy of the system is minimized. This process reflects the natural behavior of materials, which move towards configurations that minimize potential energy. As the simulation progresses over time, this rule is applied iteratively. With each time step, the evolving stress and strain fields influence subsequent decisions about where cracks are introduced, making this model a dynamic representation of crack initiation and propagation. The dual consideration of both strain energy and stress ensures that cracks are introduced in a way that reflects both the ability of cells to bear load and the physical tendency of cracks to propagate through the material along paths of highest stress.

3.6 Algorithm for the Local Rule

1. **Identify Neighboring Cells:**
The neighborhood $N(i)$ for a cell i is defined as:

$$N(i) = \{j : j \text{ is a neighbor of } i\}.$$

2. **Candidate Selection:**
For each neighboring cell $j \in N(i)$, compare the strain energy $E_s(j)$ and the area energy $E_a(j)$:

$$\text{If } E_s(j) < E_a(j), \text{ then cell } j \text{ is a candidate for crack propagation.}$$

If $E_s(j) \geq E_a(j)$, exclude the cell from further consideration.
3. **Maximum Stress Criterion:**
Among the eligible neighbors ($j \in N(i)$ where $E_s(j) < E_a(j)$), identify the cell with the maximum stress:

$$j^* = \arg \max_{j \in N(i)} \sigma(j).$$

Here, j^* is the cell with the highest stress concentration among eligible neighbors.
4. **Introduce a Crack:**
Introduce a crack in cell j^*, and set its stress to zero:

$$c_{j^*} = 1, \quad \sigma(j^*) = 0.$$

5. **Iterative Application with Precomputed Stress:**
The stress values of other cells remain unchanged, as precomputed values are used in each iteration. Repeat the process for each time step:

$$c_j(t+1) = \begin{cases} 1, & \text{if } j = j^* \text{ and } \sigma(j^*) \neq 0, \\ c_j(t), & \text{otherwise.} \end{cases}$$

4 Results and Discussion

When the load is sufficiently small, the entire matrix remains elastic. In this case, the Classical Elasticity Solution is employed for the entire matrix. The Classical Plasticity Solution, when it becomes large enough, the matrix will begin to yield from its inner surface due to the stress concentration induced by the hard particle (acting as an inclusion). As the load continues to increase, the yielded region will expand outward. From symmetry, it follows that the elastic-plastic interface in the matrix will be a spherical surface for any value of load that produces a plastic region. Let r_c be the radius of this elastic-plastic interface, and p_c be the associated interface (normal) tension under loading. Then, the matrix material in the region $a < r < r_c$ is plastically deformed, whereas the material in the region $r_c < r < b$ remains elastic under load.

To simulate the path of crack propagation through the composite material, rules are established. The cellular automata approach will be used to dynamically model how cracks initiate and propagate, based on the stress concentrations and material properties.

Crack propagation can occur in two primary ways:

Penetration of the Core. In this mechanism, the crack directly penetrates the core material. This occurs when the stress concentration of the inclusion is greater than that of the matrix.

Deviating and Separating the Interface. Alternatively, the crack can deviate and travel along the interface between the core and the matrix. This also occurs when the stress concentration of the inclusion is more significant compared to that of the matrix.

The simulation results corresponding to different cases of elasticity and plasticity solutions, using Moore and von Neumann neighborhood configurations, are presented below. These simulations illustrate the behavior of the material under various stress conditions, highlighting how cracks propagate based on the surrounding cells in different neighborhoods. Each result provides insights into how stress distribution and material properties interact in these distinct configurations (Figs. 3, 4, 5 and 7).

4.1 Classical Elasticity Solution

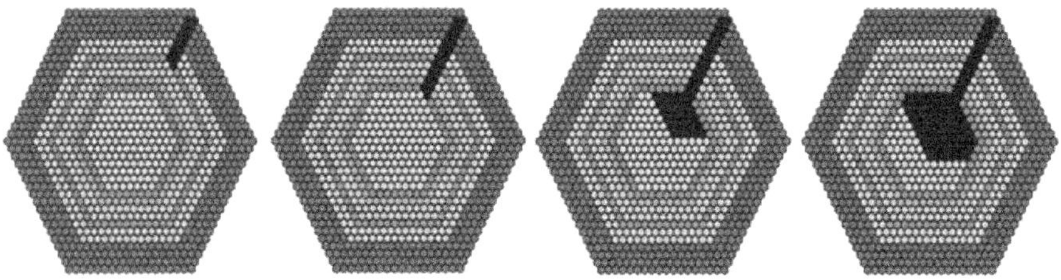

Fig. 3. Crack Propagation at time steps 6,10,15 and 18 under the Moore Neighbourhood when the stress in the inclusion is more than that in the matrix

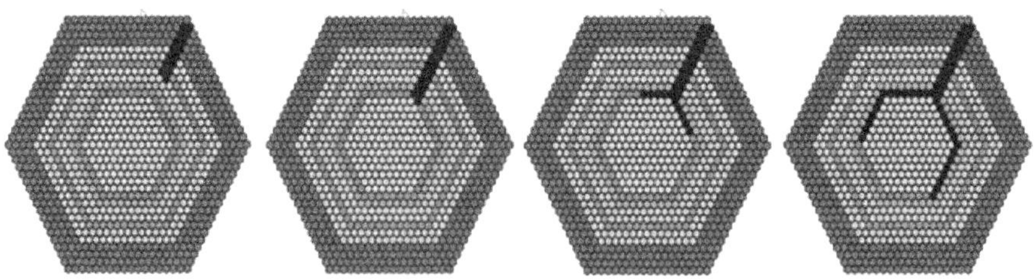

Fig. 4. Crack Propagation at time steps 6,10,14 and 25 under the Moore Neighbourhood when the stress in the inclusion is less than that in the matrix

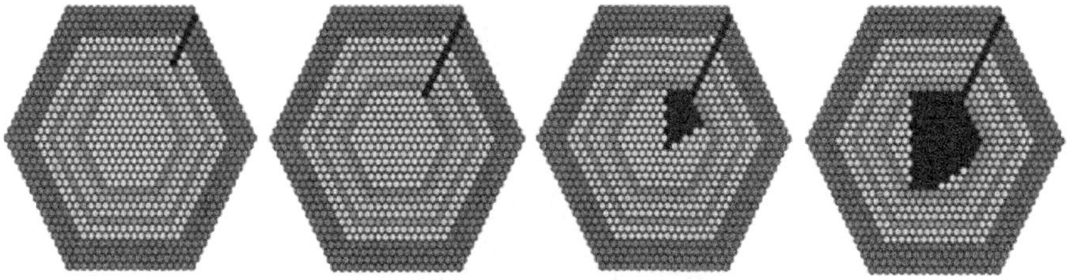

Fig. 5. Crack Propagation at time steps 6,10,16 and 23 under the von Neumann Neighbourhood when the stress in the inclusion is more than that in the matrix

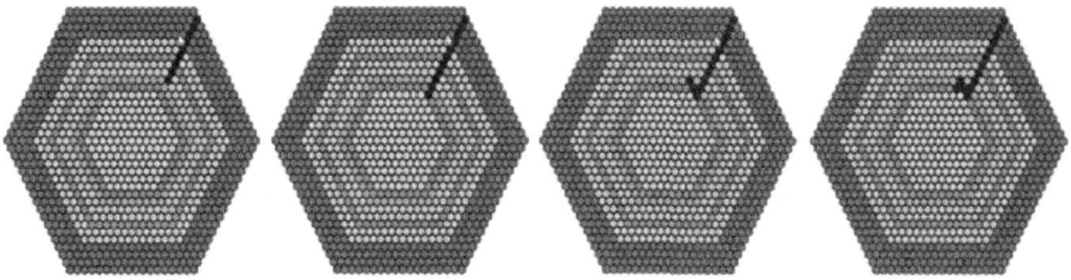

Fig. 6. Crack Propagation at time steps 8,10,12 and 13 under the von Neumann Neighbourhood when the stress in the inclusion is less than that in the matrix

4.2 Classical Plasticity Solution

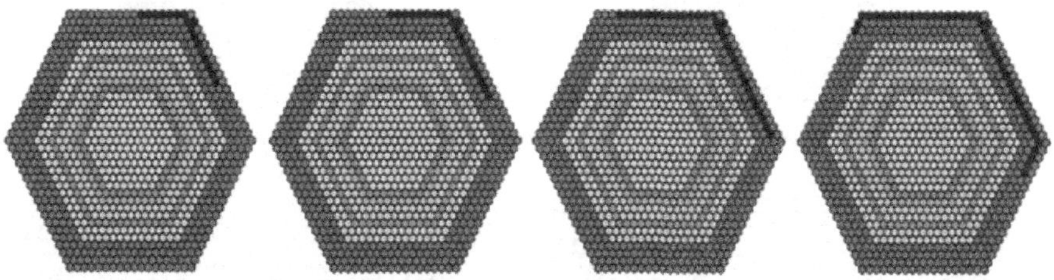

Fig. 7. Crack Propagation at time steps 9,10,15 and 20 under the Moore Neighbourhood

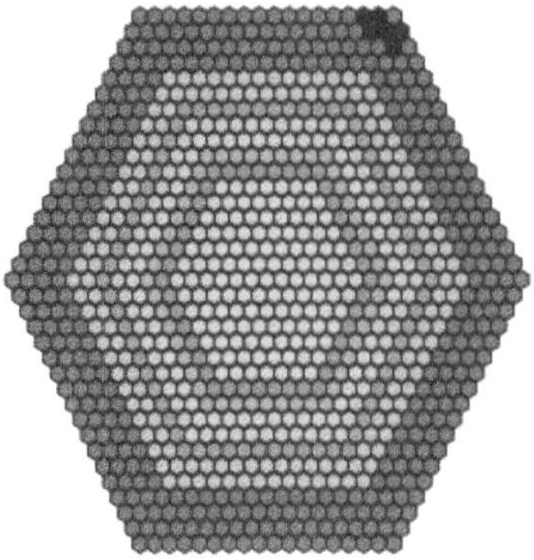

Fig. 8. Crack Propagation under the von Neumann Neighbourhood

4.3 Crack Propagation Results by Taking into Account the Comparison Between Area Energy with the Strain Energy

When the area energy of the matrix cells increases, crack propagation becomes more pronounced and extensive, particularly towards the core. This behavior occurs because, for a crack to propagate, the strain energy in neighboring cells must remain lower than the area energy. By increasing the area energy, the system provides a larger window for crack propagation, allowing cracks to form and expand more easily. In essence, the higher area energy threshold permits more stress relief through crack propagation, driving the crack deeper and more aggressively into the material (Figs. 9 and 10).

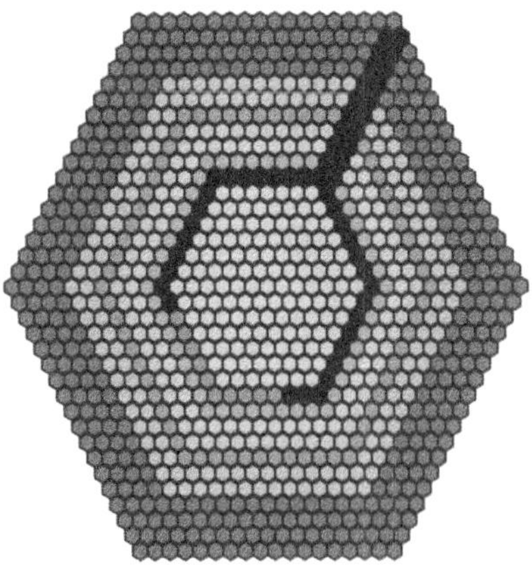

Fig. 9. Crack Propagation when the area energy is $30\text{x}10^8$

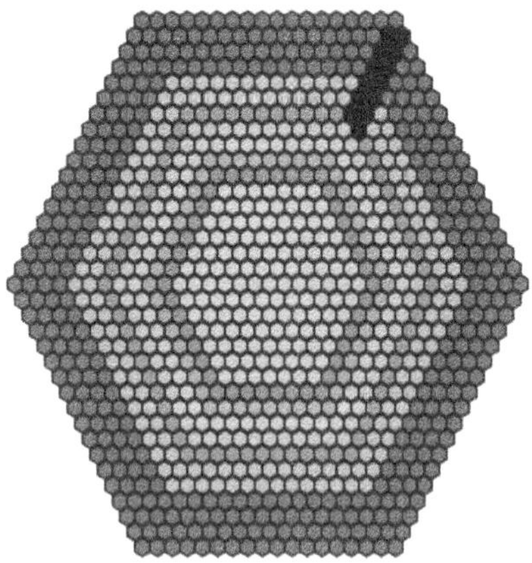

Fig. 10. Crack Propagation when the area energy is $15\text{x}10^8$

In contrast, when the area energy is relatively low, crack propagation is significantly limited. A lower area energy reduces the number of neighboring cells eligible for crack formation, as fewer cells meet the condition of having strain energy lower than the area energy. This limitation confines the cracks to smaller regions, preventing them from propagating extensively through the material. The cracks tend to remain localized, and the overall progression of the damage is more gradual and contained. Thus, by modulating the area energy of the matrix cells, it is possible to control the extent and direction of crack propagation, with higher area energies facilitating more rapid and widespread failure, while lower area energies act as a restrictive factor, minimizing crack growth.

The use of the hexagonal von Neumann neighborhood in the current model limits crack propagation to three possible directions at each step. This restriction can hinder propagation at a hexagonal corner or artificially direct it to adjacent directions, as observed in Figs. 6 and 8. Addressing this limitation will be essential to ensure realistic and unbiased crack propagation simulations.

5 Conclusion

This study developed a cellular automata model to simulate crack propagation in composite materials, using classical elasticity and plasticity solutions to analyze stress and strain distributions under different load conditions. Crack propagation was governed by stress concentration in neighboring cells, and the introduction of cracks was guided by comparing strain energy with area energy. The results demonstrated that increasing the area energy facilitated more extensive crack propagation, particularly towards the core, while lower area energy constrained cracks to more localized regions.

The presented CA model lays a foundation for simulating crack propagation in 2D materials, but extending it to three dimensions opens up opportunities for realistic modeling while presenting challenges. The computational complexity increases significantly, requiring efficient algorithms and parallel processing to manage larger cell interactions. Representing displacements, such as plastic deformation and crack paths, becomes more challenging in 3D, necessitating refined rule sets and accurate neighborhood definitions. 3D modeling enables realistic simulations of volumetric stress distributions, benefiting industries like aerospace, automotive, and civil engineering. It allows for advanced analysis of stress intensity and multi-crack interactions, guiding the development of materials resistant to crack propagation. To achieve this, algorithms must be optimized, rules refined and user-friendly tools developed. These improvements would refine the model's ability to predict material failure under more diverse and realistic conditions.

Appendix

The variables used in the equations are defined as follows: σ_0 is the applied external stress, σ_{rr} and $\sigma_{\theta\theta}$ are the radial and hoop stresses, respectively. The radial strain and hoop strain are denoted by ε_{rr} and $\varepsilon_{\theta\theta}$, respectively. p_i is the uniform normal tension acting on the inclusion surface. The material properties include Young's modulus (E) and Poisson's ratio (ν), while E^I and ν^I refer to the Young's modulus and Poisson's ratio of the inclusion. a and b are the inner and outer radii of the spherical shell, while r is the radial distance. In the plastic region, r_c is the radius of the plastic zone, σ_y is the stress at r = r_c, and n is the strain-hardening coefficient. Finally, E_s is the strain energy.

Appendix A: Classical Elasticity Solution

When σ_0 is sufficiently small, the entire matrix will remain elastic. The inclusion can be considered an elastic solid sphere of radius a, subjected to a uniform normal tension p_i on its surface.

$$p_i = \frac{\frac{3\sigma_0}{2}\frac{b^3}{b^3-a^3}}{\frac{E}{1-2\nu^I} + \frac{E^I}{1-\nu^I} + \frac{\nu}{1-\nu} + \frac{2a^3+b^3}{2(b^3-a^3)}} \tag{1}$$

In this case, the stress and strain components in the matrix for the region $a < r < b$ can be expressed as are:

$$\sigma_{rr} = \frac{\sigma_0 b^3}{b^3 - a^3}\left(1 - \frac{a^3}{r^3}\right) + \frac{p_i a^3}{b^3 - a^3}\left(\frac{b^3}{r^3} - 1\right), \tag{2}$$

$$\sigma_{\theta\theta} = \sigma_{\phi\phi} = \frac{\sigma_0 b^3}{b^3 - a^3}\left(1 + \frac{a^3}{2r^3}\right) - \frac{p_i a^3}{b^3 - a^3}\left(1 + \frac{b^3}{2r^3}\right), \tag{3}$$

$$\begin{aligned} \varepsilon_{rr} = \frac{1}{E}&\left[\frac{\sigma_0 b^3}{b^3 - a^3}\left(1 - \frac{a^3}{r^3}\right) + \frac{p_i a^3}{b^3 - a^3}\left(\frac{b^3}{r^3} - 1\right)\right] \\ -2\nu&\left[\frac{\sigma_0 b^3}{b^3 - a^3}\left(1 + \frac{a^3}{2r^3}\right) - \frac{p_i a^3}{b^3 - a^3}\left(1 + \frac{b^3}{2r^3}\right)\right] \end{aligned} \tag{4}$$

$$\begin{aligned} \varepsilon_{\theta\theta} = \varepsilon_{\phi\phi} = \frac{1}{E}&\left[(1-\nu)\left(\frac{\sigma_0 b^3}{b^3 - a^3}\left(1 + \frac{a^3}{2r^3}\right) - \frac{p_i a^3}{b^3 - a^3}\left(1 + \frac{b^3}{2r^3}\right)\right)\right] \\ -\nu&\left[\frac{\sigma_0 b^3}{b^3 - a^3}\left(1 - \frac{a^3}{r^3}\right) + \frac{p_i a^3}{b^3 - a^3}\left(\frac{b^3}{r^3} - 1\right)\right] \end{aligned} \tag{5}$$

Appendix B: Classical Plasticity Solution

This region can be considered as a thick-walled spherical shell with an inner radius r_c and an outer radius b, subjected to internal pressure p_c and external pressure σ_0. The stress at r_c is given by $\sigma_y$and the strain hardening coefficient is given by n. The solution in the plastic region of the elastic-perfectly plastic matrix can then be expressed as:

$$\sigma_{rr} = \sigma_0 - \frac{2\sigma_y}{3}\left(1 - \frac{r_c^3}{b^3}\right) + \frac{2\sigma_y}{3n}\left(1 - \frac{r_c^{3n}}{r^{3n}}\right) \tag{6}$$

$$\sigma_{\theta\theta} = \sigma_{\phi\phi} = \sigma_0 - \frac{2\sigma_y}{3}\left(1 - \frac{r_c^3}{b^3}\right) + \frac{2\sigma_y}{3n}\left[1 + \left(\frac{3n}{2} - 1\right)\left(\frac{r_c}{r}\right)^{3n}\right] \tag{7}$$

Appendix C: Strain Energy Calculation

The strain energy is computed using the formula:

$$E_s = \frac{1}{2} E \varepsilon_{rr}^2 \tag{8}$$

References

1. Gao, X.L.: Analytical solution for the stress field around a hard spherical particle in a metal matrix composite incorporating size and finite volume effects. Math. Mech. Solids **13**(4), 357–372 (2008). https://doi.org/10.1177/1081286507086906
2. Gobron, S., Chiba, N.: Crack pattern simulation based on 3d surface cellular automata. Vis. Comput. **17**, 287–309 (2001)
3. Greco, F., Lonetti, P., Nevone Blasi, P.: Crack propagation analysis in composite materials using moving mesh and multiscale techniques. Compos. Struct. **131**, 663–675 (2015)
4. Griffith, A.A.: The phenomena of rupture and flow in solids. Philos. Trans. Royal Society London A **221**, 163–198 (1921)
5. Helm, J., Lichtenberg, C., Vainas, B.: Cellular automata and their applications in computational mechanics. Comput. Mech. **68**(1), 151–169 (2022)
6. Irwin, G.R.: Analysis of stresses and strains near the end of a crack traversing a plate. J. Appl. Mech. **24**, 361–364 (1957)
7. Kari, J.: Theory of cellular automata: a survey. Theoret. Comput. Sci. **334**(1–3), 3–33 (2005). https://doi.org/10.1016/j.tcs.2004.11.021
8. Li, M., Zhongliang, R., Junlian, H.: Cellular automata to simulate rock failure. In: Proceedings of the 16th International Conference on Artificial Reality and Telexistence. pp. 35–41 (2006)
9. Sadek, K., Aour, B., Bouiadjra, B.B., Bouanani, M.F., Khelil, F.: Analysis of crack propagation by bonded composite repairs in marine structures: a numerical analysis. Int. J. Eng. Res. Afr. **35**, 175–184 (2018)
10. Yan, F., Wang, J., Zhang, S.: Discontinuous cellular automaton method for crack growth analysis without remeshing. Comput. Struct. **143**, 80–91 (2014)
11. Yan, F., Xu, S., Zhang, S.: Dynamic crack propagation analysis combined with stable scheme and continuous-discontinuous cellular automaton. Eng. Fract. Mech. **231**, 107041 (2020)

Diabetes Prediction Enhancement via Cellular Automata and Machine Learning

Deepanshu Baghel(✉) and Partha Sarathi Bishnu

Birla Institute of Technology, Mesra, Lalpur, Ranchi 834001, India
{mca40056.23,psbishnu}@bitmesra.ac.in

Abstract. This paper explores the enhancement of predictive models using Cellular Automata (CA) to create neighborhood-influenced features in machine learning models. Diabetes prediction models are crucial in early diagnosis, allowing timely intervention and management. We investigated the performance of four techniques: LightGBM, XGBoost, Random Forest, and Logistic Regression, both with and without CA-based feature engineering on a diabetes dataset. After encoding feature values into discrete states, CA principles were applied to generate new features that capture risk levels based on the influence of k closest neighbors. A comparative study of model accuracy, precision, recall, and the F1 score shows that CA-enhanced models provide better insights into neighborhood-driven predictions. This paper illustrates the potential for CA to improve traditional machine-learning techniques for classification tasks in diabetes prediction.

Keywords: Cellular Automata · Diabetes Classification · Machine Learning

1 Introduction

Diabetes is a chronic medical condition affecting millions worldwide, requiring early diagnosis and intervention to effectively manage complications. Predictive models play a significant role in aiding medical professionals in early diabetes diagnosis, with machine learning (ML) models such as LightGBM [1], XGBoost [2], Random Forest [3], and Logistic Regression [4] becoming widely adopted in this field. Despite advances, traditional ML techniques often treat features independently, potentially overlooking the influence of neighboring data points, a limitation that Cellular Automata (CA) [5] can address. Cellular Automata [6], which utilize local interactions to drive global patterns, present a promising approach to incorporating spatial relationships into feature space. When applied to ML, CA can capture neighborhood effects by encoding features into states and generating new attributes based on the influence of neighboring instances, ultimately enhancing the prediction process.

S. Kamilya et al. (Eds.): ASCAT 2025, CCIS 2499, pp. 34–43, 2025.
https://doi.org/10.1007/978-3-031-94121-4_3

This paper proposes a CA-based feature engineering method that integrates neighborhood-based influence into four well-known ML models, LightGBM, XGBoost, Random Forest and Logistic Regression, for the prediction of diabetes. The main contributions are as follows.

- We develop a method to encode continuous medical features (e.g., Glucose, BMI, Age) into discrete risk states, using CA principles to incorporate neighborhood influence through k-nearest neighbors.
- By creating CA-based features (High, Medium, and Low Risk Neighbors), we enhance ML models' sensitivity to spatially related data, offering insights into neighborhood-driven health risks.
- We evaluated and compared the effectiveness of CA-enhanced models against traditional models in performance metrics (accuracy, precision, recall, F1 score), showing CAs potential to boost predictive power in medical diagnostics.

The structure of this paper is as follows. Section 2 reviews related work, Sect. 3 details the methodology with a focus on Cellular Automata for feature engineering, Sect. 4 covers the experimental analysis, results, and performance evaluation, and Sect. 5 concludes with key findings and future directions.

2 Related Work

Cellular automata (CA) have been widely applied to enhance machine learning and predictive modeling [5,7]. Povalej et al. (2005) introduced a Classificational Cellular Automata (CCA) framework, integrating CA with decision trees and support vector machine (SVM) to improve classification accuracy on datasets like heart disease and breast cancer [8]. Kim and Kim (2020) combined CA with ANN to model traffic at Jeju Airport, effectively predicting congestion based on real-time conditions [9]. Zhang and Blcskei (2024) demonstrated that deep neural networks could learn CA rules, facilitating CA logic modeling using logic of many values [10]. Li et al. (2024) introduced Photonic Neural Cellular Automata (PNCA) for binary image classification, showcasing the noise resilience and scalability of photonic CA [11]. In manufacturing, Kaikova and Terziyan (2024) combined DNN, CA, and Petri nets to improve decision making for predictive maintenance and logistics [12]. Sun et al. (2024) developed a CA-based model for forest fire prediction, outperforming traditional models in the accuracy of the burned area [13]. Richardson et al. (2024) applied Neural Cellular Automata (NCA) to model spatio-temporal patterns, suitable for biological systems [14]. *In medical applications*, Pfeifer et al. (2008) used CA to simulate the spread of infectious diseases, providing information for pandemic planning [15]. Sree and Babu (2011) developed a Non-Linear CA (NLCA) model to predict the risk of heart attack [16], while Prez Betancourt et al. (2015) modeled disease progression under varied population densities and geographic constraints [17]. Sree and Nedunuri (2020) used CA to predict COVID-19 trends, achieving high accuracy

[18]. Valentim et al. (2023) used CA to model tumor growth dynamics, capturing valuable complex cell behaviors for cancer research [19]. Finally, Korevaar et al. (2024) applied NCA to medical image segmentation, delivering enhanced robustness across imaging domains [20]. The review of CA applications showed a research gap in diabetes prediction, encouraging us to develop a CA-enhanced feature engineering method that leverages neighbourhood influences to improve the accuracy and interpretability of predictive models for early diabetes diagnosis.

3 Methodology

In this paper, we employ *Cellular Automata (CA)* principles [7] to encode features into discrete states and incorporate neighborhood influences to create additional features. This methodology consists of *feature encoding* and *neighborhood-based feature creation*. We outline the process in Fig. 1, starting with data preprocessing, followed by the application of cellular automata for the generation of new features. These newly created features are then integrated with the existing dataset, which is subsequently used to train machine learning models for prediction purposes. Before delving into the details, refer to the Symbol Table 1 for an understanding of the symbols used throughout this paper.

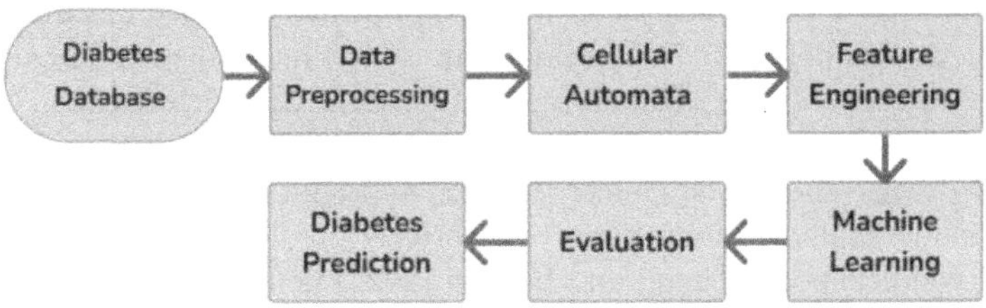

Fig. 1. Methodological Framework Diagram

Table 1. Symbol Table

Symbol	Description	Symbol	Description
X	Dataset, $\|X\| = N$	$\mathcal{F}$	Features set
$F \in \mathcal{F}$	Selected features	$\|F\| = n$	Number of selected features
T_{low}, T_{high}	Threshold values	$f \in F \subseteq \mathcal{F}$	feature
S	states {low, medium, high}	$\|S\| = r$	Number of states
$\mathcal{N}$	Neighborhood	$(\|\mathcal{N}\| =)\ k$	Number of neighbors

1. Feature Encoding: We categorize each feature into discrete states to simplify continuous features and apprehend different risk levels. Let X be a dataset with selected features F and $|F| = n$ (e.g., *Glucose*, *BMI*, and *Age*). Each feature $f \in F$ is binned into three states (S)—representing {low, medium, and high} risk—based on specified threshold values.

For a given feature f_i, the state assignment is defined as:

$$f_i^{\text{state}} = \begin{cases} 0 & \text{if } f_i < T_{\text{low}} \\ 1 & \text{if } T_{\text{low}} \leq f_i < T_{\text{high}} \\ 2 & \text{if } f_i \geq T_{\text{high}} \end{cases} \tag{1}$$

where $1 \leq i \leq N$, N is the number of records, T_{low} and T_{high} serve as thresholds defining {low, medium, and high} states. These values can be determined using domain knowledge, statistical quantiles, or optimization techniques such as grid search.

2. Neighborhood-Based Feature Creation: After encoding features, we apply *k-nearest neighbors (k-NN)* to capture neighborhood influences. For each individual i, we define a neighborhood $\mathcal{N}_i$ of size k by calculating the Euclidean distance between i and other observations based on encoded states f_i^{state}, $state \in S$.

The Euclidean distance between two individuals i and j with encoded features f^{state} is given by:

$$d(i,j) = \sqrt{\sum_{m=1}^{n} (f_{im}^{\text{state}} - f_{jm}^{\text{state}})^2} \tag{2}$$

where $1 \leq i, j \leq N, i \neq j$, n is the number of encoded features. The *k-NN* algorithm identifies the closest k neighbors based on this distance.

3. Creation of CA-Based Features: To capture risk levels in an individual's neighborhood, we create new features based on the states of the k-nearest neighbors:

- **High Risk Neighbors (HRN)**: Count of neighbors where any feature f^{state} is high ($f^{\text{state}} = 2$):

$$\text{High\_Risk\_Neighbors}_i = \sum_{j \in \mathcal{N}_i} \mathbb{I}(f_j^{\text{state}} = 2) \tag{3}$$

- **Medium Risk Neighbors (MRN)**: Count of neighbors where any feature f^{state} is medium ($f^{\text{state}} = 1$):

$$\text{Medium\_Risk\_Neighbors}_i = \sum_{j \in \mathcal{N}_i} \mathbb{I}(f_j^{\text{state}} = 1) \tag{4}$$

- **Low Risk Neighbors (LRN)**: Count of neighbors where all features f^{state} are low ($f^{\text{state}} = 0$):

$$\text{Low\_Risk\_Neighbors}_i = \sum_{j \in \mathcal{N}_i} \mathbb{I}(f_j^{\text{state}} = 0) \tag{5}$$

Here, $\mathbb{I}$ is the indicator function which returns 1 if the condition is met and 0 otherwise. Utilizing CA beyond the initial application of K-NN allows for a deeper integration of local interactions and transformations in the feature set, leading to models that can better understand and predict based on the complex, often nonlinear interactions typical in real-world data. This approach is particularly advantageous in healthcare settings, where patient outcomes can be heavily influenced by local or neighborhood effects.

4. Final Feature Set: Finally, we combine the original features X with the newly created CA-based features HRN, MRN, and LRN to form an enhanced dataset X_{combined}:

$$X_{\text{combined}} = X \cup \{\text{HRN}, \text{MRN}, \text{LRN}\}. \tag{6}$$

To provide a mathematical foundation for our proposed concept, we present the following lemmas and theorems.

Lemma 1. *Let f^{state} be a discretely encoded feature in a dataset X, where each feature value is transformed based on defined thresholds T_{low} and T_{high}. For any two individuals i and j in X, if $f_i^{state} = f_j^{state}$, then their Euclidean distance contribution from feature f is zero.*

Proof. By the definition of the Euclidean distance between two individuals i and j for feature f^{state}:

$$d(i,j) = \sqrt{\sum_{m=1}^{n} (f_{im}^{\text{state}} - f_{jm}^{\text{state}})^2}$$

If $f_i^{\text{state}} = f_j^{\text{state}}$, then $f_{im}^{\text{state}} - f_{jm}^{\text{state}} = 0$, which implies that this feature does not contribute to the overall distance $d(i,j)$. □

Theorem 1. *For a dataset X with encoded features and a neighborhood $\mathcal{N}_i$ of size k, the CA-based feature representing the count of neighbors at a specific risk level for individual i (e.g., High_Risk_Neighbors$_i$, Medium_Risk_Neighbors$_i$, or Low_Risk_Neighbors$_i$) will be maximized when all neighbors $j \in \mathcal{N}_i$ have the corresponding state value for f^{state}.*

Proof. Let f^{state} represent the encoded state of any feature at a specified risk level: $f^{\text{state}} = 2$ for high risk, $f^{\text{state}} = 1$ for medium risk, or $f^{\text{state}} = 0$ for low risk.

The CA-based feature for individual i in the neighborhood $\mathcal{N}_i$ is calculated as:

$$\text{Risk\_Neighbors}_i = \sum_{j \in \mathcal{N}_i} \mathbb{I}(f_j^{\text{state}} = r)$$

where r is the risk level (0 for low, 1 for medium or 2 for high).

If each $j \in \mathcal{N}_i$ satisfies $f_j^{\text{state}} = r$, then $\mathbb{I}(f_j^{\text{state}} = r) = 1$ for all j. Therefore, the sum Risk_Neighbors$_i$ reaches its maximum value of k, as each indicator contributes 1 to the total count. □

Lemma 2. *If threshold values T_{low} and T_{high} are set based on statistical quantiles, then for any feature f in the dataset X, approximately 50% of the data points will fall into the medium risk category $f^{state} = 1$ under the assumption of a normal distribution.*

Proof. Setting T_{low} and T_{high} to statistical quantiles (e.g., the 25th and 75th percentiles) divides the data approximately into three equal parts for a normal distribution. Thus, under this assumption, the interval $T_{\text{low}} \leq f_i < T_{\text{high}}$ will contain approximately 50% of the data points, categorizing them into the medium-risk state $f^{\text{state}} = 1$. □

Theorem 2. *For a given individual i with neighborhood $\mathcal{N}_i$ of size k, the CA-based features HRN_i, MRN_i, and LRN_i provide a comprehensive local representation of risk levels that enhances predictive accuracy when used as additional features in a machine learning model.*

Proof. The features HRN_i, MRN_i, and LRN_i capture the number of neighbors in each risk category (high, medium, and low) based on f^{state}. These features incorporate neighborhood risk levels, which add valuable context to X and contribute to a more robust predictive model by capturing spatial dependencies and local variation in risk, thus enhancing model accuracy. □

4 Experiments

This section discusses the dataset, evaluation metrics, parameter settings, and result analysis. For comparison purposes, we implemented four machine learning techniques: XGBoost [2], LightGBM [1], Random Forest [3], and Logistic Regression [4] - with and without Cellular Automata (CA) enhancements, using Python programming. Before that, we describe the dataset as follows.

4.1 Dataset

This study utilizes Pima Indians Diabetes Database [21] containing 768 records, each representing a unique individual. The dataset includes eight features: *Pregnancies*, *Glucose*, *Blood Pressure*, *Skin Thickness*, *Insulin*, *BMI*, *Diabetes Pedigree Function*, and *Age*. These attributes provide relevant medical and genetic information, with the target variable, *Outcome*, indicating diabetes status (1 for diabetic, 0 for non-diabetic). Data preprocessing, such as handling missing values and feature scaling, was applied to improve model performance.

4.2 Evaluation Metrics

For evaluating the performance of the diabetes prediction model, we use common classification metrics: *Accuracy*, *Precision*, *Recall*, and *F1 Score* (Table 2) to provide a comprehensive assessment. To obtain all parameters TN, TP, FN and FP, we may refer to Confusion Matrix Table 3 as follows:

Table 2. Evaluation Metrics for Model Performance

Metric	Formula
Accuracy	$\frac{TP+TN}{TP+TN+FP+FN}$
Precision	$\frac{TP}{TP+FP}$
Recall	$\frac{TP}{TP+FN}$
F1 Score	$2 \cdot \frac{Precision \times Recall}{Precision+Recall}$

Table 3. Confusion Matrix for Diabetes Prediction

	Predicted Non-Diabetic	Predicted Diabetic
Actual Non-Diabetic	True Negatives (TN)	False Positives (FP)
Actual Diabetic	False Negatives (FN)	True Positives (TP)

4.3 Parameter Settings

The parameter settings for this research are described below. We selected three key characteristics, glucose, BMI and age, for feature encoding. The feature encoding thresholds categorize continuous values into discrete states: for glucose, the thresholds are $\{100, 140\}$; for BMI, $\{25, 30\}$; and for age, $\{40, 60\}$, producing low, medium and high states. The size of the neighborhood, $k = 5$, determines the number of nearest neighbors used in the CA-based feature engineering process, with k-NN defining the neighborhood for each individual based on their encoded states. The encoded states, glucose state, BMI state, and age state, represent risk levels, forming the basis for neighborhood analysis and providing context for each individual relative to their neighbors. Lastly, the CA-based features-HRN, MRN, and LRN-indicate the count of neighbors falling into high, medium, or low-related risk categories, offering additional contextual information to support the CA-enhanced feature engineering.

In this study, we first split the dataset into training and testing sets with an 80-20 ratio, using a random state of 42 for reproducibility. For model training, we applied several machine learning algorithms. In LightGBM, a binary LightGBM model was trained using gradient boosting (GBDT) on an unbalanced dataset with 100 boosting rounds to predict binary outcomes on the test set. In XGBoost, we set a binary logistic objective with log loss evaluation and disabled label encoding, training the model on the training data to produce binary predictions on the test data. For Random Forest, a model with 100 estimators and a random state of 42 was trained in the training set and used for binary prediction on the test set. Lastly, in Logistic Regression, a model with up to 1000 iterations and a random state of 42 was trained on the training data to classify test outcomes.

In this study, our software environment was set up on a Microsoft Windows 10 operating system. We conducted our programming using Python within a Jupyter Notebook environment on an HP computer model 15g-br0xx. Hardware features include an Intel Core i5-7200U CPU, 16 GB of RAM, and a 64-bit system architecture.

4.4 Results and Analysis

Table 4 provides a comparative analysis of performance metrics: accuracy, precision, recall, and F1 score between models with and without the integration of Cellular Automata (CA). Overall, models incorporating CA demonstrate slight performance improvements, particularly in recall and F1 score, critical metrics for unbalanced datasets. The *LightGBM* model with CA achieves a marginal improvement in all metrics compared to its standard version, the F1 score increasing from 0.60 to 0.63, indicating a better balance between precision and recall. Similarly, *XGBoost with CA* shows slight enhancements in recall and F1 score (0.67 and 0.62, respectively) compared to its standard counterpart, although overall gains are minimal. *RandomForest with CA* displays a more noticeable improvement in accuracy and precision, where accuracy increases from 0.78 to 0.80 and the F1 score from 0.62 to 0.65, underscoring CA's positive influence on the model's ability to identify both classes correctly. *Logistic Regression* sees the highest increase in accuracy with CA, from 0.80 to 0.82, indicating that CA's neighborhood-based feature engineering complements simpler models, enhancing their predictive capacity. Therefore, CA-based feature engineering benefits all models, with more pronounced effects in RandomForest and Logistic Regression. This suggests that CA may provide additional predictive value, especially in models that benefit from neighborhood-based feature enhancement.

Table 4. Comparison of Model Performance Metrics with and without CA

Model	Accuracy	Precision	Recall	F1 Score
LightGBM	0.76	0.56	0.65	0.60
LightGBM with CA	0.77	0.58	0.69	0.63
XGBoost	0.76	0.58	0.65	0.61
XGBoost with CA	0.76	0.58	0.67	0.62
RandomForest	0.78	0.61	0.62	0.62
RandomForest with CA	0.80	0.64	0.65	0.65
Logistic Regression	0.80	0.64	0.67	0.65
Logistic Regression with CA	0.82	0.66	0.64	0.65

5 Conclusions

This study explored the use of Cellular Automata (CA) principles in feature engineering to improve binary classification. By encoding features into discrete states and integrating neighbourhood influences, CA-based features improved accuracy, precision, recall, and F1 scores across multiple models. The approach added valuable spatial relationships and context, with notable gains in the RandomForest and Logistic Regression models. The results show that CA-enhanced models can capture interactions within the data, improving predictive accuracy. Future work could apply CA-based methods to other domains and models to assess broader applicability.

Acknowledgments. We would like to express our sincere gratitude to Birla Institute of Technology, Mesra, Ranchi, Jharkhand, for their support and for providing us with a conducive research environment. We also extend our thanks to all anonymous reviewers for their valuable feedback and insights.

Disclosure of Interests. The authors have no competing interests to declare that are relevant to the content of this article.

References

1. Ke, G., et al.: LightGBM: a highly efficient gradient boosting decision tree. Adv. Neural Inform. Process. Syst. **30**, 3146–3154. Curran Associates, Inc. (2017)
2. Chen, T., Guestrin, C.: XGBoost: a scalable tree boosting system. In: Proceedings of the 22nd ACM SIGKDD International Conference on Knowledge Discovery and Data Mining, pp. 785–794. ACM, New York (2016). https://doi.org/10.1145/2939672.2939785
3. Breiman, L.: Random forests. Mach. Learn. **45**(1), 5–32 (2001). https://doi.org/10.1023/A:1010933404324
4. Cox, D.R.: The regression analysis of binary sequences. J. R. Stat. Soc. Ser. B **20**(2), 215–242 (1958)
5. Wongthanavasu, S., Ponkaew, J.: A cellular automata-based learning method for classification. Expert Syst. Appl. **49**, 99–111 (2016). https://doi.org/10.1016/j.eswa.2015.12.003
6. Wolfram, S.: Theory and Appllications of Cellular Automata. World Scientific, Singapore (1986)
7. Djouama, I., Kadache, N., Seghir, R.: Machine learning-driving cellular automata: application in urban engineering. In: Novel and Intelligent Digital Systems: Proceedings of the 4th International Conference (NiDS 2024). Lecture Notes in Networks and Systems, vol. 1170, pp. 507–511. Springer, Cham (2024). https://doi.org/10.1007/978-3-031-73344-4_43
8. Povalej, P., Kokol, P., Welzer Družovec, T., Stiglic, B.: Machine-learning with cellular automata. In: Famili, A.F., et al. (eds.) Lecture Notes in Computer Science, vol. 3646, pp. 305–315. Springer-Verlag (2005). https://doi.org/10.1007/11552253
9. Kim, B.S., Kim, T.G.: Modeling and simulation using artificial neural network-embedded cellular automata. IEEE Access **8**, 24056–24061 (2020). https://doi.org/10.1109/ACCESS.2020.2970547
10. Zhang, Y., Bölcskei, H.: Cellular automata, many-valued logic, and deep neural networks. arXiv preprint arXiv:2404.05259 (2024)
11. Li, G.H.Y., Leefmans, C.R., Williams, J., Gray, R.M., Parto, M., Marandi, A.: Deep learning with photonic neural cellular automata. Light: Sci. Appl. **13**(283) (2024). https://doi.org/10.1038/s41377-024-01651-7

12. Kaikova, O., Terziyan, V.: Deep neural networks, cellular automata and petri nets: useful hybrids for smart manufacturing. Procedia Comput. Sci. **232**, 2334–2346 (2024). https://doi.org/10.1016/j.procs.2024.02.052
13. Sun, X., et al.: A forest fire prediction model based on cellular automata and machine learning. IEEE Access **12**, 55389–55402 (2024). https://doi.org/10.1109/ACCESS.2024.3389035
14. Richardson, A.D., Antal, T., Blythe, R.A., Schumacher, L.J.: Learning spatio-temporal patterns with neural cellular automata. PLoS Comput. Biol. **20**(4), e1011589 (2024). https://doi.org/10.1371/journal.pcbi.1011589
15. Pfeifer, B., et al.: A cellular automaton framework for infectious disease spread simulation. Open Med. Inform. J. **2**, 70–81 (2008). https://doi.org/10.2174/1874431100802010070
16. Sree, P.K., Babu, I.R.: Non linear cellular automata in predicting heart attack. Int. J. Hybrid Inform. Technol. **4**(1), 33–40 (2011)
17. Pérez Betancourt, Y.G., Rodríguez Puente, R., Mufeti, K.: Cellular automata and its applications in modeling and simulating the evolution of diseases. In: Presented at the International Conference on Epidemiology, Havana, Cuba (2015). https://www.researchgate.net/publication/282249343
18. Sree, P.K., Nedunuri, S.S.S.N.U.: A novel cellular automata classifier for COVID-19 trend prediction. J. Health Sci. **10**(1), 34–38 (2020). https://doi.org/10.17532/jhsci.2020.907
19. Valentim, C.A., Rabi, J.A., David, S.A.: Cellular-automaton model for tumor growth dynamics: Virtualization of different scenarios. Comput. Biol. Med. **153**, 106481 (2023). https://doi.org/10.1016/j.compbiomed.2022.106481
20. Korevaar, S., Tennakoon, R., Bab-Hadiashar, A.: Generalization capabilities of neural cellular automata for medical image segmentation: a robust and lightweight approach. In: Proceedings of the International Conference on Medical Imaging, pp. 1–10 (2024)
21. Dua, D., Graff, C.: UCI Machine Learning Repository (2019). https://www.kaggle.com/datasets/uciml/pima-indians-diabetes-database Accessed 30 Oct 2024

Empirical Study on Reversibility of Rotation-Symmetric Rules of 2D Cellular Automata-Finite Triangular Lattice

A. Bhavadharani and D. K. Sheena Christy(✉)

Department of Mathematics, Faculty of Engineering and Technology, SRM Institute of Science and Technology, Kattankulathur, Chengalpattu District, Tamil Nadu 603203, India
{ba7497,sheenac}@srmist.edu.in

Abstract. This study presents the rotation-symmetric rules of finite two-dimensional cellular automata. The lattice considered is the finite triangular lattice with von Neumann neighborhood of radius one under null boundary conditions. There exist 256 rotation-symmetric rules on two-dimensional finite triangular lattice cellular automata. This study reveals the fractal-generating capability of a rotation-symmetric rule with self-replicating patterns. This study also aims to experimentally discuss the reversibility of rotation-symmetric rules in two-dimensional (2D) Cellular Automata on the finite triangular lattice with von Neumann neighborhood of radius one under null boundary.

Keywords: 2D Cellular Automata · Finite Triangular Lattice · Null boundary · von Neumann Neighborhood

1 Introduction

In 1966, John von Neumann [1] unveiled the concept of cellular automata (CA), setting the stage for a groundbreaking field focused on the intricate modeling of self-reproduction in biological systems. The unique feature of CA is their capacity to produce complex behaviors from simple local rules. The behavior of the automaton is significantly affected by its neighborhood consideration. Von Neumann began his work with 2D CA, considering a five-cell neighborhood which included the center cell and its four orthogonal neighbor cells. Moore [2] extended this approach to a nine-cell neighborhood, incorporating the center cell, its orthogonal neighbor cells, and the diagonal neighbor cells.

The neighborhood of a CA is dependent on its dimension, where the von Neumann and Moore neighborhood are specifically associated with the 2D CA.

This work is carried out as a project in the Indian Summer School on Cellular Automata 2024.

S. Kamilya et al. (Eds.): ASCAT 2025, CCIS 2499, pp. 44–56, 2025.
https://doi.org/10.1007/978-3-031-94121-4_4

The 2D CA are focused in their works by Packard and Wolfram [3], Kari [4], Durand [5] and so on. After introducing the basic concepts of neighborhood and dimension, the next key aspect to consider is the lattice structure. Early research in this domain predominantly concentrated on the finite or infinite square lattice. A difference in the shape of lattice results in different neighborhood dependencies. In this way, Morita [6], Imai [7], and Zawidzki [8] introduced the idea of CA on the triangular lattice. In particular, Morita [6] addressed the concept of partitioned CA. In this work, we deal with the finite 2D triangular lattice CA with von Neumann neighborhood under null boundary.

A significant number of researchers in CA have focused primarily on square lattices, compared to other lattice configurations such as triangular and hexagonal lattices. While the algebraic characterization and reversibility of linear rules on hexagonal [9] and triangular (in [10] and [11]) lattices have been examined, further investigation into the non-linear rule space enhance research in these lattices. A 2D triangular lattice CA with a von Neumann neighborhood of radius one generates a vast rule space of 65,536 rules. Exploring such a large space is a challenging task.

To narrow the vast rule space, we focus on rotation-symmetric rules. Rotation symmetry in CA ensures that the rules governing state transitions remain invariant under rotations of the neighborhood configuration, making them consistent across all orientations. It enhances the application of CA in studying natural phenomena with rotational invariance. The rotation symmetry in CA rules has been studied specifically for CA on the traditional square lattice. Kornyak [12], Imai et al. [13] and Dzedzej et al. [14] investigated the concept of rotation symmetry in CA on square lattices.

In this work, we consider the 256 rotation-symmetric rules, and perform the experimental analysis on their reversibility. Kari [15] presented the fundamental classical results on reversible CA. Although the reversibility of 2D CA is undecidable, this study experimentally confirms the reversibility or irreversibility of these rotation-symmetric rules, offering valuable insights into future research on triangular lattice CA.

Section 2 outlines the pre-requisites. Section 3 introduces the rotation-symm etric rules for 2D von Neumann CA on a triangular lattice and demonstrates the evolution of a fractal-forming rule. Section 4 presents the reversibility test which is applied to the rotation-symmetric rules. Section 5 discusses the conclusion and future scope of our work.

2 Pre-requisites

Before delving into the work, we provide some foundational concepts and methodologies essential for understanding the subsequent analysis.

Definition 1. *[16] A cellular automaton (CA) is stated as a quadruple*
$\mathbb{A} = (\mathbb{L}, \mathbb{Q}, \mathbb{N}, \mathbb{R})$, *where*
$\mathbb{L} \subseteq Z^m$ *is the m-dimensional lattice,*

$\mathbb{Q}$ is the finite set of states $\{0, 1\}$,
$\mathbb{N} = (\mathrm{n}_1, \mathrm{n}_2, ..., \mathrm{n}_v)$ be the set of neighborhood vectors of all distinct states of $\mathbb{L}$,
$\mathbb{R} : \mathbb{Q}^v \to \mathbb{Q}$ is the local transition rule where $\mathbb{R} = \{f | f : \mathbb{Q}^v \to \mathbb{Q}\}$.

In traditional square lattice, the neighborhood cell combination is uniform. But in triangular lattice we have to deal with two types of neighborhood configurations due to the presence of two types of triangles in the lattice. The two types are upright triangles and inverted triangles. **Fig.** 1a and **Fig.** 1b represents the von Neumann neighborhood configuration for a state of the CA to be inside the inverted and upright triangle. **Fig.** 1c and **Fig.** 1d represents the Moore neighborhood configuration for a CA state to be inside the inverted and upright triangle.

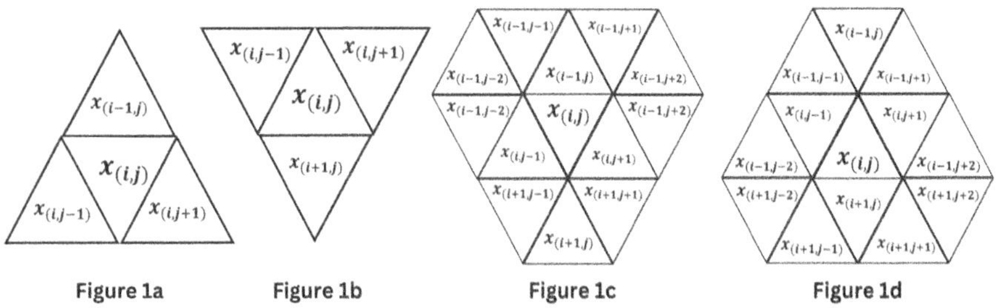

Fig. 1. Neighborhood for upright and inverted triangle

Definition 2. *[17] A configuration of an m-dimensional cellular automaton on a finite lattice with state set $\mathbb{Q}$ is defined by the mapping $\mathscr{C} : Z^m \to \mathbb{Q}$ where Z^m is a m-dimensional finite lattice and $\mathbb{Q}$ is a finite set of states.*

The state of each cell in the *m-dimensional finite lattice is mapped to a state in Q*. The configuration of a 2D CA is represented by a matrix of order $m \times n$. The configuration at time t is denoted by $\mathscr{C}^{(t)}$ where

$$\mathscr{C}^{(t)} = \begin{pmatrix} \wp_{1,1} & \wp_{1,2} & \cdots & \wp_{1,n} \\ \wp_{2,1} & \wp_{2,2} & \cdots & \wp_{2,n} \\ \cdot & \cdot & \cdots & \cdot \\ \cdot & \cdot & \cdots & \cdot \\ \cdot & \cdot & \cdots & \cdot \\ \wp_{m,1} & \wp_{m,2} & \cdots & \wp_{m,n} \end{pmatrix}_{m \times n}$$

and $\wp_{i,j}$ represents the state of each cell $x_{i,j}$ at time t.

In a triangular lattice CA, each cell has four neighbors as in **Fig.** 1a and **Fig.** 1b (von Neumann neighborhood of radius 1). We consider the states as 0 or 1. There exists 16 (i.e.,2^4) possible neighborhood configurations. Each neighborhood configuration is mapped to a state in Q using the local transition rule. The rule is in the form of a 16-bit binary string and is represented by its decimal equivalent.

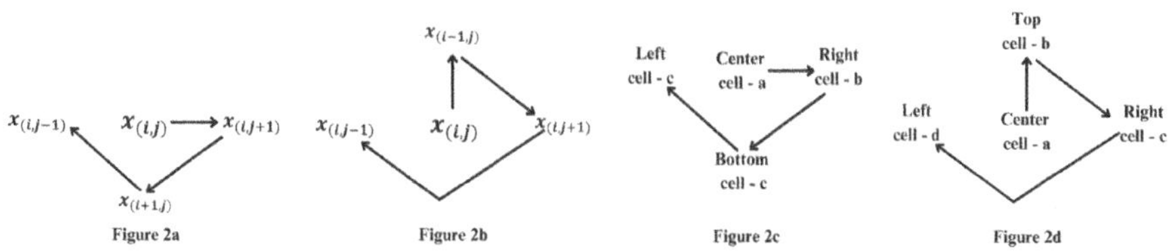

Fig. 2. Triangular lattice neighborhood orientation for Example 1

Table 1. Rule 26985

1111	1110	1101	1100	1011	1010	1001	1000	0111	0110	0101	0100	0011	0010	0001	0000
0	1	1	0	1	0	0	1	0	1	1	0	1	0	0	1

Example 1. Let us consider a 2D CA rule 26985(decimal form) with binary states and the neighborhood configuration of length 4. The binary expansion of 26985 is 0110100101101001. Each possible neighborhood configuration is mapped to a value in the string sequentially starting from 1111 to 0000 as depicted in *table* 1. The combination of binary values of length 4(For example *abcd*) is taken as a neighborhood configuration in two ways for the two types of triangles namely the upright and inverted triangles. **Fig.** 2a and **Fig.** 2c depicts the orientation of neighborhood states for a state to be inside an upright triangle. Similarly, **Fig.** 2b and **Fig.** 2d correspond to a state inside an inverted triangle.Let the initial configuration of the proposed CA at time t i.e. $\mathscr{C}^{(t)}$ be represented by the matrix of order 4×5

$$\mathscr{C}^t = \begin{pmatrix} 0\,0\,1\,0\,1 \\ 0\,1\,1\,0\,1 \\ 1\,1\,0\,1\,1 \\ 0\,0\,1\,1\,1 \end{pmatrix}$$

The states in the matrix are updated using the rule 26985.

- The cell in the position (i, j) of the matrix is considered to be inside the upright triangle when $i + j = 2n$ where n is a finite positive integer. It is updated using the rule 26985 with the neighborhood orientation mentioned in **Fig.** 2a. For x_{11}, the next state is $f(0000) = 1$ using the Rule 26985.
- The cell in the position (i, j) of the matrix is considered to be inside the inverted triangle when $i + j = 2n + 1$ where n is a finite positive integer. It is updated using the rule 26985 with the neighborhood orientation mentioned in **Fig.** 2a. For x_{12}, the next state is $f(0010) = 0$ using the Rule 26985.

Similarly the cells are updated using the rule 26985 and the resulting next step configuration at time $t + 1$ is

$$\mathscr{C}^{t+1} = \begin{pmatrix} 1\,0\,0\,1\,0 \\ 0\,1\,1\,0\,0 \\ 0\,1\,0\,0\,1 \\ 0\,0\,0\,1\,1 \end{pmatrix}$$

Fig. 3. Triangular lattice 2D CA evolution for Example 1

The space time diagram evolution of this CA is shown in **Fig.** 3.

Note: Throughout this work the CA always starts with upright triangle. (i.e. $x_{1,1}$ denotes the upright triangle)

The following definition explains the rotation-symmetric rule on the traditional square lattice (Table 1).

Definition 3. *[13] A von Neumann neighborhood cellular automaton* $\mathbb{A} = (\mathbb{L}, \mathbb{Q}, \mathbb{N}, \mathbb{R})$ *is rotation symmetric if its local transition function f is invariant under rotation. For every* $\mathtt{c}, \mathtt{u}, \mathtt{r}, \mathtt{d}, \mathtt{l} \in \mathbb{Q}$,

$$f(c, u, r, d, l) = f(c, r, d, l, u)$$

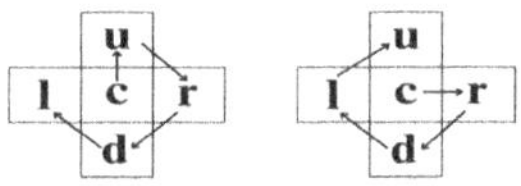

Fig. 4. Neighborhood orientation of 5-neighborhood CA on the Square lattice

Example 2. Let us consider a 2D CA rule with binary states and neighborhood configuration of length 4 for the finite square lattice. The definition 4 implies the 5-neighborhood dependency. Let the states are considered from $\{0, 1\}$. Then the transition function for all the rules satisfies the following conditions: $f(00000) = f(00000), f(00001) = f(00010), f(00010) = f(00100), f(00011) = f(00110)$ and so on. The orientation of the neighborhood configuration is depicted in **Fig.** 4.

3 Rotation Symmetry

Definition 4. *A two-dimensional von Neumann neighborhood Cellular Automaton* $\mathbb{A} = (\mathbb{L}, \mathbb{Q}, \mathbb{N}, \mathbb{R})$*, on the finite triangular lattice is said to be rotation-symmetric if its local transition function f is invariant under rotation. For every* $\mathtt{a}, \mathtt{b}, \mathtt{c}, \mathtt{d} \in \mathbb{Q}$,

$$f(\mathtt{abcd}) = f(\mathtt{acdb}) = f(\mathtt{adbc})$$

Example 3. Consider the 2D von Neumann neighborhood CA on the finite triangular lattice. The orientation of the neighborhood configuration of a rotation-symmetric CA on triangular lattice is depicted in **Fig.** 5. Consider the *Rule*

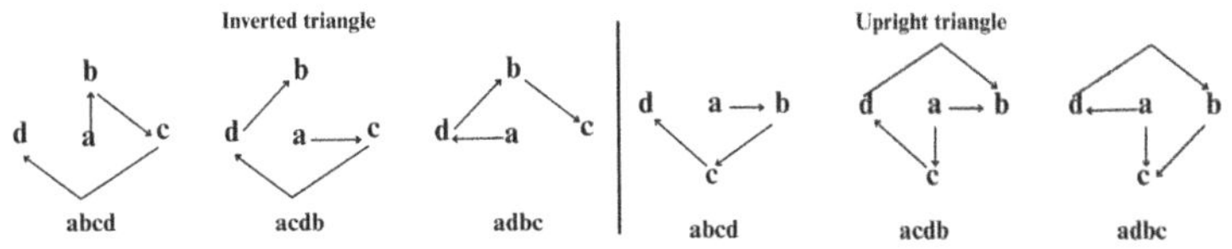

Fig. 5. Neighborhood orientation of rotation-symmetric CA on the triangular lattice

26985 from *example 1*. By applying the transition rule from *example 1* with the neighborhood dependency as in **Fig.** 2a and **Fig.** 2b, the transition function satisfies the following conditions.

$f(0001) = f(0010) = f(0100) = 0$, $f(0011) = f(0110) = f(0101) = 1$
$f(1001) = f(1010) = f(1100) = 1$, $f(1011) = f(1101) = f(1110) = 1$.
which satisfies the equality $f(abcd) = f(acdb) = f(adbc)$
Therefore, the *rule 26985* is rotation-symmetric for the above mentioned CA.

In 2D von Neumann neighborhood CA over the finite triangular lattice, the rule is in the form of a 16-bit binary string. There exists a total of 65,536 (i.e. 2^{2^4}) rules for the finite triangular lattice CA with von Neumann neighborhood of radius one. Then in the rotation-symmetric rules, out of all 16 bits in the string, only eight bits are independent and the other eight bits are dependent.
Dependency of the rule:

Consider the Wolfram number: 15 14 13 12 11 10 9 8 7 6 5 4 3 2 1 0

The positions 2 and 4 depends on 1; 5 and 6 depends on 3; 10 and 12 depends on 9; 13 and 14 depends on 11. The rules which satisfy the condition of this dependency(as mentioned above) are computed. A list of 256 rules are obtained satsifying the condition to be rotation-symmetric by considering the different combinations of the 8 independent bits. The resulting binary strings are also converted to its decimal equivalent. Table 2 represents the list of 256 rotation-symmetric rules denoted by (R_s) and are numbered from Rule 0 to Rule 255.

Through an exploration of the space time evolution of 256 rotation-symmetric rules for finite 2D CA on a triangular lattice with a von Neumann neighborhood, a novel property has been identified: the emergence of fractal (self-similar) structures under one specific rule. Notably, this is the first instance of fractal formation observed on a triangular lattice, as such behavior has previously been reported only on square lattices. This discovery is not a direct consequence of the rotation symmetry constraints but rather emerged from the systematic investigation of all such rules. The identified rule expands the understanding of fractal-generating mechanisms in CA, demonstrating that triangular lattice geometries can also serve as a substrate for these complex self-similar pattern replications. This unique behavior is illustrated in **Fig.** 6.

4 Reversibility of Rotation-Symmetric Rules

Definition 5. *[15] A cellular automaton* $\mathbb{A} = (\mathbb{L}, \mathbb{Q}, \mathbb{N}, \mathbb{R})$ *is reversible if and only if* $\mathbb{R} = \{f | f : \mathbb{Q}^v \rightarrow \mathbb{Q}\}$ *is bijective.*

Table 2. Rotation-symmetric rules(R_s)

Rule	R_s	Rule	R_s	Rule	R_s	Rule	R_s	Rule	R_s	Rule	R_s
0	0	1	1	2	22	3	23	4	104	5	105
6	126	7	127	8	128	9	129	10	150	11	151
12	232	13	233	14	254	15	255	16	256	17	257
18	278	19	279	20	360	21	361	22	382	23	383
24	384	25	385	26	406	27	407	28	488	29	489
30	510	31	511	32	5632	33	5633	34	5654	35	5655
36	5736	37	5737	38	5758	39	5759	40	5760	41	5761
42	5782	43	5783	44	5864	45	5865	46	5886	47	5887
48	5888	49	5889	50	5910	51	5911	52	5992	53	5993
54	6014	55	6015	56	6016	57	6017	58	6038	59	6039
60	6120	61	6121	62	6142	63	6143	64	26624	65	26625
66	26646	67	26647	68	26728	69	26729	70	26750	71	26751
72	26752	73	26753	74	26774	75	26775	76	26856	77	26857
78	26878	79	26879	80	26880	81	26881	82	26902	83	26903
84	26984	85	26985	86	27006	87	27007	88	27008	89	27009
90	27030	91	27031	92	27112	93	27113	94	27134	95	27135
96	32256	97	32257	98	32278	99	32279	100	32360	101	32361
102	32382	103	32383	104	32384	105	32385	106	32406	107	32407
108	32488	109	32489	110	32510	111	32511	112	32512	113	32513
114	32534	115	32535	116	32616	117	32617	118	32638	119	32639
120	32640	121	32641	122	32662	123	32663	124	32744	125	32745
126	32766	127	32767	128	32768	129	32769	130	32790	131	32791
132	32872	133	32873	134	32894	135	32895	136	32896	137	32897
138	32918	139	32919	140	33000	141	33001	142	33022	143	33023
144	33024	145	33025	146	33046	147	33047	148	33128	149	33129
150	33150	151	33151	152	33152	153	33153	154	33174	155	33175
156	33256	157	33257	158	33278	159	33279	160	38400	161	38401
162	38422	163	38423	164	38504	165	38505	166	38526	167	38527
168	38528	169	38529	170	38550	171	38551	172	38632	173	38633
174	38654	175	38655	176	38656	177	38657	178	38678	179	38679
180	38760	181	38761	182	38782	183	38783	184	38784	185	38785
186	38806	187	38807	188	38888	189	38889	190	38910	191	38911
192	59392	193	59393	194	59414	195	59415	196	59496	197	59497
198	59518	199	59519	200	59520	201	59521	202	59542	203	59543
204	59624	205	59625	206	59646	207	59647	208	59648	209	59649
210	59670	211	59671	212	59752	213	59753	214	59774	215	59775
216	59776	217	59777	218	59798	219	59799	220	59880	221	59881
222	59902	223	59903	224	65024	225	65025	226	65046	227	65047
228	65128	229	65129	230	65150	231	65151	232	65152	233	65153
234	65174	235	65175	236	65256	237	65257	238	65278	239	65279
240	65280	241	65281	242	65302	243	65303	244	65384	245	65385
246	65406	247	65407	248	65408	249	65409	250	65430	251	65431
252	65512	253	65513	254	65534	255	65535				

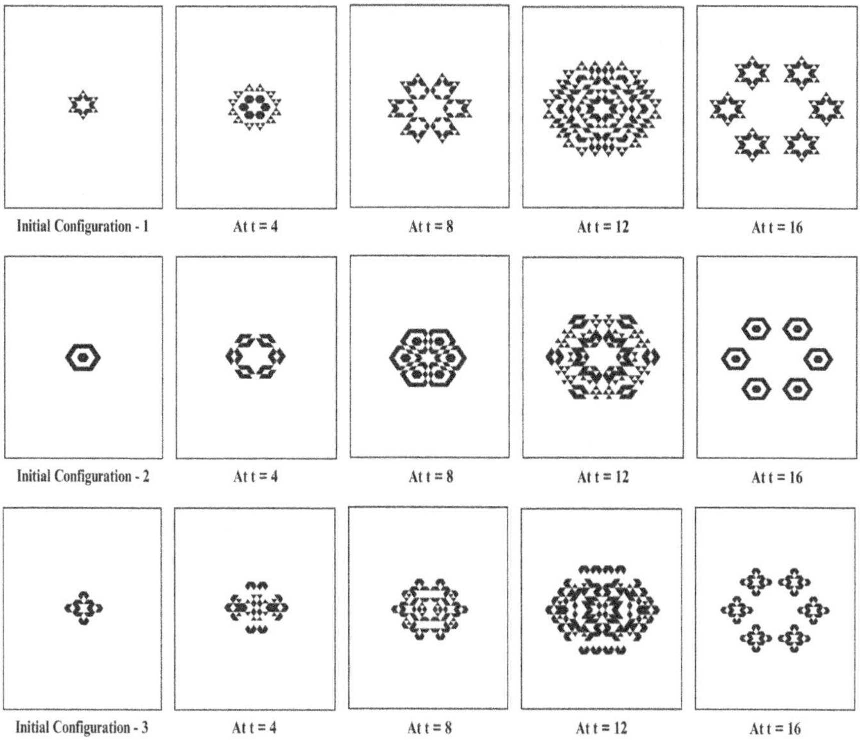

Fig. 6. Fractal formation of Rule 90 ($R_s = 27030$)

If the transition diagram is cyclic, each configuration is mapped to a unique configuration. The CA will be injective. For a 2D CA of order $m \times n$, there exist $2^{m \cdot n}$ possible configurations. For all possible configurations of the specified order, the existence of a previous configuration implies surjectivity. The rotation-symmetric rules are classified into two types in order to discuss the reversibility. The two types are linear rotation-symmetric rules and non-linear rotation-symmetric rules. $0, 90, 170, 240$ are the 4 linear rotation-symmetric rules. The rest are non-linear rotation-symmetric rules.

The reversibility of linear rules for a 2D CA on the finite triangular lattice with null boundary conditions is characterized using matrix algebra by Selman Uguz et al. [11]. A CA is said to be reversible if each configuration is mapped to a unique configuration. Let us consider a 2D CA on finite triangular lattice. The configuration of each CA can be written in a matrix form. The transition diagram is constructed for any rule and any order of the proposed CA in this work by using a method.

Steps involved in the construction of 2D CA transition diagram:

Step 1: Any 2D CA on the finite triangular lattice can be represented by a matrix of order $m \times n$. The matrix of order $m \times n$ of 2D is converted to a one-dimensional string of order $1 \times mn$.

$$\begin{pmatrix} x_{11} & x_{12} & \cdots & x_{1n} \\ x_{21} & x_{22} & \cdots & x_{2n} \\ \vdots & \vdots & \cdots & \vdots \\ x_{m1} & x_{m2} & \cdots & x_{mn} \end{pmatrix} \longrightarrow \left(x_{11} \cdots x_{1n}\, x_{21} \cdots x_{2n}\, x_{31} \cdots x_{mn}\right)$$

Step 2: The 1D matrix is then considered as a 1D string. As we use the states 0 and 1, the resulting string is a binary string. The decimal equivalent of the binary strings are computed.

Step 3: For a $m \times n$ CA there exists $2^{m \cdot n}$ possible configurations. Each possible configuration is converted using steps 1 and 2 to their decimal equivalent. The decimal equivalents are considered as the nodes. If a configuration $\mathscr{C}^t$ updates to configuration $\mathscr{C}^{(t+1)}$, an edge passes from the decimal equivalent of $\mathscr{C}^t$ to the decimal equivalent of $\mathscr{C}^{(t+1)}$. The transition diagram is constructed for a $m \times n$ CA involving all the possible configurations.

Note: A 2×2 CA have 16 possible configurations. The transition diagram will have 16 nodes and the edges are drawn based on the transition rule. For a $m \times n$ CA, there exists $2^{m \cdot n}$ possible configurations. The transition diagram of this CA must have $2^{m \cdot n}$ nodes.

Example 4 explains the construction of nodes in the transition diagram from to 2D finite configuration.

Example 4. Let us consider a 2D finite triangular lattice CA with null boundary of order 2×2. The states of the CA be represented in the form of a matrix of order 2×2. Let the matrix be (where each entry a_{ij} denotes the entry of each cell x_{ij} of the CA.

$$\begin{pmatrix} 0 & 0 \\ 0 & 1 \end{pmatrix}$$

This 2×2 matrix is converted to a one-dimensional matrix of order 1×4.

$$\begin{pmatrix} 0 & 0 \\ 0 & 1 \end{pmatrix} \longrightarrow (0\,0\,0\,1)$$

This one-dimensional matrix is converted to a one-dimensional string 0001. The decimal equivalent of 0001 is 1 which is one of the nodes of transition diagram.

In this way, the transition diagram is constructed by considering all the possible combinations. The transiton diagrams for linear rotation-symmetric rules are given below.

Rule 90

Figure 7 depicts the transition diagram for all possible 2×2 configurations by applying the rule $90(\mathtt{R_s} = 27030)$.

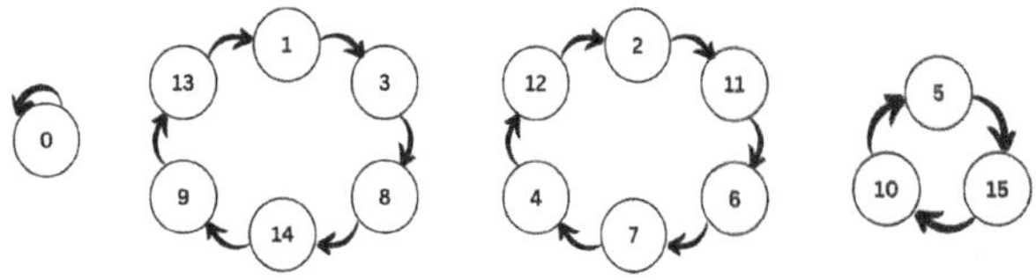

Fig. 7. State-transition diagram for 2×2 configurations for Rule 90

Rule 170

Figure 8 depicts the transition diagram for all possible 2×2 configurations by applying rule $170(\mathtt{R_s} = 38550)$.

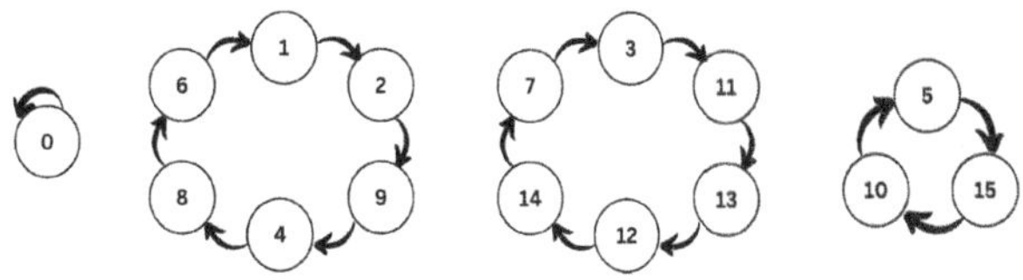

Fig. 8. State-transition diagram for 2×2 configurations for Rule 170

The transition diagram is cyclic for all $m \times n$ triangular lattice CA when n is even. Each configuration is mapped to a unique configuration out of all the possible configurations. This concept paves way for the reversibility of finite triangular lattice CA.

On the other side, if we consider a non-linear rule and the order of the CA as 2×3, the transition diagram is as follows(**Fig.** 9). The linear rules due to their reversible nature have a cyclic transition diagram. The non-linear rules will have several branches converging into a single node and nodes without pre-images.

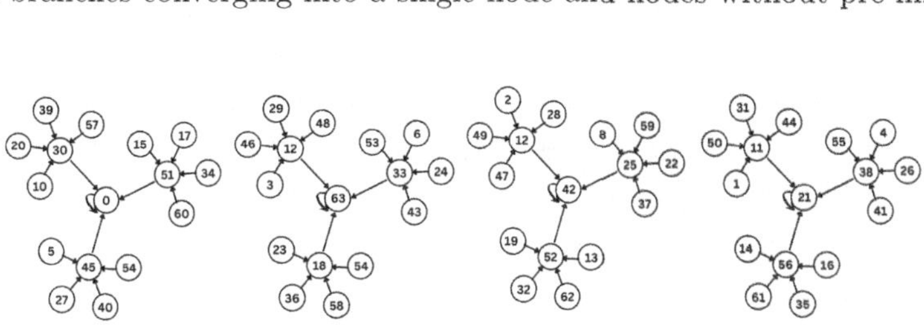

Fig. 9. State-transition diagram of 2×3 CA for *Rule* 90

4.1 On the Reversibility of Non-linear Rotation-Symmetric Rules

The non-linear rotation-symmetric rules were analyzed to determine their reversibility, starting with a 2×2 CA. Each rule was initially tested for bijectivity in the 2×2 CA. If a rule does not satisfy this bijectivity test, it will be irreversible by default. Rules that completely satisfied the bijectivity test for a 2×2 CA were further evaluated for bijectivity in 2×3, 3×2, and 3×3 configurations. The rules that fulfilled the respective tests are presented in detail (Table 3).

The linear rotation-symmetric rules are reversible for a special case only when the value of n is even for a $m \times n$ CA on the finite triangular lattice. The transition diagram (**Fig.** 9) is not cyclic and more than one configurations have same next state configuration which is an example for a non-reversible linear CA because of the order $m \times n$ where n is odd.

Table 3. Bijectivity of rotation-symmetric rules (R_s)

2×2 Bijective CA	127, 255, 5654, 5737, 5782, 26902, 32512, 32640, 32895, 33023, 38422, 38505, 38633, 59670, 59753, 59798, 59881, 65408
2×3 Bijective CA	127, 255, 32512, 32640, 32895, 33023
3×2 Bijective CA	127, 255, 32512, 32640, 32895, 33023, 38422, 38505, 38633, 59670, 59798 59881, 65408
3×3 Bijective CA	255,

Apart from linear rotation-symmetric CA rules, the only non-linear rotation-symmetric CA rule which passed all the conducted tests is the rule 255. This analysis is currently limited to configurations up to 3×3 CA. Further testing and verification are required for larger grids or higher orders to confirm if Rule 255 retain its bijective and reversible properties universally or not. Also *Definition 6* presents a function to find the previous configuration of the rule 255.

Definition 6. *In a cellular automaton with grid size $m \times n$ with $2^{m \cdot n}$ possible configurations, let z be the decimal representation of the configuration at time $t+1$. The previous configuration at time t is computed as:*

$$f^{-1}(z) = (2^{m \cdot n} - 1) - z.$$

The conversion of a matrix configuration and its decimal equivalent can be referred from *Example 4*. However, the reasons why certain rules fail to exhibit reversibility under these orders remain unclear. Developing a theoretical framework is necessary to explain this phenomenon. In future, we plan to focus on the theoretical aspects of reversibility for these non-linear, rotation-symmetric rules.

5 Conclusion

This study has provided the list of 256 rotation-symmetric rules for two-dimensional CA on the finite triangular lattice. The reversibility of these 256 rotation-symmetric rules is discussed based on the experimental results. Also presented the property of a rotation-symmetric rule to generate fractals with self-replicating patterns. Future study aims to develop a theoretical proof for this through a theoretical study of the proposed rotation-symmetric rules of 2D CA on the finite triangular lattice.

Disclosure of Interests. The authors have no competing interests to declare that are relevant to the content of this article.

References

1. Von Neumann, J., Burks, A.W.: Theory of self-reproducing automata. IEEE Trans. Neural Netw. **5**(1), 3–14 (1966)
2. Moore, E.F.: Machine models of self-reproduction. In: Proceedings of Symposia in Applied Mathematics, vol. 14, no. 5, pp. 17–33 (1962)
3. Packard, N.H., Wolfram, S.: Two-dimensional cellular automata. J. Stat. Phys. **38**(5), 901–46 (1985)
4. Kari, J.: Reversibility of 2D cellular automata is undecidable. Physica D **45**(1–3), 379–85 (1990)
5. Durand, B.: Global properties of 2D cellular automata: some complexity results. In: Borzyszkowski, A.M., Sokołowski, S. (eds.) MFCS 1993. LNCS, vol. 711, pp. 433–441. Springer, Heidelberg (1993). https://doi.org/10.1007/3-540-57182-5_35
6. Morita, K.: Universality of 8-state reversible and conservative triangular partitioned cellular automata. In: El Yacoubi, S., Was, J., Bandini, S. (eds.) ACRI 2016. LNCS, vol. 9863, pp. 45–54. Springer, Cham (2016). https://doi.org/10.1007/978-3-319-44365-2_5
7. Imai, K., Morita, K.: A computation-universal two-dimensional 8-state triangular reversible cellular automaton. Theoret. Comput. Sci. **231**(2), 181–91 (2000)
8. Zawidzki, M.: Application of semitotalistic 2D cellular automata on a triangulated 3D surface. Int. J. Des. Nat. Ecodyn. **6**(1), 34–51 (2011)
9. Siap, I., Akin, H., Uguz, S.: Structure and reversibility of 2D hexagonal cellular automata. Comput. Math. Appl. **62**(11), 4161–9 (2011)
10. Uguz, S., Acar, E., Redjepov, S.: 2D triangular von neumann cellular automata with periodic boundary. Int. J. Bifurcat. Chaos **29**(03), 1950029 (2019)
11. Uguz, S., Redjepov, S., Acar, E., Akin, H.: Structure and reversibility of 2D von neumann cellular automata over triangular lattice. Int. J. Bifurcat. Chaos **27**(06), 1750083 (2017)
12. Kornyak, V.V.: Cellular automata with symmetric local rules. In: International Workshop on Computer Algebra in Scientific Computing 2006 Sep 11, pp. 240–250 Springer Berlin Heidelberg, Berlin, Heidelberg (2006)
13. Imai, K., Ishizaka, H., Poupet, V.: 5-state rotation-symmetric number-conserving cellular automata are not strongly universal. In: Cellular Automata and Discrete Complex Systems: 20th International Workshop, AUTOMATA, pp. 31–43, Himeji, Japan, July 7-9, 2014, Revised Selected Papers 20 2015 . Springer International Publishing (2014)

14. Dzedzej, A., Wolnik, B., Nenca, A., Baetens, J.M., De Baets, B.: Two-dimensional rotation-symmetric number-conserving cellular automata. Inf. Sci. **577**, 599–621 (2021)
15. Kari, J.: Reversible cellular automata: from fundamental classical results to recent developments. N. Gener. Comput. **36**(3), 145–72 (2018)
16. Bhattacharjee, K., Naskar, N., Roy, S., Das, S.: A survey of cellular automata: types, dynamics, non-uniformity and applications. Nat. Comput. **19**, 433–61 (2020)
17. Sahin, U., Uguz, S., Akin, H., Siap, I.: Three-state von neumann cellular automata and pattern generation. Appl. Math. Model. **39**(7), 2003–2024 (2015)

An Elementary Cellular Automata Based Two-Class Data Imbalance Problem: Initial Study and Observations

Neha Kumari(✉), Suvendu Kanungo, and Jit Mukherjee

Department of Computer Science and Engineering, Birla Institute of Technology, Mesra, Ranchi, Jharkhand, India
neha92.ranchi@gmail.com

Abstract. Real-life data may exhibit skewed data distributions across various classes, which detrimentally influences the learning paradigm. In the two class problems, the number of samples in the majority class substantially exceeds the number of samples from the minority class. It may lead to the marginalization of the minority class by machine learning algorithms, i.e. creating a bias favoring the majority class. To mitigate this issue, the formulation of a strategy to balance them, is vital. Oversampling and undersampling represent two of the principal methodologies for achieving a balanced distribution. This paper proposes a novel approach using Elementary Cellular Automata (*ECA*) rules to address the data imbalance problem through oversampling of the minority points. Cellular Automata (*CA*) is a versatile and adaptable method for modeling complex patterns using simple rules and synthetic minority points are generated strategically. CA rule exhibits diverse behaviour, which can be classified into four categories. One representative rule chosen from each category is studied here. This paper shows that it can effectively address imbalances across large and complex datasets. Further, this paper discusses the strengths and limitations of these rules. The proposed technique exhibits accuracy comparable to that of classical techniques, in particular Rule 255 achieves $> 90\%$ accuracy, whereas Rules 110 and 90 achieve 90% accuracy.

Keywords: Data Imbalance · Cellular Automata Rules · Minority Class · Majority Class

1 Introduction

Class imbalance is a critical problem in machine learning, especially with binary classes where, class label distribution is substantially skewed i.e. when the majority class significantly exceed the number of samples in the minority class. Such imbalances cause classifiers to be biased towards the majority class, often overlooking the minority instances [20]. The class imbalance problem is typically of greater importance in different real-world applications such as fraud detection [1],

S. Kamilya et al. (Eds.): ASCAT 2025, CCIS 2499, pp. 57–68, 2025.
https://doi.org/10.1007/978-3-031-94121-4_5

medical diagnostics [12], sentimental analysis [24], and anomaly detection [17] where true negative cases are rare.

The work illustrated in [5] highlights the significant challenges of class imbalance by systemically analyzing data resampling methods and cost-sensitive learning. The study in [19] observed challenges such as algorithmic bias, inadequate evaluation metrics, and limited generalizability remain in data imbalance. Hence, there is still a dire requirement for developing robust algorithms for data imbalance. There can be various directions, such as exploring novel data synthesis methods, enhancing model interpretability, and developing adaptive learning algorithms to balance the classes better [7]. Specific machine learning models such as ensemble methods and resampling techniques are also finding prominence in the literature [14].

Data imbalance typically have two different avenues: undersampling and oversampling. Undersampling effectively reduces the number of instances in the majority class to match the number of instances in the minority class [6]. However, critical information often gets lost. The oversampling technique increases the size of the minority class by introducing new synthetic points. While random oversampling helps to achieve a more equal representation of classes. It may increase the risk of overfitting by replicating pre-existing positive instances by learning too much from the duplicated samples and perform poorly on new and unseen data [3]. Synthetic Minority Oversampling Technique (*SMOTE*) is a popular oversampling technique, where new synthetic minority class instances are generated by interpolating between existing minority class samples [14]. Yet, it tends to amplify the existing noise in the data [10]. Different algorithms try to overcome this limitation. Borderline SMOTE focuses on generating synthetic samples for the minority class near the decision boundary between the minority and majority classes, which are more likely to be misclassified [13,21]. In Borderline SMOTE, all minority class samples have three groups: noise (inaccurate samples near the majority class), danger (samples on class boundaries and overlap with the majority class samples), and safe (true minority class) [21]. Adaptive synthetic sampling (*ADASYN*) generates synthetic samples based on data density [11]. Oversampling techniques like *SMOTE* and *ADASYN* can reduce bias against minority classes but may add noise [18]. The work demonstrated in [21] indicates that areas of feature space with low minority class density generate more synthetic data.

Although these techniques are effective, they have several drawbacks that can limit the learning performance. These limitations necessitate exploring novel techniques to balance datasets effectively. This paper proposed a model to resolve the class imbalance problem using Elementary Cellular Automata (*ECA*). Cellular Automata (*CA*) is a simple but powerful computational model that evolved over discrete time steps based on local interactions between neighboring cells. It has profound capabilities in simulating complex systems, solving computational problems, and representing natural processes [4,22]. Its adaptability and inherent parallelism make it well-suited to various domains incuding machine learning [8]. In recent advances, *CA* has been used in various machine learning tasks such as feature generation [25], classification [23], as convolutional neural network [9], and many others. However, there is a research gap of using *CA* for the data imbalance problem. Hence, this paper focused on one-dimensional elementary *CA* Rules to generate synthetic samples for the minority class in a two-class imbalanced data. To the best of the authors knowledge, elementary cellular automata rules have not been applied in data imbalance problem in the literature.

A critical aspect of this paper is not just developing a working algorithm, but understanding the strengths, limitations, and underlying reasons of elementary cellular automata rules for the data imbalance problem. The proposed technique provides a bridge between elementary cellular automata and theoretical machine learning through which, it generates a robust technique where one dimensional elementary cellular automata rules can be applied to multi-dimensional spaces of machine learning.

2 Dataset

We have generated synthetic datasets with a specified number of total points and the imbalance ratio. The minority points are sampled from a normal distribution centered at 0 with a standard deviation of 1. The majority points are generated from a normal distribution that is shifted from the origin, typically centered around a point 2 with the same standard deviation, but adjusted by a shift of 0.5. This shift in the centre of distribution creates a separation between the majority and minority classes in the feature space, with the majority point forming a cluster in a different region. This arrangement reflects an imbalanced data set. Similarly, we have also experimented with doughnut and spiral shape overlaps. The other dataset is a real-world dataset, namely the spambase dataset, which is collected from UCI Machine Learning Repository. It consists of 39.4% of spam emails, and non-spam emails for 60.6%. It contains 58 attributes in total, with 57 continuous attributes and one class label.

3 Preliminaries: Elementary Cellular Automata

A Cellular Automata (CA) is an autonomous finite state machine (FSM) that operates in a discrete time and space. Each cell in a CA updates its state at each time step, based on a defined set of rules [22]. *CA* rules exhibit diverse behaviors, broadly classified into four categories , i.e. Class 1, Class 2, Class 3 and Class 4. They offer insight into how different CA rules influence system dynamics and pattern formation [16]. In class 1, CAs progress towards a stable, homogeneous state where all cells ultimately take on the same value, such as all 0*s* or all 1*s*. As an example, Rule 255 produces a uniform final configuration with no internal variation. Rules in Class 2, generate repeating patterns that evolve in predictable cycles. They forms stable, oscillatory structures that repeat over time such as Rules 90 and 50. In Class 3, *CA*s display chaotic or random behavior. Rules 30 and 22 fall under this category. In Class 4, *CA*s operate at the edge of chaos or complex, producing a mix of ordered and chaotic behavior. Rule 110 and the Game of Life, generate such complex structures. They are known for their ability to model and mimic complex systems due to their balance between stability and unpredictability. In this work, we utilize four distinct CA rules from each class - Rule 30, Rule 90, Rule 110, and Rule 255 - to study their pattern and limitations in data imbalance problems.

4 Methodology

The rules described above can be applied to one dimensional cellular automata.[1] One of the contributions of this work is the process of using these rules in the multidimensional space of machine learning to generate new points. These rules decide whether a new synthetic point should be created or not at a certain position. Let the majority class be denoted as 0 (off) and the minority class be described as 1 (on). Here, we are considering the ideal balance as 0.9, i.e. the target ratio of the minority and majority class. The proposed technique goes through all the points having the marker 1. For each such point, the algorithm computes the k nearest neighbors of a minority point in the Euclidean space. Here, $k = 8$. Let, this set of points be N. N can have different markers such as 0 or 1. The algorithm accumulates the number of 1, i.e. the points of minority class in the neighborhood. Let it be Ω. For each neighbor $n \in N$ the accumulation is computed as shown in Eq 1.

$$\Omega = \sum_{n \in N} t(n = 1) \tag{1}$$

Here $t(n = 1)$ is an indicator function which provides 1 when $n = 1$, and 0 when $n = 0$. Thus, it defines the distributions of the minority class points in the neighborhood. A minority class is an outlier if $\Omega = 0$, i.e. it does not have any minority point neighbor. Hence, generating new synthetic points in this position may hamper the distribution of the minority class. Thus, new synthetic points are generated based on CA rules when $1 \leq \Omega \leq 8$. Ω is the count of neighbors in the neighborhood that belong to the minority class. The position of an elementary one dimensional cellular automata rule can be represented by $S = b_2 b_1 b_0 | b_2, b_1, b_0 \in [0, 1]$. Every rule has a resultant value of either 0 or 1 in each combination of $b_2 b_1 b_0$. We further map each value of Ω to $b_2 b_1 b_0$ such that the cellular automata rules can be applied in the multidimensional space.

$$\Omega = 4 \times b_2 + 2 \times b_1 + b_0 + 1 \quad \text{when} \quad \Omega \neq 0 \tag{2}$$

Let f be a function which provides 0 or 1 based on a CA rule. New synthetic points are generated when the CA rule provides 1 given the mapping of Ω.

$$\text{Synthetic Points} = \begin{cases} \text{Generate} & \text{if } f(\Omega - 1) = 1, \\ \text{Don't Generate} & \text{if } f(\Omega - 1) = 0. \end{cases} \tag{3}$$

[1] A segment of this work is a part of a project done in Indian Summer School on Cellular Automata 2024.

This provides an opportunity to generate synthetic points based on any elementary CA rule. As an example, the Rule 90 can be defined as shown in Table 1. The first row is the representation of $\Omega - 1$. The second and third rows define Rule 90. We have computed the 8 nearest neighbors in the Euclidean space for each minority point. Among them we accumulate the number of minority points. If $\Omega = 3$, three neighbors belong to the minority class while the rest five neighbors belong to the majority class. It denotes the binary expression of Ω -1, i.e. 2. We map the current state in binary representation $(b_2 b_1 b_0)$, such as the binary representation of 2 - 010. Further, Rule 90 is applied. In Rule, the next state is defined as 0 as per the rule table. Hence, no synthetic points will be generated. If it is 1, synthetic points will be generated.

Table 1. Truth Table for Rule 90: for each 3-cell neighborhood configuration

$\Omega - 1$	7	6	5	4	3	2	1	0
Binary $(b_2 b_1 b_0)$	111	110	101	100	011	010	001	000
Next State	0	1	0	1	1	0	1	0

Normal, Binomial, and Chi-square distributions are used to generate synthetic points. Let X_a be the synthetic point selected based on the *CA* rule. A new synthetic instance X_{new} is generated around X_a by applying normal, binomial or chi-square distributions. First, the average and variance from the minority point and their minority neighbors are computed. Using normal distribution $\mathcal{N}(\mu, \sigma^2)$, m number of random points are generated such as $X_{new} \sim \mathcal{N}(\mu, \sigma^2)$. Here, m is defined as $m = \Omega \times 5$. Similarly, m number of synthetic points are generated from chi-square distribution by $X_{new} \sim \chi^2(\delta)$, where degree of freedom (δ) is defined as 2. Likewise, for binomial distribution, number of trials (t) and probability of success (p) are chosen as 10, and 0.5, respectively. Further, m random points are generated by $X_{new} \sim \text{Binomial}(t, p)$.

5 Result and Discussion

In this work, rules from *ECA* are considered. *CA* rules can be classified into four categories. One representative rule is chosen from each category as detailed in Sect. 3. They are applied to the minority data points to study their characteristics. We employ random forest as the base classifier for our experiment due to its efficiency and reliability. In the spambase dataset, the proposed technique provides a similar outcome, i.e. 95%, with SMOTE, Borderline SMOTE and ADASYN. In the Synthetic Dataset, initially the majority to minority ratio is taken as 0.25 for experimentation. The performance of the proposed approach with different elementary *CA* rules and different probability distributions are shown in Fig. 1a. The accuracy of traditional techniques such as *SMOTE*, and borderline *SMOTE* are shown in Fig. 1b.

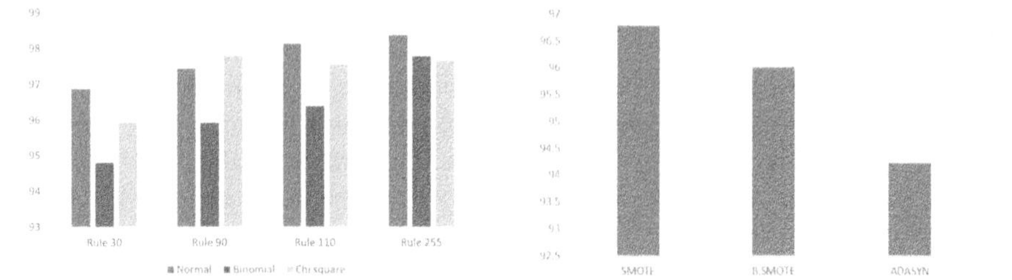

(a) Accuracy of the Proposed Approach in Different Rules and Distributions

(b) Accuracy by Traditional Approaches: SMOTE, Borderline SMOTE, ADASYN

Fig. 1. Accuracy Comparison

- Rule 30 with normal, binomial and chi-square distribution framework, achieved accuracy of $96.87\%, 94.79\%$, and 95.94% respectively.
- Rule 90 employing the framework of distributions, achieved accuracy of $97.44\%, 95.92\%$, and 97.79% respectively.
- Rule 110 with the distribution framework, achieved accuracy of $98.13\%, 96.38\%$, and 97.55% respectively.
- The assessment of rule 255 achieved accuracy of $98.37\%, 97.78\%$, and 97.67%.

To further comprehend the outcome of different rules, scatter plots are studied as shown in Fig. 2. Figure 2a shows the initial distribution of the majority and minority class. It can be observed that the majority and minority classes have a overlap. Hence, a clear decision boundary is difficult to fathom. There are a few points of each class which belong to the distribution of other class, i.e. outliers. They do not have any nearest neighbor of the same class in their vicinity. Such minority point where $\Omega = 0$, is not considered (Sect. 4). Hence, a minority point, which has at least one minority point neighbor, will be used for synthetic point generation. This guarantees that synthetic points are not generated by an outlier minority point or noise. Synthetic points will be generated for a minority point having only one minority neighbor if the rule is odd. Rule 90 extends the minority points when it has $2, 4, 5$, or 7 numbers of minority neighbors (Eq: 2). Similarly, Rule 110 extends the minority points when $\Omega = 2, 3, 4, 6$, or 7 (Eq: 2). These rules have a balanced combination of point generation, where a minority point is effectively included. Hence, these rules extend the minority class in the decision boundary and also at those points where density of minority points are higher. Rule 255 generates new synthetic points for every minority point if it has at least one minority neighbor. It provides better outcome among all the rules. However, it may create more density where the distribution of the minority points can be easily modeled by any machine learning technique, whereas it may ignore decision boundary region. In an imbalance problem, where weightage at the decision boundary is required for better outcome of a model, Rule 255 may not be found adequate. Further, as shown in Fig. 2d, it also creates synthetic points at the outer boundary of the minority class. As it considers

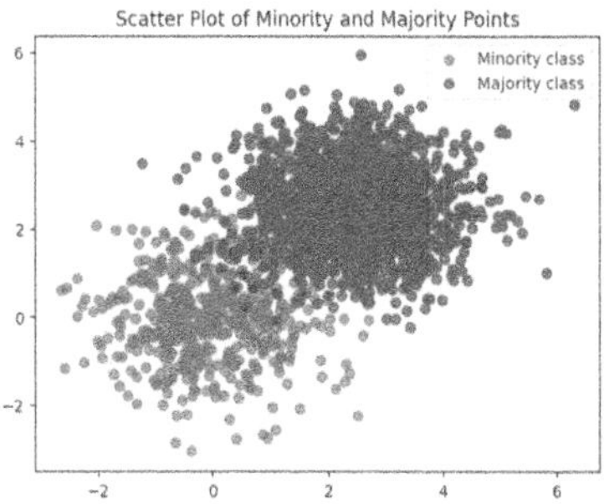

(a) Scatter Plot of Initial Majority (Blue) and Minority (Red) Points

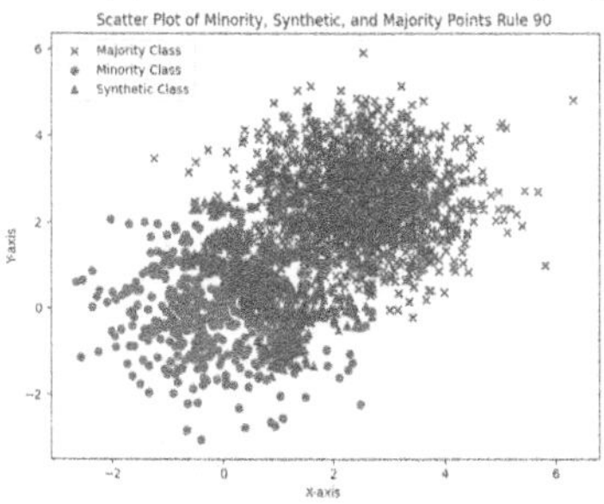

(b) Rule 90 in Normal distribution

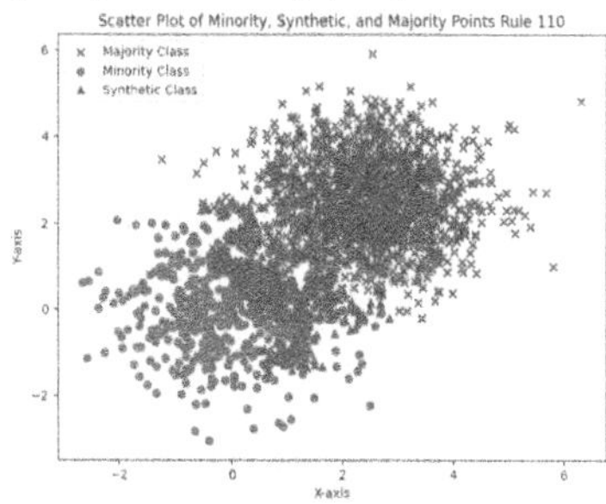

(c) Rule 110 in Normal distribution

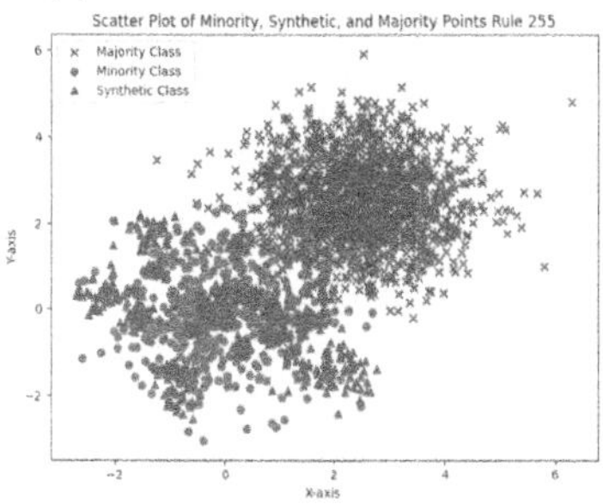

(d) Rule 255 in Normal distribution

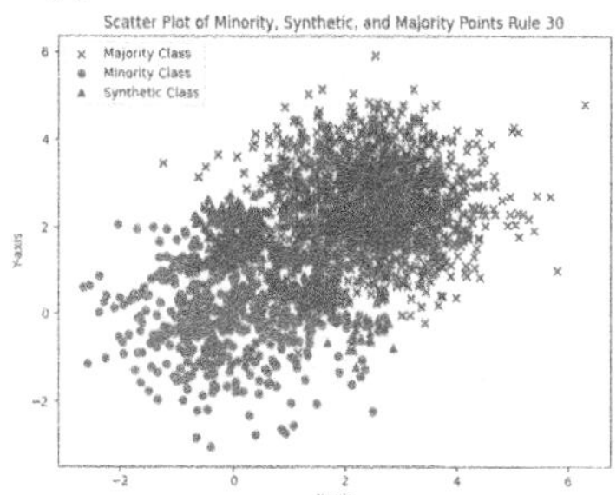

(e) Rule 30 in Normal distribution

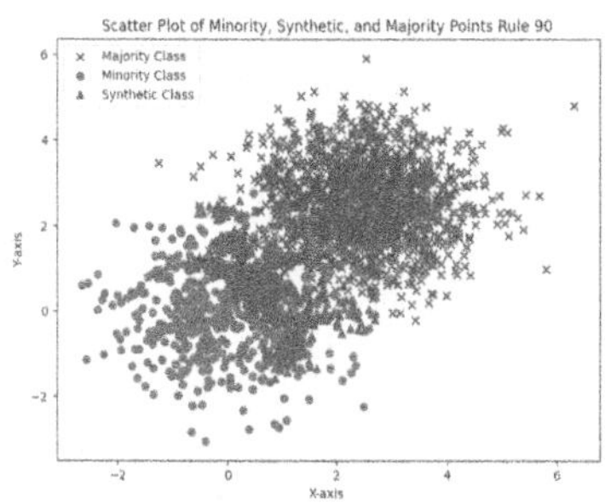

(f) Rule 30 in Chi-Square distribution

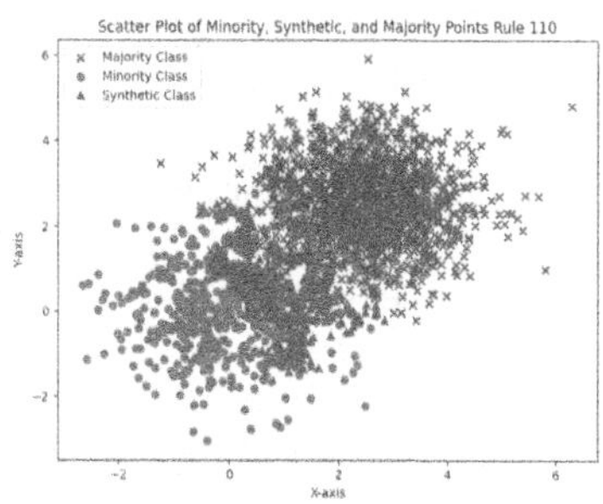

(g) Rule 30 in Binomial distribution

Fig. 2. Scatter Plots; Blue: Majority, Red: Minority, and Green: Synthetic Points. (Color figure online)

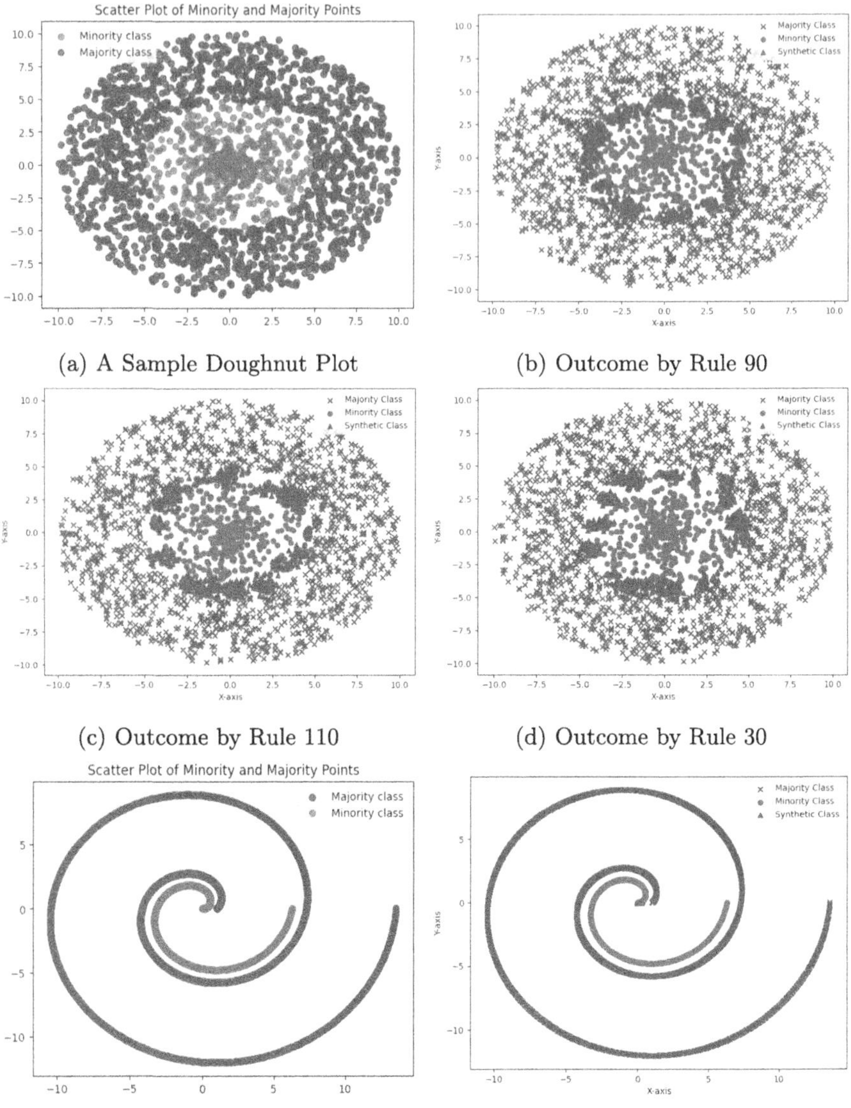

(a) A Sample Doughnut Plot

(b) Outcome by Rule 90

(c) Outcome by Rule 110

(d) Outcome by Rule 30

(e) Spiral Distribution

(f) Outcome by Rule 255

Fig. 3. Scatter Plot With Different Overlaps of Minority and Majority Classes

most of the neighbors to create new synthetic points, Rule 255 or any higher rules will be faster. New synthetic points will be generated using Rule 30, only when $\Omega = [2, 5]$ (Eq. 2). Hence, it requires at least a few majority point in the neighborhood to create a synthetic point. Hence, new synthetic points are created at the borderline of the majority and minority class (Fig. 2e). No syn-

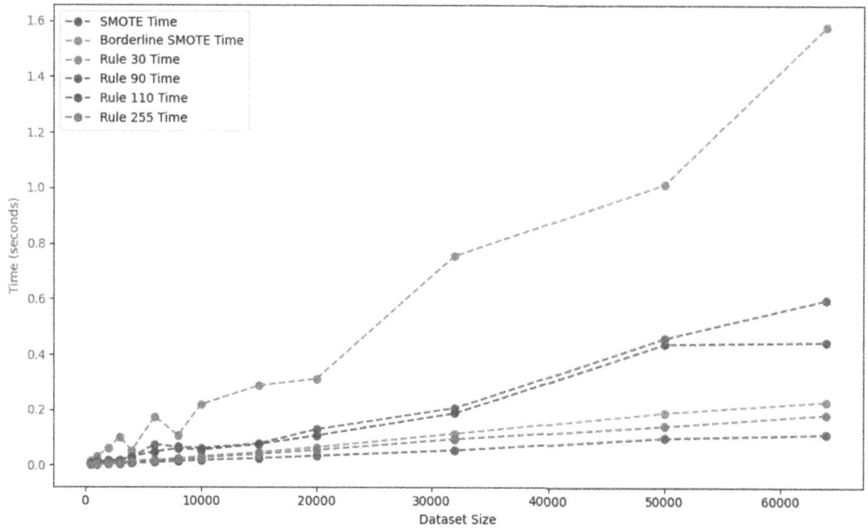

Fig. 4. Execution Time for Propose Method and Traditional Techniques

thetic points are generated where the density of minority class is higher. The behavior may closely reciprocate the behavior of Borderline SMOTE at the decision boundary. Further experimentation regarding this, is considered as a future work. We are using iterative process until the desired ratio of 0.9 is achieved. As lower rules such as Rule 30 considers less number of minority points to generate synthetic points and new synthetic points are included in every iteration, there is high chance that the new synthetic points may intrude into the distribution of majority class and take more time to achieve the balance. (Fig. 2e). As Rule 90 and 110 has more generating points, it generates new synthetic points at the decision boundary, with less intrusion towards the majority distribution as shown in Fig. 2b and Fig. 2c . However, the collection of points changes based on different distributions as shown in Fig. 2f and 2g. It is shown in Fig. 1a that chi-square and binomial distribution in Rule 30 have shown poorer outcome. The normal distribution is used for evenly spread data around a central mean, suitable for datasets with a balanced spread. Whereas, binomial distribution applies discrete, clustered data points, useful for datasets with inherent grouping tendencies [15]. The chi-square distribution is skewed, capturing minority points concentrated near lower values [2]. Hence, they may provide marginally poorer outcome than normal distribution. It needs further experimentation which is treated as a future work.

Figure 3 shows outcome of Rule 90 and 110 with different shape of overlaps in majority and minority class. It corroborates with our previous observations (Fig. 3b, 3c, 3d). The proposed *CA* rule based techniques shows similar or marginal improvement in accuracy than *SMOTE*, and Borderline *SMOTE* for the overlap shown in Fig. 3a. Further, the study also finds that lower rules may intrude into the majority class due to the iterative nature of the algorithm. Furthermore, a rule with very high number may produce better outcome, however, they may generate more synthetic point where a model can easily understand the pattern. Hence, an even rule with an optimal number of generating point such as Rule 90 or 110 can be more suitable for the two class data imbalance problem. However, these rules may never reach to the target ratio of 0.9 for some special overlap of minority and majority point such as Fig. 3e. In such a case, higher rules (> 200) may become better choice as shown in Fig. 3f. Further experimentation regarding this is considered as a future work. Next, the execution time of the traditional techiques and proposed method are also checked for different sizes of data (Fig. 4). It can be observed that higher rules, Rule 110 and 255 take less execution time. In a few cases, it performs better than borderline *SMOTE*. However, as it should be, lower rules take more time. Here, we are generating new data based on minority points as $\Omega \times 5$. If we increase the value of 5 it will enhance the data generation process. It is observed that higher *CA* rules perform with similar and sometimes better execution time than the traditional techniques. The proposed technique tries to take decision based on the local distribution of minority classes, which increases the execution time. Any technique that studies the decision boundary, like borderline *SMOTE* has higher times (Fig. 4, [13]).

6 Conclusion

Class imbalance is a critical issue for obtaining a balanced dataset. Robust Strategies need to be implemented to address the imbalance and to enhance the model performance. This work presents a novel technique by which the elementary cellular automata rule can be applied to the multidimensional feature space to resolve the class imbalance. This paper uses synthetic and real imbalanced datasets with the traditional methods. It can be found that the proposed methodology effectively works with the even number cellular automata rule in contrast to the odd number rule. It also found that lower rules may intrude into the majority class due to the iterative nature of the algorithm, whereas higher rules may produce better outcomes. However, they may generate more synthetic points where a model can easily understand the pattern. Hence, rules 90 and 110 can be more suitable for the class imbalance problem. The proposed methodology highlights several areas for further exploration. For instance, robust experimentation with different distributions and analysis of their performance may provide significant insights. The multi-class classification problem involves categorizing data into more than two classes. For instance, in a multi-class problem with three classes i.e. 0, 1 and 2, where class 0 is the majority, class 1 is the minority and class 2 is moderately imbalanced. This scenario will be explored using *CA* rules in future. Future research could explore the application of *CA* rules in dynamic datasets, where features evolve over time, or the integration of *CA*-based synthetic sampling techniques with deep learning models.

References

1. Ali, H., et al.: A review on data preprocessing methods for class imbalance problem. Int. J. Eng. Technol. **8**(3), 390–397 (2019)
2. Arnold, B.C., Lin, G.D.: Characterizations of the skew-normal and generalized chi distributions. Sankhyā Indian J. Stat. 593–606 (2004)
3. Batista, G.E., Prati, R.C., Monard, M.C.: A study of the behavior of several methods for balancing machine learning training data. ACM SIGKDD Explor. Newsl. **6**(1), 20–29 (2004)
4. Bhattacharjee, K., Naskar, N., Roy, S., Das, S.: A survey of cellular automata: types, dynamics, non-uniformity and applications. Nat. Comput. **19**, 433–461 (2020)
5. Chen, W., Yang, K., Yu, Z., Shi, Y., Chen, C.: A survey on imbalanced learning: latest research, applications and future directions. Artif. Intell. Rev. **57**(6), 1–51 (2024)
6. Das, S.: A new technique for classification method with imbalanced training data. Int. J. Inf. Technol. **16**(4), 2177–2185 (2024)
7. Dube, L., Verster, T.: Enhancing classification performance in imbalanced datasets: a comparative analysis of machine learning models. Data Sci. Finance Econ. **3**(4), 354–379 (2023)
8. Ghosh, M., Kumar, R., Saha, M., Sikdar, B.K.: Cellular automata and its applications. In: 2018 IEEE International Conference on Automatic Control and Intelligent Systems (I2CACIS). pp. 52–56. IEEE (2018)
9. Gilpin, W.: Cellular automata as convolutional neural networks. Phys. Rev. E **100**(3), 032402 (2019)
10. Gosain, A., Sardana, S.: Handling class imbalance problem using oversampling techniques: a review. In: 2017 International Conference on Advances in Computing, Communications and Informatics (ICACCI). pp. 79–85. IEEE (2017)
11. He, H., Bai, Y., Garcia, E.A., Li, S.: Adasyn: adaptive synthetic sampling approach for imbalanced learning. In: 2008 IEEE International Joint Conference on Neural Networks (IEEE world congress on computational intelligence). pp. 1322–1328. IEEE (2008)
12. He, H., Garcia, E.A.: Learning from imbalanced data. IEEE Trans. Knowl. Data Eng. **21**(9), 1263–1284 (2009)
13. Japkowicz, N., Stephen, S.: The class imbalance problem: a systematic study. Intell. Data Anal. **6**(5), 429–449 (2002)
14. Li, J., Zhu, Q., Wu, Q., Fan, Z.: A novel oversampling technique for class-imbalanced learning based on smote and natural neighbors. Inf. Sci. **565**, 438–455 (2021)
15. Malakar, I.: Theorizing probability distribution in applied statistics. J. Popul. Dev. **1**(1), 79–95 (2020)
16. Mitchell, M., et al.: Computation in cellular automata: a selected review. Non-Stan. Comput. 95–140 (2005)
17. Ogwueleka, F.N.: Data mining application in credit card fraud detection system. J. Eng. Sci. Technol. **6**(3), 311–322 (2011)
18. Piyadasa, T.D., Gunawardana, K.: A review on oversampling techniques for solving the data imbalance problem in classification. Int. J. Adv. ICT for Emerg. Reg. **16**(1) (2023)
19. Rekha, G., Tyagi, A.K., Sreenath, N., Mishra, S.: Class imbalanced data: open issues and future research directions. In: 2021 International Conference on Computer Communication and Informatics (ICCCI). pp. 1–6. IEEE (2021)

20. Rezvani, S., Wang, X.: A broad review on class imbalance learning techniques. Appl. Soft Comput. **143**, 110415 (2023)
21. Sun, Y., et al.: Borderline smote algorithm and feature selection-based network anomalies detection strategy. Energies **15**(13), 4751 (2022)
22. Wolfram, S.: Statistical mechanics of cellular automata. Rev. Mod. Phys. **55**(3), 601 (1983)
23. Wongthanavasu, S., Ponkaew, J.: A cellular automata-based learning method for classification. Expert Syst. Appl. **49**, 99–111 (2016)
24. Wosiak, A., Karbowiak, S.: Preprocessing compensation techniques for improved classification of imbalanced medical datasets. In: 2017 Federated Conference on Computer Science and Information Systems (FedCSIS). pp. 203–211. IEEE (2017)
25. Yilmaz, O.: Machine learning using cellular automata based feature expansion and reservoir computing. J. Cell. Automata **10** (2015)

Search of 3-State Cellular Automata for Clustering

Sukanya Mukherjee(✉) and Arka Prabha Acharyya

Department of Computer Science and Engineering, Institute of Engineering and Management, Kolkata, University of Engineering and Management, Kolkata, Kolkata 700091, West Bengal, India
sukanyaiem@gmail.com

Abstract. This work finds a special category of 3-state 3-neighborhood cellular automata that produce only *multi length* cycles and are named as *strictly multi length cycle* cellular automata. These automata show their efficacy for *clustering* problem by a comparative study with existing benchmark clustering algorithms on real datasets. In this work, *rule min terms* (RMTs) play an instrumental role for the following purposes: i) to generate strictly multi length cycle cellular automata, this work figures out that the *rules* should maintain a *minimum* proportion of *non self-replicating* RMTs; ii) a *valid* set of non self-replicating RMTs ensures the generation of strictly multi length cycle cellular automaton only and iii) the rules of strictly multi length cycle cellular automata maintain *different* state values for almost all *sibling* RMTs and *equivalent* RMTs to generate the *optimal* number of clusters (cycles) and maintain less *intra-cluster* distance (*close* configurations are in same cluster) which are the key properties for clustering. Additionally, this work finds *non-strictly multi length cycle* cellular automaton which guarantees the generation of *at least one* multi length cycle with *minimum* number of non self-replicating RMTs.

Keywords: cellular automata (CAs) · multi length cycle · strictly multi length cycle CA · non self-replicating RMTs · clustering

1 Introduction

Cellular Automaton (CA) is a discrete mathematical tool that consists of a grid of *cells*. The *state* of each cell evolves based on the states of its *neighbors* according to the given *rule*. A *finite* CA consisting of a finite number of cells always forms a *cycle* of finite *length*. The length of the cycle is the number of configuration(s) present in it. A cycle involving only one configuration is called *single length cycle*; otherwise the cycle is *multi length* cycle. A CA which generates only single length cycle(s) is called *single length cycle cellular automaton*.

This work is carried out as a project in the Indian Summer School on Cellular Automata 2024.

S. Kamilya et al. (Eds.): ASCAT 2025, CCIS 2499, pp. 69–85, 2025.
https://doi.org/10.1007/978-3-031-94121-4_6

CA cycles have immense contributions in technological advancements [3, 6,17,19]; for example, in machine learning, CAs with single length cycles are used as pattern classifier [10,17,21]. Specially designed CAs, with *large length* cycles which ensures *optimal* number of *clusters* (CA cycles) and designed by the rules with moderate number of *self-replicating rule min terms* for maintaining less intra-cluster distance [1,11–14,16,20], are useful for clustering problem. CAs with single length cycles are used for *convergence* property analysis [2,7,15,18,19,23]. In general, single length cycle CAs are *irreversible* (with the exception of one scenario) where some configurations are *acyclic* (not belonging to any cycle). *Reversible* CAs (where all configurations belong to some cycles) find application in clustering [1,11,13,14,16] due to the fact that every cycle maintains *closeness* among the respective *reachable* configurations and obviously, an *effective* clustering maintains *close* objects in the same cluster. Clearly, large length cycles, which are also multi length cycles, are effective for clustering. Our objective in this paper is to search for rules that generate only multi length cycles for every arbitrary size of automaton. We call such rules *strictly multi length cycle* rules and the corresponding CAs are *strictly multi length cycle* CAs. Note that such CAs can be either reversible or irreversible. The present paper generalizes the work of applying reversible CAs for clustering [13] by extending the state count from 2 to 3 and considering strictly multi length cycle CAs.

The transition between any two consecutive configurations in a CA is governed by a *sequence* of rule min terms (RMTs). An RMT is the combination of the present states of a cell and its neighbor(s). For 3-state 3-neighborhood CA, there are $3^3 = 27$ possible RMTs. The configuration belonging to a single length cycle transits to itself; this indicates the presence of an RMT sequence where every RMT is *self-replicating*. An RMT xyz is self-replicating if there is no change in the next state i.e., next state remains y. Recall that our objective is to generate 3-state CA with no single length cycle, therefore, the rule needs to be designed in such a manner that for every possible RMT sequence, at least *one* RMT is *non self-replicating*. A trivial solution is to ensure that all the 27 RMTs are non self-replicating. Thus we can get $2^9.2^9.2^9 = 2^{27}$ strictly multi length cycle rules; however, these rules are not useful for clustering due to generation of large number of cycles leading to similar objects put in different clusters. On the other hand, if a negligible amount of impurity is added (for example, only RMT 000 is self-replicating), then the generation of only multi length is not guaranteed in a CA (configuration 0^n forms single length cycle where n is the number of cells in a CA). The following observations are also interesting: the rule 022011100100101201202201201 with 52% non self-replicating RMTs does not generate strictly multi length cycle CA whereas rule 102202222002101201001111220 with 41% non self-replicating RMTs generates. These observations lead to the following questions: do there exist *valid* sets of non self-replicating RMTs which ensure the generation of strictly multi length cycle CAs? If so, what is the *minimal* set? Also, for clustering, we need to find a set of 3-state 3-neighborhood rules that generates large length cycles. The study also raises question on the design strategy of rules that ensure the existence of a multi length cycle in a non-strictly multi length cycle CA.

2 3-State 3-Neighborhood Cellular Automata

This work uses 1-dimensional *finite* cellular automaton which is represented as an array of $n \in \mathbb{N}$ cells where the leftmost and rightmost cells are numbered as 0 and $n-1$ respectively. This n signifies the *size* of the CA and the corresponding automaton is called as n-cell CA. Each cell has a *state* and as this work uses 3-state CA, therefore, the cells' states are from the set $S = \{0, 1, 2\}$. The sequence of states from cell 0 to cell $n-1$ refers to a *configuration* of an n-cell CA. Suppose $x_0, x_1, \cdots, x_{n-1}$ be the states of cells $0, 1, \cdots, n-1$ respectively at a particular time step; therefore, $(x_0 x_1 \cdots x_{n-1})$ refers to a *configuration*. Thus, a configuration is essentially a *ternary* string of length n. Here the CA maintains *synchronous update*; meaning all CA cells together update their state values at every time step and finds the *next configuration* from the current configuration.

As this work uses 3-neighborhood architecture, therefore, the state of every cell is influenced by its *three* neighbors: the cell *itself* (cell i), its *left* neighbor (cell $i-1$) and its *right* neighbor (cell $i+1$). Let x_{i-1}, x_i and x_{i+1} be the *present* states of cells $i-1$, i and $i+1$ respectively. Therefore, y_i which is the *next state* of cell i is evaluated as follows: $y_i = \mathcal{R}(x_{i-1}, x_i, x_{i+1})$ where $\mathcal{R}$ is the *local transition* function. In this work, $\mathcal{R} : \{0, 1, 2\}^3 \rightarrow \{0, 1, 2\}$ and we get 3^{27} next state functions. These functions are called as *rules* also. The sequence of present states of a cell's neighbors is called as *Rule Min Term* (RMT); an RMT can also be represented as the *decimal* equivalent of that sequence. Let r be an RMT; in this work, $r = x_{i-1}x_i x_{i+1}$ is a ternary string of length 3 (can alternatively be $3^2.x_{i-1} + 3^1.x_i + 3^0.x_{i+1}$). Therefore, we get $3^3 = 27$ RMTs. Let $\mathcal{R}(x_{i-1}, x_i, x_{i+1})$ or $\mathcal{R}[r]$ denote the *value* at RMT r and it can be either 0 or 1 or 2. Table 1 shows the RMTs and (a very few) rules of 3-state 3-neighborhood cellular automata. For example, the value at RMT 212 (23) is 2 for the rule at row 2 of Table 1. Without abuse of notation, we also represent $\mathcal{R}$ as a ternary string of length 27 given by $\mathcal{R}(2, 2, 2)\mathcal{R}(2, 2, 1) \cdots \mathcal{R}(0, 0, 0)$. The left and right neighbors of cell 0 and $n-1$ respectively should be assigned with some *boundary values*. In this study, both *null* and *periodic* boundary CAs are considered. For periodic boundary CA, $x_{-1} = x_{n-1}$ and $x_n = x_0$. Therefore, every cell of 3-state 3-neighborhood periodic boundary CA considers all *twenty* seven RMTs as *valid*. In null boundary CA, $x_{-1} = x_{n-1} = 0$. Therefore, cells 0 and $n-1$ consider two different sets of *valid* RMTs: cell 0 uses RMTs from $\{000(0), 001(1), 002(2), 010(3), 011(4), 012(5), 020(6), 021(7), 022(8)\}$ whereas $\{0\ 00(0), 010(3), 020(6), 100(9), 110(12), 120(15), 200(18), 210(21), 220(24)\}$ is considered by cell $n-1$.

This work uses n-cell *uniform* CA where the same rule $\mathcal{R}$ is used by every cell. A cycle is a set of configurations where the members are reachable from one another using state transitions. Any configuration on a cycle is called *cyclic* configuration; otherwise the configuration is *acyclic*. A cycle is called *single length* cycle when its length is one; otherwise, it is *multi length* cycle. A cycle is said

Table 1. In this table, RMTs are represented in both the forms: 3-bit ternary strings and their decimal equivalents. NS is the *next state* or *value* at the RMT.

RMT	222	221	220	212	211	210	202	201	200	122	121	120	112	111	110	102	101	100	022	021	020	012	011	010	002	001	000
	(26)	(25)	(24)	(23)	(22)	(21)	(20)	(19)	(18)	(17)	(16)	(15)	(14)	(13)	(12)	(11)	(10)	(9)	(8)	(7)	(6)	(5)	(4)	(3)	(2)	(1)	(0)
	0	2	2	0	1	1	1	0	0	1	0	0	1	0	1	2	0	1	2	0	2	2	0	1	2	0	1
NS	1	0	2	2	0	2	2	2	2	0	0	2	1	0	1	2	0	1	0	0	1	1	1	1	2	2	0
	2	2	2	1	1	1	0	0	0	2	0	2	1	1	1	0	2	0	2	2	2	1	1	1	0	0	0
	2	2	2	1	1	1	0	0	0	2	0	2	1	1	1	0	0	0	2	2	2	1	2	1	0	0	0

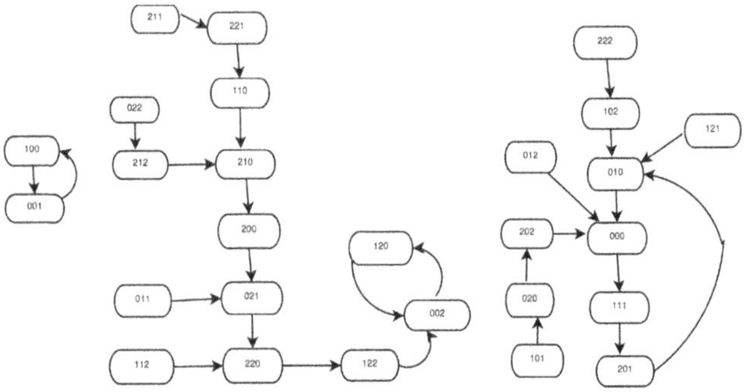

Fig. 1. A 3-state 3-neighborhood CA with only multi length cycles.

to be reachable from a configuration if any member of the cycle is reachable from that configuration. The *cyclic space* represents number of cycles, lengths of the cycles, cyclic and acyclic (reachable and non-reachable both) configurations. When all the configurations are cyclic, then the CA is *reversible*. An *irreversible* CA contains cyclic and acyclic configurations both.

Let us explain the above mentioned terminologies using an 3-cell 3-state 3-neighborhood null boundary CA. 012120012010201120120020201 is the candidate rule applied to all *three* cells of the CA. The cyclic space contains *three* multi length cycles; the lengths of two cycles are 2 and the length of another is 4. One cycle of length 2 contains no acyclic configuration. Another cycle of length 2 is reachable from 12 acyclic configurations whereas the cycle of length 4 is reachable from 7 acyclic configurations. The CA contains 8 non-reachable configurations. Obviously, the current CA is irreversible.

As the change of cell's state is governed by an RMT, there exists an *RMT sequence* in between the transition of any two consecutive configurations. Let us consider a 4-cell periodic boundary uniform CA with rule $\mathcal{R} = 22211100020211$ 1020222111000. Suppose the current configuration is 2121. Its next configuration is computed as follows: $\mathcal{R}(1,2,1)\mathcal{R}(2,1,2)\mathcal{R}(1,2,1)\mathcal{R}(2,1,2) = \mathcal{R}[16]\mathcal{R}[23]$ $\mathcal{R}[16]\mathcal{R}[23] = 0101$ (see row 3 of Table 1). Here, $(16, 23, 16, 23)$ is an RMT sequence of 4-cell periodic boundary 3-state 3-neighborhood CA. Let $rs = (r_0, r_1,$ $\cdots, r_{n-1})$ be an RMT sequence of an n-cell CA and let us explain the process of generation of such RMT sequence. Let $r_i = xyz$ be the RMT of cell i; therefore, r_{i+1} is selected from $\{yz0, yz1, yz2\}$. In null boundary CA, r_{n-1} is always $yz0$

where $r_{n-2} = xyz$. Let rs be the RMT sequence in between configurations x and y such that y is the next configuration of x. Now, rs is the *associated RMT sequence* of x and y is the *associated configuration* of rs. Therefore, $(16, 23, 16, 23)$ is the associated RMT sequence of configuration 2121 and 0101 is the associated configuration of RMT sequence $(16, 23, 16, 23)$. When the value at RMT xyz $(x, y, z \in S)$ of a rule $\mathcal{R}$ is y, then such RMT is called as *self-replicating RMT*. For example, RMT 23 (212) of rule 222111000202111020222111000 is self-replicating RMT as value at RMT 23 is 1 whereas RMT 23 is *non* self-replicating for rule 102202222002101201001111220 as value at RMT 23 (212) is 2 (see rows 3 and 2 respectively of Table 1). A configuration forms a *single length cycle* if *all* the RMTs of the associated RMT sequence are self-replicating. Let us consider configuration 2222 of the periodic boundary CA constituted by the rule 222111000202111020222111000. As RMT 26 (222) is self-replicating (see row 3 of Table 1), therefore every RMT of the associated RMT sequence $(26, 26, 26, 26)$ of configuration 2222 is self-replicating which makes the associated configuration of $(26, 26, 26, 26)$ as 2222 also. Thus configuration 2222 forms single length cycle. Obviously, the associated RMT sequence of every configuration of a multi length cycle maintains *at least one* non self-replicating RMT. For example, in the same CA, configurations, 1010 and 1212 form a *multi* length cycle and the respective associated RMT sequences $(3, 10, 3, 10)$ and $(23, 16, 23, 16)$ maintain RMTs 10 and 16 as non self-replicating (at position 2 and 4). Next, we introduce strictly multi length cycle CA.

Definition 1. *Let us consider a rule $\mathcal{R}$ for designing an n-cell CA. If the CA always generate multi length cycle(s) only for an arbitrary n, then $\mathcal{R}$ is called as strictly multi length cycle rule. A CA constructed by strictly multi length cycle rule, is called as strictly multi length cycle CA.*

$\mathcal{R} = 012120012010201120120020201$ is strictly multi length rule and any CA designed by $\mathcal{R}$ is strictly multi length CA (see Fig. 1). In 3-state 3-neighborhood CAs, a set of *sibling* RMTs is represented in the given manner: $\{xy0, xy1, xy2\}$ where $xy \in \{00, 01, 02, 10, 11, 12, 20, 21, 22\}$. Thus, we get *nine* different sets of sibling RMTs. Similarly, we get *nine* more sets of *equivalent* RMTs where every set follows $\{0yz, 1yz, 2yz\}$ where $yz \in \{00, 01, 02, 10, 11, 12, 20, 21, 22\}$.

3 Strictly Multi Length Cycle Cellular Automata

In this section and hereafter, 3-state 3-neighborhood CA will be alternatively called CA or 3-state CA. The first objective of this study is to figure out those 3-state rules which form CAs with multi length cycles only for any arbitrary CA size. As we get 3^n configurations in an n-cell CA, obviously, there exist 3^n (associated) RMT sequences.

Lemma 1. *An n-cell CA cannot generate a single length cycle if there does not exist any RMT sequence $(r_0, r_1, \cdots, r_{n-1})$ where every r_i is self-replicating [15].*

Lemma 1 implies that every RMT sequence of n-cell CA is traversed to check whether there exists at least one non self-replicating RMT in that sequence of n-length for deciding whether the given CA is strictly multi length cycle CA or not. Obviously, this process is computationally demanding. Thus we are motivated to figure out some intrinsic properties of RMTs in terms of strictly multi length cycle rules. The proofs of these are deferred to the full version of the paper due to space constraints.

Property 1. If all 27 RMTs of a rule are non self-replicating in a rule $\mathcal{R}$, then $\mathcal{R}$ forms both null and periodic boundary strictly multi length cycle CAs.

Property 2. Let $\mathcal{A}$ be the set of 26 non self-replicating RMTs in rule $\mathcal{R}$. Then $\mathcal{R}$ forms strictly multi length cycle CAs with boundary condition (i) null if $\mathcal{A}$ contains RMT 0 and (ii) periodic if $\mathcal{A}$ contains RMTs 0, 13 and 26.

Property 3. Let $\mathcal{A}$ be the set of 25 non self-replicating RMTs in rule $\mathcal{R}$. Then $\mathcal{R}$ forms strictly multi length cycle CAs with boundary condition

- null if $\mathcal{A}$ contains RMT 0, at least *one* RMT from $\{3, 10\}$ and at least *one* RMT from $\{6, 20\}$,
- periodic if $\mathcal{A}$ contains $\{0, 13, 26\}$, at least *one* RMT from $\{3, 10\}$, at least *one* RMT from $\{6, 20\}$ and at least *one* RMT from $\{16, 23\}$.

Property 4. A rule $\mathcal{R}$ forms strictly multi length cycle CA under null boundary condition if any one of the followings holds:

- Every RMT of $\{0, 1, 2, 3, 4, 5, 6, 7, 8\}$ is non self-replicating,
- Every RMT of $\{0, 3, 6, 9, 12, 15, 18, 21, 24\}$ is non self-replicating.

These *four* properties identifies a proper subset of the 3-state strictly multi length cycle rules. Property 1 detects that all 2^{27} target rules form strictly multi length cycle CAs for both null and periodic boundary CAs. Using Property 2, out of 27.2^{26} CAs, there exist 26.2^{26} null boundary and 24.2^{26} periodic boundary strictly multi length cycle CAs. Property 3 figures out at most 24.2^{26} strictly multi length cycle CAs under null boundary condition. For periodic boundary condition CAs, Property 3 detects at most 21.2^{26} strictly multi length cycle CAs. Property 4 finds two sets of strictly multi length cycle rules where every set contains $2^9.3^{18}$.

Although the above mentioned properties detect a set of strictly multi length cycle CAs, but the detection process is time consuming. For example, when Property 3 is used, the rules need to maintain 25 non self-replicating RMTs and 2 RMTs of those rules are self-replicating. Now, the question is which those *two* RMTs are? one pair or more? To answer this, we need to figure out from $\binom{27}{2}$. When the count of self-replicating RMTs is *more* (upto 13), the possible combinations are also more.

To deal this computational issue, we use *de Bruijn Graph* [4,22] which is one of the characterization tools of 1-dimensional cellular automata. In this graph, all possible *elementary cycles* and *elementary paths* are used to recognize the strictly multi length cycle CAs. The most interesting point is that the count of elementary paths and cycles are *constant* for d-state k-neighborhood CAs for a fixed d and a fixed k of any n size CA.

3.1 3-State CA: De Bruijn Graph

Let us explain de Bruijn Graph, which is a directed graph in which any two consecutive vertices shares common information.

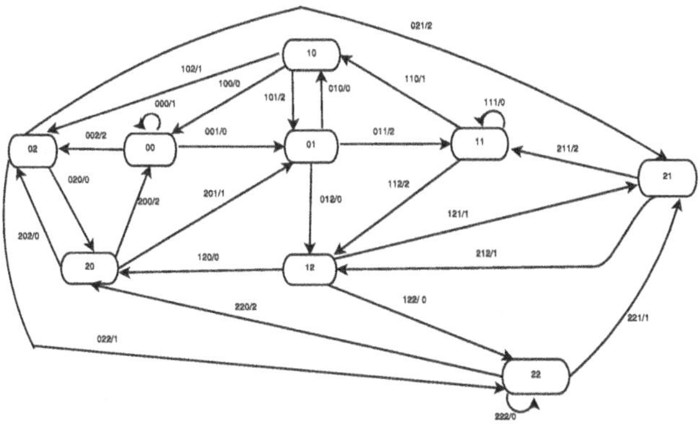

Fig. 2. de Bruijn Graph of CA 012120012010201120120020201: A 3-state 3-neighborhood CA.

Definition 2. *Let Σ be a set of symbols and $s \geq 1$ be a number. Then de Bruijn Graph is $B(s, \Sigma) = (V, E)$, where $V = \Sigma^s$ is the set of vertices and $E = \{(ax, xb) | a, b \in \Sigma, x \in \Sigma^{s-1}\}$ is the set of edges [4].*

Let us consider a de Bruijn Graph $B(s, \Sigma)$ where $s = 2$ and $\Sigma = \{0, 1, 2\}$. Therefore, $V = \{0, 1, 2\}^2 = \{00, 01, 02, 10, 11, 12, 20, 21, 22\}$ and $E = \{(ax, xb)\}$ where $a, x, b \in \{0, 1, 2\}$. In this directed graph, every vertex maintains equal in-degree and out-degree i.e., 3. Now, a de Bruijn graph can be mapped to an 1-dimensional d-state m-neighborhood cellular automaton. This work represents de Bruijn graph in-terms of 3-state 3-neighborhood CA.

A de Bruijn graph $B(s, \Sigma)$ can represent a 3-state 3-neighborhood CA as $s = m - 1 = 3 - 1$ and $\Sigma = \{0, 1, 2\} = S$ where m and S be respectively the neighborhood size and the set of cell states of a CA. Here, edge (ax, xb) of $B(s, \Sigma)$ represents the connection between overlap vertices ax and xb. In d-state k-neighborhood CA, $ax, xb \in S^{m-1}$ and $axb \in S^m$ where the length of the sequence of the present states represented by x is $m - 2$. Therefore, the respective de Bruijn graph of a 3-state 3-neighborhood CA confirms that axb is an RMT which is a ternary string of length *three*. In this graph, ax represents the sequence of length *two* which is the present states of cells $i - 1$ and i; similarly, xb represents the sequence of present states of cell i and $i + 1$. The edge between ax and xb explicitly denotes the corresponding RMT axb along with the value at axb. Let us see the de Bruijn graph of rule $\mathcal{R} = 012120012010201120120020201$ (Fig. 2). Here, the edge between vertices 00 and 01 represents RMT 001 and its value ($\mathcal{R}(0, 0, 1) = \mathcal{R}[1] = 0$) in the following way: 001/0. The outgoing edges of any vertex represents a collection of sibling RMTs; for example, the outgoing edges of 20 refers to sibling RMTs $\{200, 201, 202\}$. The incoming edges of a vertex refers to the equivalent RMTs (for example, the incoming edges of vertex 20 refers to equivalent RMTs 020, 021 and 022). As every edge of the graph holds an unique RMT and the value at that RMT which is a *state* also, therefore, a sequence of edges of finite length refers to an RMT sequence and its associated configuration. For example, $(001, 010, 100) = (1, 3, 9)$ is an RMT sequence of 3-cell CA and the associated configuration is 010 as 0, 1 and 0 are present on the following edges respectively $(00, 01)$, $(01, 10)$ and $(10, 00)$. Obviously, the sequence of edges cannot be arbitrary. The reason behind is that there exist valid combinations of starting and the ending vertices based on the periodic and null boundary conditions.

Property 5. Every RMT sequence of an n-cell periodic boundary CA is represented by a *cycle* of length n of the de Bruijn graph.

Property 6. Every RMT sequence of an n-cell null boundary CA is represented by any one of the following:

1. a *cycle* of length n where the starting and ending vertices are 00.
2. a *path* of length n where the starting and ending vertices are distinct and are respectively from $\{00, 01, 02\}$ and $\{00, 10, 20\}$.

For example, $(3, 9, 1, 3, 9, 1)$ is an RMT sequence of 6-cell periodic boundary cellular automaton as the starting and ending vertices are same i.e. 01; whereas $(1, 3, 10, 5, 15)$ is an RMT sequence of an 5-cell null boundary CA provided that the starting vertex is 00 and the ending vertex is 20. The transition of configurations can be explained by a de Bruijn graph. Let 012012 be a configuration of an 6-cell CA by rule 222111000222111000222111001; therefore, its associated RMT sequence is $(1, 5, 15, 19, 5, 15)$ and next configuration is also 012012. Thus, configuration 012012 forms a single length cycle as every RMT of $(1, 5, 15, 19, 5, 15)$ is self-replicating. As our target is get a CA without any single length cycle, therefore, we need make at least one RMT as non self-replicating in every RMT sequence. If we look carefully, every cycle of de Bruijn graph (which is an RMT sequence of an n-cell CA) contains at least one *elementary cycle*. In analogous manner, an path (which is an RMT sequence for an n-cell null boundary CA) contains at least one *elementary path* from $0x_0$ vertex to $x_{n-1}0$ vertex where $x_0 \neq x_{n-1}$. For example, $(3, 9, 1, 3, 9, 1)$ contains elementary cycle $(3, 9, 1)$ and $(1, 3, 10, 5, 15)$ contains elementary path $(1, 5, 15)$. Prior arts also use the concepts of elementary cycles [3,8] for studying different properties of CAs. In those work, the RMT sets of elementary cycles are called as *primary RMT sets* [3]. Besides elementary cycles, as this work uses elementary paths also; now onward, an "elementary path" in the De bruijn graph of CA is elementary path with $0x_0$ and $x_{n-1}0$ as starting and destination vertices provided $x_0 \neq x_{n-1}$. Obviously, the elementary cycles and paths are also the RMT sequences but the length is *restricted*. For elementary cycles, the lengths of RMT sequences are 1 to 9 and the lengths of paths vary from 1 to 8, obviously for 3-state 3-neighborhood CAs.

Property 7. The de Bruijn graphs of strictly multi length cycle CAs maintain the following properties: (i).Every elementary cycle contains *at least one* non self-replicating RMT when periodic boundary condition is considered. (ii).Each elementary path or cycle (where starting and ending vertices are 00) contains *at least one* non self-replicating RMT under null boundary condition.

Let us now focus on the *minimum* proportion and *valid* sets of non self-replicating RMTs in a rule for designing strictly multi length cycle cellular automata.

3.2 Strictly Multi Length Cycle CA: Significant Non Self-replicating RMTs

Now, the first task is to figure out the *count* of elementary cycles and elementary paths in the de Bruijn graph of 3-state 3-neighborhood CA. All the elementary paths and cycles can be found using the existing efficient algorithms, such as [9]. In this paper, we use *networkx* package in python for obtaining elementary paths and elementary cycles in the de Bruijn graph.

Remark 1. The de Bruijn graph of an n-cell (where n is arbitrary) 3-state 3-neighborhood cellular automaton contains

- 148 elementary cycles when boundary condition is periodic,
- 407 elementary cycles and paths when boundary condition is null.

Table 2 refers to the details of the counts of elementary paths and elementary cycles for both null and periodic boundary cellular automata. The *first* column of the table refers to the category of boundary condition maintained by the given CA. If periodic, then the graph generates only elementary cycles (see *second* column of first row). Each of columns *three* to *eleven* refers to the count of sets (elementary cycles) based on the count of RMTs. For example, column *five* of first row refers to the count of elementary cycles where every such cycle contains 8 RMTs. In analogous manner, the counts of elementary cycles and elementary paths are depicted in Table 2 based on the counts of RMTs. For detecting whether a rule $\mathcal{R}$ forms strictly multi length cycle CA, the following steps are to be followed.

1. Let $\mathcal{A}$ denote the non self-replicating RMTs of $\mathcal{R}$. Let $\mathcal{D}_j$ denote elementary cycle or elementary path where $\mathbb{D}$ refers to a collection of 407 or 148 such cycles (and paths) based on boundary condition.
2. Let $r \in \mathcal{A}$. Every $\mathcal{D}_j$ is checked whether r is contained in $\mathcal{D}_j$. If *yes*, then discard the respective $\mathcal{D}_j$s from $\mathbb{D}$ and update $\mathbb{D}$.
3. Repeat previous step (2) for $|\mathcal{A}|$. After that if $\mathbb{D} = \phi$, then $\mathcal{R}$ is strictly multi length cycle rule; otherwise, $\mathcal{R}$ is non-strictly multi length cycle rule.

Obviously, the above mentioned process is computationally less costly as it does not depend on the number of RMT sequences which increases exponentially

Table 2. Counts of sets based on the length of the elementary cycles and elementary paths. Here, the length refers to the count of RMTs present in the cycle or path.

Types of Boundary Condition	Elementary Cycles/ Paths	# of 1 RMT	# of 2 RMTs	# of 3 RMTs	# of 4 RMTs	# of 5 RMTs	# of 6 RMTs	# of 7 RMTs	# of 8 RMTs	# of 9 RMTs
Periodic	# of Elementary Cycles	3	3	8	12	17	21	24	26	24
Null	# of Elementary Cycles	1	-	2	4	8	12	25	11	24
	# of Elementary Paths	2	8	16	30	40	56	96	72	-

Table 3. Valid and minimal sets of non self-replicating RMTs for null boundary strictly multi length cycle cellular automata

Set of RMTs	Count of non self Replicating RMTs	Count of strictly multi length cycle rules
{0, 3, 6, 15, 24, 12, 21}	7	$2^7 \cdot 3^{20}$
{0, 3, 6, 7, 8, 4, 5}	7	$2^7 \cdot 3^{20}$
{0, 3, 6, 7, 24, 25, 4, 5}	8	$2^8 \cdot 3^{19}$
{0, 3, 6, 7, 14, 8, 12, 5}	8	$2^8 \cdot 3^{19}$
{0, 3, 6, 22, 15, 24, 4, 21}	8	$2^8 \cdot 3^{19}$
{0, 3, 6, 15, 17, 8, 12, 21}	8	$2^8 \cdot 3^{19}$
{0, 3, 6, 22, 24, 23, 4, 21, 5}	9	$2^9 \cdot 3^{18}$
{0, 3, 6, 14, 24, 23, 12, 21, 5}	9	$2^9 \cdot 3^{18}$
{0, 3, 6, 22, 23, 8, 4, 21, 5}	9	$2^9 \cdot 3^{18}$
{0, 3, 6, 14, 23, 8, 12, 21, 5}	9	$2^9 \cdot 3^{18}$
{0, 3, 6, 7, 14, 24, 25, 12, 5}	9	$2^9 \cdot 3^{18}$
{0, 3, 6, 7, 15, 24, 25, 12, 16}	9	$2^9 \cdot 3^{18}$
{0, 3, 6, 7, 15, 24, 25, 4, 16}	9	$2^9 \cdot 3^{18}$
{0, 3, 6, 7, 15, 17, 8, 4, 16}	9	$2^9 \cdot 3^{18}$
{0, 3, 6, 7, 15, 17, 8, 12, 16}	9	$2^9 \cdot 3^{18}$
{0, 3, 6, 22, 15, 17, 8, 4, 21}	9	$2^9 \cdot 3^{18}$

with respect to CA size n. Now, even if we use this process, we may get a reasonable number of non-strictly multi length cycle rules from a pool of rules. This motivates us to find a *valid* set of RMTs useful for the elementary cycles and elementary paths.

Definition 3. *A set of non self-replicating RMTs is said to be valid for strictly multi length cycle CA if at least one member of every set $\mathcal{D}_j$ of $\mathbb{D}$ is present in that set of non self-replicating RMTs.*

This problem can be addressed by the problem of *hitting set*. Next, we figure out the *minimal* valid sets; for that we use *sat* package in python.

Remark 2. To generate strictly multi length cycle cellular automata under *periodic* boundary condition, a *valid* set contains at least *eleven* RMTs. For *null* boundary CA, the cardinality of *valid* set is at least *seven.*

There are 408 *minimal* valid sets under periodic boundary condition whereas null boundary condition uses only 14 *minimal* valid sets. In case of periodic boundary condition, out of 408, 84 valid sets contain only *eleven* which is the *minimum* number of non self-replicating RMTs; remaining 300 valid sets contain *twelve* and 24 valid sets contain *thirteen* non self-replicating RMTs. For null boundary condition, 2 valid sets contain the minimum number of non self-replicating RMTs and that is *seven.* The counts of non self-replicating RMTs in other minimal sets are *eight* and *nine.* There are 4 valid sets of cardinality *eight* and 10 valid sets of cardinality *nine.*

Property 8. A rule can never generate a strictly multi length cycle CA if (i) $|\mathcal{A}| \leq 6$ and (ii) $|\mathcal{A}| \leq 10$ under *null* and *periodic* boundary conditions respectively.

Property 9. There exist at least (i) $2^7 \cdot 3^{20}$ and (ii) $2^{11} \cdot 3^{16}$ strictly multi length cycle rules when the boundary conditions are respectively *null* and *periodic.*

This section can give a figure of counts of the strictly multi length cycle cellular automata using *valid* and *minimal* set. See Table 3 for null boundary strictly multi length cycles rules. The *third column* of the table represents the count of the rules based on every minimal and valid set of non self-replicating RMTs.

4 Usefulness of 3-State Cellular Automata for Clustering

As this work aims to use the strictly multi length cycle CAs as clustering tool, let us explain clustering in terms of *multi length cycles.* The prior arts on CA based clustering used reversible cellular automata (already mentioned in Sect. 1). A reversible CA can have multi length cycles also. In such case, the multi length cycle is *reachable* only from *cyclic* configurations. The irreversible CAs *relax* that limitation: a cycle can be reachable from acyclic and cyclic configurations both. For, example, a cycle of length 6 of the CA 012120012010201120120020201 is reachable from 111 (cyclic configuration), 102 (acyclic: reachable configuration) and 101 (acyclic: non reachable configuration) (see Fig. 1). This concept of *reachability of a cycle from a configuration* (irrespective of cyclic or acyclic) leads us to use multi length cycles of any CA for clustering. Traditionally, every cluster holds those objects who are similar with respect to some features. In this work, obviously, a cluster means a multi length cycle; two configurations can be kept in same cluster if both of them have the *reachability* via some multi length cycle.

Definition 4. *Two configurations of a CA belong to the same* **cluster** *if there exists a multi length cycle which is reachable from both the configurations.*

If we consider the prior example in this section, configurations 111, 102 and 101 belong to the same *cluster*. Thus according to Definition 4, a pair of acyclic configurations or cyclic configurations and even a pair of cyclic and acyclic configurations can also be part of same cycle. Our next focus is on the identification of effective strictly multi length cycle rules for clustering. In this section, we call a multi length cycle as "cluster" afterwards.

4.1 Effective Rules for Clustering

From the prior work, null boundary cellular automata have proven their efficacy in clustering. This work also uses null boundary CA for clustering. For choosing the rules for clustering, the following properties are to be considered.

1. Limited number of clusters which can be obtained with a cycle reachable from *large* number of configurations.
2. Moderate number of non self-replicating RMTs.

To fulfil the *first* property, the target rule should be designed in such a manner that the rule maintains *more* dependency on the neighbors. The reason is explained as follows: if a change in one neighbor (right or left) *affects* the state value of a cell, then the *information propagation* is more. Let us explain with an example. If $\mathcal{R}(x_{i-1}, x_i, 0) = \mathcal{R}(x_{i-1}, x_i, 1) = \mathcal{R}(x_{i-1}, x_i, 2)$ for every i, there is less change in cell's state values in consecutive configurations. Thus a cycle is formed of small length; also there is a tendency to form more number cycles which are not suitable for clustering. Actually, here, a constraint is imposed on *right* neighbor and make the rule as *right independent* as there is no change in the next states of a set of sibling RMTs $\{xy0, xy1, xy2\}$. In analogous manner, the same restriction can be raised in *left* neighbor. Therefore, we need to make right and left neighbors more proactive by affecting the state changes of the sibling RMTs as well as the equivalent RMTs. The best way of doing that is all the possible sibling RMTs and equivalent RMTs should have different values ($\forall a, b, c$, $\mathcal{R}[a] \neq \mathcal{R}[b] \neq \mathcal{R}[c]$ and $\forall r, s, t$ $\mathcal{R}[r] \neq \mathcal{R}[s] \neq \mathcal{R}[t]$ where a, b, c refers to sibling and r, s, t refers to equivalent RMTs); in this case, obviously, one RMT from the set of sibling (resp. equivalent) RMTs must be self-replicating. In this way, $\frac{1}{3}$ RMTs are self-replicating and maintain the *second* property of designing rule aims to perform good clustering. Now, this feature makes conflict in the design of strictly multi length cycle rules. Let us look at Table 3. Here, out of 14, 12 valid sets are violating the feature directly as at least one set of either sibling or equivalent RMTs are non self-replicating. Even if another two sets (rows 11 and 16) also violate as two sibling RMTs from same set and their respective equivalent RMTs also belong to the same valid set and it fails to design strictly multi length cycle rule. Therefore, we need to design our target rule using a slight variant of the above strategy.

1. Generate a pool $\mathbb{R}$ of 3-state 3-neighborhood rules where $\forall a, b, c$, $\mathcal{R}[a] \neq \mathcal{R}[b] \neq \mathcal{R}[c]$ and $\forall r, s, t$ $\mathcal{R}[r] \neq \mathcal{R}[s] \neq \mathcal{R}[t]$ where a, b, c refers to sibling and r, s, t refers to equivalent RMTs. In this way, we get $12^3 = 1728$ rules.

2. Select every rule from $\mathbb{R}$ and take *two* sets of sibling RMTs (from $\binom{9}{2}$) arbitrarily. For these two sets, make all the RMTs as non self-replicating in given manner $\mathcal{R}[a] = \mathcal{R}[b] \neq \mathcal{R}[c]$. In this way, we get 6 different combinations of values at the RMTs in one set: xxy, xyy, xyx, yxx, yxy and yyx where y and x are different. Analogously, we get 6 more possibilities for another set.
3. Thus we get $1728 * 6 * 6 = 62208$ and $\mathbb{R}$ is updated.
4. From $\mathbb{R}$, only 5504 ($\approx 10\%$) are strictly multi length cycle rules.

4.2 Results

This section shows the *effectiveness* of the *useful* 5504 3-state 3-neighborhood CA rules for clustering on three real datasets: Iris, User Knowledge Modeling, and BuddyMove(https://archive.ics.uci.edu/). Two metrics - *Silhouette Score* and *Dunn Index* [5] are reported. For a given dataset, a CA is said to be the *best* for clustering among all the useful CAs if any one of the followings is satisfied: (i) *highest* silhouette value and (ii) *highest* value of dunn index. The number of features and datapoints for these datasets are presented in Table 4 (columns 2 and 3). All the features for each of these datasets are quantitative.

First we need to *map* the target objects into *CA configuration* which are ternary string. The length of this string is decided by the counts of qualitative and quantitative attributes. For this mapping, we use the same *encoding* technique, which is mentioned in [13]. This work uses datasets having quantitative attributes only for which we use *frequency encoding*. We sort the values of a feature in ascending (resp. descending) order and distribute them in w bins *equally*. In this work, $w = 3$. The elements of the first, second and third bins are encoded respectively to 0, 1 and 2. For example, if a dataset has 6 quantitative attributes, each attribute value is converted to 0, 1 and 2 and finally each object encoded as a member of $\{0, 1, 2\}^6$ which is a configuration of a 6-cell 3-state CA. Note that two different objects may be mapped to the same configuration.

Now, for every dataset, we consider the useful configurations space and distribute them among k clusters. In this work, k is the *optimal* number of clusters for a given dataset based on CA based clustering for which the silhouette and dunn index scores are the *maximum* among a number of clusters. Let us look at the rows corresponding to Strictly Multi Length Cycle CA in Table 4. Here, all 5504 strictly multi length cycles CAs have been used for each of the datasets. The objects of each dataset are targeted to distribute among 2 to 6 clusters. The *seventh* column refers to the *best* CA and the *sixth* column shows the count of optimal clusters. Interestingly, for every dataset, a fixed CA is called as *best*. This paper demonstrates the ability of natural clustering of strictly multi length cycle CAs. A comparative analysis shows that the best CAs are *effective* for clustering. See Table 4 for the explanation. Here, as benchmark algorithms, *hierarchical*, k-means and PAM clustering algorithms [5] are taken into consideration.

Table 4. Comparative Analysis of 3-state strictly multi length cycle based clustering and some benchmarks algorithms [5]

Dataset	# Features	# Objects	Algorithm	k	n	$\mathcal{R}$	Silhouette Score	Dunn Index
Iris	4	150	Hierarchical	2			0.6326	0.3597
			Kmeans				0.6279	0.0765
			PAM				0.6318	0.0811
			Strictly Multi Length Cycle CA		4	012120012010201120120020201	0.6803	0.0413
BuddyMove	6	249	Hierarchical	4			0.3247	0.1432
			Kmeans				0.3423	0.0687
			PAM				0.3310	0.0208
			Strictly Multi Length Cycle CA		6	210012102021201210010020021	0.1147	0.0158
User Knowledge Modeling	5	403	Hierarchical	2			0.2926	0.1979
			Kmeans				0.2230	0.1396
			PAM				0.2161	0.0738
			Strictly Multi Length Cycle CA		5	201102120012021012010020201	0.01458	0.0698

5 Non-strictly Multi Length Cycle Cellular Automata

3-state null boundary strictly multi length cycle CA has already shown its efficacy for clustering. Now, to be a strictly multi length cycle rule, at least 7 RMTs need to be non self-replicating for null boundary CAs. Next our target is to observe the cyclic behaviour of those CAs which are *null* boundary and designed by the rules which have *less* than 7 non self-replicating RMTs. Now, do these CAs generate only single length cycles or single and multi length cycles both? One category generates only single length cycles and such CAs are called as *strictly single length cycle* cellular automata [15]. The trivial strictly single length cycle CA is that CA when the corresponding rule maintains all 27 RMTs as self-replicating. Now, if we increase the impurity in terms of non self-replicating RMT(s), then what is the behaviour of CA? Let us consider the situation where number of non self-replicating RMT is *one*; if an RMT sequence rs_i holds such non self-replicating RMT at position p, then the associated configuration can never repeat, so not be *cyclic* also.

Lemma 2. *The exist* $(d-1).d^m$ *strictly single length cycle cellular automata when the corresponding rules maintain only one non self-replicating RMT.*

Lemma 2 verifies that there exist $2.27 = 54$ 3-state 3-neighborhood strictly single length cycle CAs for the rules with only one non self-replicating RMT. Now, if we get rules with *two* non self-replicating RMTs, then we can get different types of behaviour in CAs. Let us consider the following scenario. (i) Suppose a CA constructed by rule 222111000202111020222111000 (where RMTs 16 (121) and 10 (101) are non self-replicating) guarantees the existence of multi length cycle is of length *two*. The reason is as follows: when an RMT sequence rs_1 holds RMT 16 at some position p, then in the next RMT sequence rs_2, at the same position, we get the RMT in the form of $x0z$ as the value at RMT 16 is 0. Now, RMTs at $p-1$ and $p+1$ positions in rs_1 are from $x12$ and $21z$ respectively. As all RMTs in the form $x1z$ are self-replicating, therefore, rs_2 contains RMT 10 (101) at position p. As the value at RMT 10 (101) is 0 and the RMTs at $p-1$ and $p+1$ contains RMTs from $x1z$, therefore, RMT 16 will be obtained at position p in rs_3 and obviously, $rs_3 = rs_1$. Thus a cycle of two length is created. Let take a look in the following example. Consider the following situation. In rule 222111000202111000222121000, all RMTs in the form $x0z$ are self-replicating. All the RMTs in the forms of $x1z$ and $x2z$ are self-replicating excluding RMTs 16 (121) and 4 (011). The values at RMTs 16 and 4 are 0 and 2 respectively. Now, an RMT sequence rs_1 holds RMT 16 (121) at position p and as value at RMT 16 is 0 here, therefore, the next RMT sequence rs_2 holds RMT $x0z$ at position p which is self-replicating. Thus in the consecutive RMT sequences of rs_1, RMT 16 can never be obtained. So, no multi length cycle will be generated. If we consider any RMT sequence with RMT 4 (011) also, we can never get any multi length cycle. Let us explain the reason. If an RMT sequence rs_1 holds RMT 4 at position p and as value at that RMT is 2, we get RMT $x2z$ at position p in the next RMT sequence rs_2. As RMTs at $p-1$ and $p+1$ of rs_1 are in the form of $x01$ and $11z$ respectively and both of them maintain self-replicating property. Therefore, RMT at position in rs_2 is 021 or 7 which is self-replicating also. So, there is no way of getting RMT 4 in the consecutive RMT sequences of rs_1.

The above scenarios raise the question whether a particular pair of non self-replicating RMTs in a rule ensures the generation of a cycle of length *two*. For 3-state 3-neighborhood CAs, there exist $\binom{27}{2} \cdot 2^2 = 1404$ rules which maintains only *two* non self-replicating RMTs.

Property 10. Let $\mathcal{R}$ be the rule where r_p^1 (xy_1z) and r_p^2 (xy_2z) be the only non self-replicating RMTs and $y_1 \neq y_2$. If $\mathcal{R}[r_p^1] = y_2$ and $\mathcal{R}[r_p^2] = y_1$, then a CA designed by $\mathcal{R}$ of an arbitrary size n generates a cycle of length two.

Property 10 identifies the rules which ensure the generation of multi length cycle with *two* non self-replicating RMTs. Let us distribute the 27 RMTs in 9 sets where each of them represents 3 RMTs in the following form: $B_{xz} = \{x0z, x1z, x2z\}$ for the fixed x and z. For example, $B_{10} = \{100(9), 110(12), 120$ $(15)\}$. Now, from B_{10}, we get *three* possibilities ($\{9, 12\}$, $\{9, 15\}$ and $\{12, 15\}$) which satisfy Property 10. Thus we get $\binom{9}{1} \cdot \binom{3}{1} \cdot 1 = 27$ rules which forms at least *one* multi length cycle and the cycle length is *two* always. Thus out of all rules with two non self-replicating RMTs, 98% are strictly single length cycle CAs and only 2% contribute in generation of multi length cycle. The interesting point is that besides multi length cycles, these 2% CAs generate single length cycles also as at least one of the RMT sequence $(0, 0, \cdots, 0)$ (configuration 0^n) or $(4, 13, 13, \cdots, 12)$ (configuration 1^n) or $(8, 26, 26, \cdots, 24)$ (configuration 2^n) maintains self-replicating property for each of its RMT.

6 Conclusion

This work finds a collection 3-state 3-neighborhood null boundary strictly multi length cycle cellular automata which are capable of generating *limited* number of cycles and a large number of configurations are attracted by some cycles of those CAs. This category of CAs also shows the effectiveness of its ability of natural clustering. Besides, this work identifies a rule set which are non-strictly multi length cycle rules but ensure the generation of multi length cycles using the *minimum* number of non self-replicating RMTs.

References

1. Abhishek, S., Dharwish, M., Das, A., Bhattacharjee, K.: A cellular automata-based clustering technique for high-dimensional data. In: Proceedings of Second Asian Symposium on Cellular Automata Technology. pp. 37–51. Springer Nature Singapore, Singapore (2023)
2. Adak, S., Naskar, N., Maji, P., Das, S.: On synthesis of non-uniform cellular automata having only point attractors. J. Cell. Autom. **12**(1–2), 81–100 (2016)
3. Bhattacharjee, K.: Cellular Automata : Reversibility, Semi-reversibility and Randomness. Ph.D. thesis, Indian Institute of Engineering Science and Technology, Shibpur (2019)
4. Bhattacharjee, K., Naskar, N., Roy, S., Das, S.: A survey of cellular automata: types, dynamics, non-uniformity and applications. Nat. Comput. **19**(2), 433–461 (2020)
5. Brock, G., Pihur, V., Datta, S., Datta, S.: clValid: An R package for cluster validation. J. Stat. Softw. **25**(4), 1–22 (2008)
6. Das, S.: Theory and Applications of Nonlinear Cellular Automata In VLSI Design. Ph.D. thesis, Bengal Engineering and Science University, Shibpur, India (2007)
7. Das, S., Mukherjee, S., Naskar, N., Sikdar, B.K.: Characterization of single cycle CA and its application in pattern classification. Electron. Notes Theor. Comput. Sci. **252**, 181–203 (2009)
8. Hazari, R., Das, S.: Number conservation property of elementary cellular automata under asynchronous update. Complex Syst. **23**(2), 177–195 (2014)
9. Johnson, D.B.: Finding all the elementary circuits of a directed graph. SIAM J. Comput. **4**(1), 77–84 (1975)
10. Maji, P.: Cellular Automata Evolution for Pattern Recognition. Ph.D. thesis, Jadavpur University, Kolkata, India (2005)
11. Manoranjan, V., Sneha Rao, G., Vaidhianathan, S.V., Bhattacharjee, K.: Optimized reversible cellular automata based clustering. In: Cellular Automata and Discrete Complex Systems. pp. 74–89. Springer Nature Switzerland, Cham (2023)
12. Moere, A.V., Clayden, J.J., Dong, A.: Data clustering and visualization using cellular automata ants. In: Australasian Joint Conference on Artificial Intelligence. pp. 826–836. Springer (2006)
13. Mukherjee, S., Bhattacharjee, K., Das, S.: Clustering using cyclic spaces of reversible cellular automata. Complex Syst. **30**(2), 205–237 (2021)
14. Mukherjee, S., Bhattacharjee, K., Das, S.: Reversible cellular automata: a natural clustering technique. J. Cell. Automata **16** (2021)
15. Mukherjee, S., Sindhuja, J.: Characterization of 3-state cellular automata with single length cycles. In: Cellular Automata Technology. pp. 137–152. Springer Nature Switzerland, Cham (2024)
16. Narodia Parth, P., Bhattacharjee, K.: Gödel number based encoding technique for effective clustering. In: Pattern Recognition and Machine Intelligence. pp. 49–58. Springer Nature Switzerland, Cham (2023)
17. Naskar, N.: Characterization and Synthesis of Non-Uniform Cellular Automata with Point State Attractors. Ph.D. thesis, Indian Institute of Engineering Science and Technology, Shibpur, India (2015). http://it.becs.ac.in

18. Naskar, N., Das, S., Sikdar, B.K.: Characterization of nonlinear cellular automata having only single length cycle attractors. J. Cell. Autom. **7**(5–6), 431–453 (2012)
19. Pal Chaudhuri, P., Roy Chowdhury, D., Nandi, S., Chatterjee, S.: Additive Cellular Automata – Theory and Applications, vol. 1. IEEE Computer Society Press, USA, ISBN 0-8186-7717-1 (1997)
20. Saha, S., Maji, P., Ganguly, N., Roy, S., Chaudhuri, P.P.: Cellular automata based model for pattern clustering. In: Proceedings of International Conference on Advances in Pattern Recognition. pp. 122–126 (2003)
21. Sethi, B., Roy, S., Das, S.: Asynchronous cellular automata and pattern classification. Complexity **21**(S1), 370–386 (2016)
22. Sutner, K.: De Bruijin graphs and linear cellular automata. Complex Syst. **5**(1), 19–30 (1991)
23. Wolfram, S.: A new kind of science. Inc, Wolfram Media (2002)

Use of 5-Neighborhood Cellular Automata for Pattern Classification

Sukanya Mukherjee(✉) and Priya Kumari

Department of Computer Science and Engineering, Institute of Engineering and Management, University of Engineering and Management, Kolkata 700091, West Bengal, India
sukanyaiem@gmail.com

Abstract. The convergence property of cellular automata makes them effective for *pattern classification* and elementary cellular automata have already proven their efficacy. This work uses *binary* cellular automata with 5-neighborhood architecture. In this work, the candidate cellular automata for classification obey the following properties: (i) cycles of length *one* only, (ii) *stable cyclic behavior* and (iii) moderate number of cycles. These three properties are investigated on *balanced* and *unbalanced* rules both. In case of unbalanced category, rules with at most 25% 0 and rules with at most 12.5% 1 are investigated for a set of CA sizes. Finally, the model is trained and tested using the rules, some of which perform at par with the existing classifiers by exhibiting optimal classification performance. This makes to the observation that the 5-neighborhood rules can be used as efficient classifiers. Experimental results establish that balanced and unbalanced both rules perform efficiently, and interestingly, rule 1114110 shows its efficacy as classifier on three different real data sets.

1 Introduction

Classification is a process in machine learning where the aim is to predict the category, class or label of an object based on its attributes. In classification tasks, a model is trained using data that have known labels, and the model learns to recognize patterns in the data to predict labels for new, unseen data. A cellular automaton (CA) can be used as the classification model [2].

A cellular automaton is a grid of *cells* where each cell presents a *state* value. A collection of states of all cells together is called as *configuration*. CA evolves in time and transition of configurations occurs at every time step using some *rules* [15]. Finally, CA reaches to a configuration or collection of configurations which forms *cycle*. Actually, the real objects is *mapped* to CA configurations (using an *encoding* technique) and finally reaches to some cycles. A configuration which represents a real object is called as *useful configuration* and the

This work is carried out as a project in the Indian Summer School on Cellular Automata 2024.

S. Kamilya et al. (Eds.): ASCAT 2025, CCIS 2499, pp. 86–100, 2025.
https://doi.org/10.1007/978-3-031-94121-4_7

corresponding cycle (where the configuration reaches finally) is called as *useful attractor*, as cycles are alternatively called as *attractors* in CA terminologies. Let a CA be used for 2-class useful attractor problem (class 0 and class 1). If an useful attractor represents *more* useful configurations from class 1, then such attractor is *trained* as class 1. In *testing* phase, that attractor is capable of recognizing the useful configuration *correctly* if it is from class 1 only. Thus the *cycles* of cellular automata (CAs) enable them to play the role of *classifiers*. Now the question is whether any arbitrary CA is *successful* for pattern classification. Prior art says that CA with *point state attractors* is the most successful CA model [4,7,8,11,12,14]. In this category, a collection of the configurations converge to an attractor (obviously a configuration) and finally the system turns into a stable state. A CA with point state attractor means a CA with *single length cycle* i.e., the *length* of the cycle is *one*. If a CA contains only single length cycles but the count of the cycle is *more than one*, such CA is called a multi attractor cellular automaton (MACA). Actually, MACA is *useful* for classifier problem due to its cyclic behavior. MACA is extensively studied for theoretical aspects and pattern classification both [1–3,6,11,13]. Even MACA may not be suitable for pattern classification if it generates exponential number of cycles (with respect *CA size*). Thus, one of the challenges is to figure out *effective* MACA which maintains *moderate* number of cycles. As single length cycle generation depends on *self replicating RMTs* (more self replicating RMTs in rule, generates CA with more single length cycles), traditionally, MACA generation is influenced by this type of RMTs. The current work focuses on the *stability* of the cyclic behavior of CAs along with the contribution of self replicating RMTs in the rule for designing the classifier. If a CA generates μ_1 number cycles of length l_1, μ_2 number cycles of length l_2 and so on, then the *cycle structure* of the CA is $[\mu_1(l_1), \mu_2(l_2), \cdots]$. CAs of different sizes can be designed by a fixed rule. If those CAs maintain *approximately* same cycle structures, then they maintain *stable* cyclic behavior. As MACA is useful for pattern classification, therefore, it needs to hold the stability in cyclic behavior. To design MACA, here we use *strictly single length cycle cellular automata rules* [10]; a CA of an arbitrary size designed by this type of rule always generates single length cycles only. Prior arts of CA based pattern classification mainly concentrates on *elementary cellular automata* where the CAs maintain 2-states per cell and 3-neighborhood architecture. Recently, [10] presents the efficacy of CAs with *extended* states (3-state 3-neighborhood CAs) in pattern classification. This motivates us to see the performance of binary CAs in pattern classification by extending the *neighbors*. This works uses 5-neighborhood architecture in the following way: *two* neighbors at *left* side of a cell, *two* neighbors at the *right* side of that cell and finally the cell *itself*.

The useful CA terminologies for this work are explained in 2. This work uses the traditional CA based classification (Sect. 3). The only variant is the way of rule selection. This work figures out 2-state 5-neighborhood strictly single length cycle CAs (i) with approximately same cycle structures and (ii) maintain a *linear* relationship between the number of cycles and the CA sizes in (Sect. 3). Here, as candidate CAs, the balanced rules (where the number of 0 s and 1 s are same) as well as the unbalanced rules are also used by varying the number of states 0 and 1 (up to 25%) in the rules. Section 4 reports the results of the classification model on the real datasets. The comparative analysis with the existing classification models indicates that the performance of our CA-based classifier technique is at least as good as the best known classifier algorithms that exist today.

2 Useful Terminologies

Cellular Automaton deals with the gradual *evolution* of discrete grids of *cells* using some *rules* and *neighborhood* properties. This work uses 1-dimensional CA with *finite* number of cells. Let us consider n be that number of cells; therefore, n is called as *CA size* and the CA having such n cells, is called as *n-cell* CA. In CA, each cell can reside in one of a finite number of *states*. For current work, the *states* of cells can be binary states where each cell can be assigned with either value 0 or value 1 only. The collection of all cells' states together at a particular time step is called as *configuration*. Traditionally, the cells are numbered as 0 to $n-1$ and let x_i be a state of cell i. Therefore, a sequence $(x_0x_1 \cdots x_{n-1})$ of n length is called as *configuration* of an n-cell CA; the decimal equivalent of this n-length binary string is also called as configuration.

Now, every cell *updates* its state based on the CA's *neighborhood* structure. This work uses 5-neighborhood architecture. Let us consider cell i. In this work, these *five* neighbors are selected in the following way: cells $i+1$, $i+2$ are the *right* neighbors of cell i and cells $i-1$, $i-2$ are the *left* neighbors; cell i also acted as the neighbor of itself. Let a binary string $x_{i-2}x_{i-1}x_ix_{i+1}x_{i+2}$ of length 5 represents the *present* states of cells $i-2$ to $i+1$ (for a given cell i). Therefore, the *next* state y_i of cell i is computed as follows: $y_i = \mathcal{R}(x_{i-2}, x_{i-1}, x_i, x_{i+1}, x_{i+2})$ where $\mathcal{R}$ is the *local transition* function. In this work, $\mathcal{R} : \{0,1\}^5 \rightarrow \{0,1\}$. Each of these functions are called as *rule*; thus we get 2^{32} rules when CAs with 2-state 5-neighborhood architecture are considered. Rules can alternatively be represented in their decimal equivalents. This work supports both the representation format. Now, each $x_{i-2}x_{i-1}x_ix_{i+1}x_{i+2}$ or its decimal equivalent is called as *Rule Min Term* (RMT). For example, 11100 or 28 is an RMT of 2-state 5-neighborhood CA. Thus, we get $2^5 = 32$ (0 to 31) RMTs. Table 1 refers to RMTs and rules of 5-neighborhood *binary* CAs. The *next* state of RMT or *value* at RMT is denoted as $\mathcal{R}(x_{i-2}, x_{i-1}, x_i, x_{i+1}, x_{i+2})$ or $\mathcal{R}[r]$ where r is the decimal equivalent of $x_{i-2}x_{i-1}x_ix_{i+1}x_{i+2}$. Let us consider a rule $\mathcal{R}$ = 10101010110100110111001101001010. Then the value at (or next state of) RMT 11101 (29) in the $\mathcal{R}$ is 1 (see row 2 of Table 1). We can distribute the 32 RMTs among 2 sets based on their *middle* bits (of their 5-bit representation). Now, S_0 and S_1 denote the sets of RMTs whose middle bits are 0 and 1 respectively: $S_0 = \{0,1,2,3,8,9,10,11,16,17,18,19,24,25,26,27\}$ and $S_1 = \{4,5,6,7,12,13,14,15,20,21,22,23,28,29,30,31\}$. An RMT can be called as *self replicating* if the value at that RMT is its middle bit. For example, RMT 31 (11111) of $\mathcal{R}$ = 10101010110100110111001101001010 is self replicating as

Table 1. In this table, RMTs are represented in both the forms: 5-bit binary strings and their decimal equivalents. NS is the *next state* or *value* at the RMT.

RMT	11111 (31)	11110 (30)	11101 (29)	11100 (28)	11011 (27)	11010 (26)	11001 (25)	11000 (24)	10111 (23)	10110 (22)	10101 (21)	10100 (20)	10011 (19)	10010 (18)	10001 (17)
	1	0	1	0	1	1	0	0	1	1	0	0	1	0	1
NS	1	0	1	0	1	0	1	0	1	1	0	1	0	0	1
	0	1	1	0	1	1	0	1	0	1	0	1	1	0	0

RMT	10000 (16)	01111 (15)	01110 (14)	01101 (13)	01100 (12)	01011 (11)	01010 (10)	01001 (9)	01000 (8)	00111 (7)	00110 (6)	00101 (5)	00100 (4)	00011 (3)	00010 (2)	00001 (1)	00000 (0)
	1	0	1	1	0	1	0	1	1	0	0	1	0	1	0	1	0
NS	1	0	1	1	1	0	0	1	1	0	1	0	0	1	0	1	0
	1	1	1	0	0	1	1	0	0	1	1	0	1	1	1	0	1

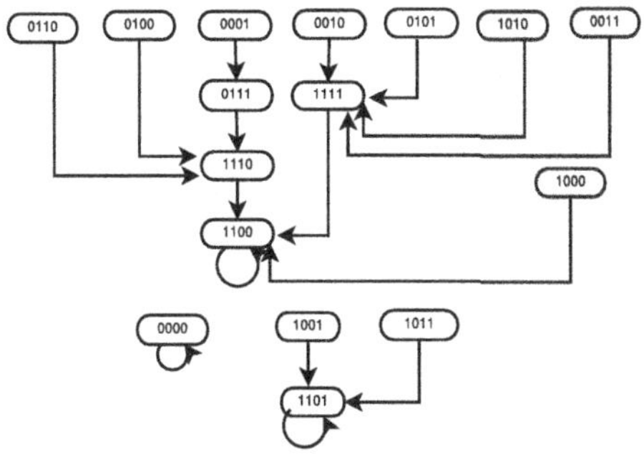

Fig. 1. A 2-state 5-neighborhood 4-cell irreversible CA with only single length cycles

$\mathcal{R} = [31] = 1$. As this work uses CAs of finite cells, to compute the next state of cell 0, *two* binary states need to be concatenated at the left neighbors of cell 0. In analogous manner, *two* binary states are to be concatenated at the right neighbor cells of cell $n-1$. Thus, $x_{-2} = x_{-1} = x_n = x_{n+1} = 0$. Therefore, we get $2^3 = 8$ RMTs which are *valid* for cell 0. These RMTs are 0 to 7 ($00xyz$ where $x, y, z \in \{0, 1\}$ for every x, y, z). Similarly, we get another set of 8 RMTs $\{0, 4, 8, 12, 16, 20, 24, 28\}$ for cell $n-1$; obviously, these 8 RMTs are considered as valid RMTs only for cell $n-1$.

This work uses *uniform* CA where all cells of a CA follow a *fixed* rule. As cells update their states in every time step, transition of configurations also occurs. This transition is governed a *global* transition function. Let $\mathtt{x} = (x_0x_1 \cdots x_{n-1})$ and $\mathtt{y} = (y_0y_1 \cdots y_{n-1})$ be the configurations. Let $G_n : \{0,1\}^n \rightarrow \{0,1\}^n$ where G_n is the global transition function. Therefore, $G_n(\mathtt{x}) = \mathtt{y}$ and every y_i is computed by same $\mathcal{R}$ as we consider uniform CA. Here, $\mathtt{y}$ is *reachable* from $\mathtt{x}$. If any configuration is reachable from *itself*, then such configuration is called as *cyclic* configuration; otherwise, the configuration is *acyclic*. An acyclic configuration is either reachable or *non-reachable* (which are not reachable from any configuration). A cyclic configuration or collection of cyclic configurations is called as *cycle*. The cycle will have some *length* depending on how many configurations are trapped into it. A cycle have length *one* when *only* one configuration is trapped and that cycle is called as either *single length cycle* or *point state attractor*. When a cycle has more than one configuration, we call that cycle as *multi length cycle*. This work targets CAs with single length cycle only. A CA with only single length cycles is always irreversible CA (excluding only one exception). Let us see Fig. 1. The 4-cell 2-state 5-neighborhood null boundary uniform CA designed by rule $\mathcal{R} = 1114110$ generates only single length cycles and the count is 3 (1100, 0000 and 1101). Here, for example, 0011 is non-reachable acyclic and 0111 is reachable acyclic configuration. Next, we introduce *cycle structure* of an n-cell CA. The count of cycles having *same* length is represented by the cycle structure of a CA. Thus the cycle structure of 4-cell CA designed by 1114110 is represented as [3(1)].

A transition of consecutive configurations is governed by a *RMT sequence*. For example, let us consider a configuration 1110 of an 4-cell null boundary CA. The corresponding RMT sequence is $(7, 14, 28, 24)$. The next configuration can be computed as $\mathcal{R}[7]\mathcal{R}[14]\mathcal{R}[28]\mathcal{R}[24] = 0100$ as $\mathcal{R} =$ 10101010110100110111001101001010 (see row 2 Table 1). Here, 0100 is called as the *associated* configuration of $(7, 14, 28, 24)$. If *all* the RMTs of a RMT sequence are self replicating, then the associated configuration forms single length cycle.

3 Pattern Classification: Useful 5-Neighborhood Rules

For 2-state 5-neighborhood CAs, the total number of possible rules are $2^{32} = 4,294,967,296$. Out of total 2^{32}, some always form CAs with only single length cycles, some with multi length cycles and other form single length as well as multi length cycles. As stated in the introduction, the main objective of our work is to identify the rules out of 2^{32} that exhibit *stable* cyclic behavior and produce only *single length cycles* for *all* possible CA sizes, which is a challenging work. A rule $\mathcal{R}$ which creates an n-cell CA with only single length cycle(s) for any $n \in \mathbb{N}$, then $\mathcal{R}$ is called as *strictly single length cycle* rule and any CA designed by $\mathcal{R}$ is called as *strictly single length cycle* CA [10].

3.1 Significant 5-Neighborhood Rules

Here, first target is to get strictly single length cycle rules. Since the rule search space is reasonably large, we proceed by the following strategy. Initially, a rule is considered where the values at all the 32 RMTs are same i.e. either 0 or 1. Let we start with rule $\mathcal{R} =$ 00000000000000000000000000000000. Obviously, this $\mathcal{R}$ is strictly single length cycle rule and every CA designed by $\mathcal{R}$ maintains the cycle structure [1(1)]. The reason is as follows: the associated configurations of all the RMT sequences in any CA constructed by this $\mathcal{R}$ are 0000 as all RMTs hold 0 as their next states. Moreover, every RMT of $(0, 0, \cdots, 0)$ is self replicating; thus only configuration 0^n (for any n) always form single length cycle. Another interesting observation is that the cycle structures remain *same* of the CAs which are designed by $\mathcal{R}$ while the CA sizes *increase*. Thus $\mathcal{R} =$ 00000000000000000000000000000000 maintains *stable* cyclic behavior. Though $\mathcal{R}$ follows the *two* key properties for classifier, but it fails to perform as *classifier* because the count of cycle is *one*.

Property 1. A strictly single length cycle CA can be a classifier if it generates at least *two* cycles.

Table 2. Number of strictly single length cycle rules with *majority* 0s where the count of cycles are at least *two*

# of 1s in rule	# of 0s in rule	Total possible rules	# of strictly single length cycle rules
1	31	32	1
2	30	496	32
3	29	4960	482
4	28	35960	4640
5	27	201376	30189

Next we want to observe strictly single length cycle rules if we add *impurity* in the rules i.e., in rule, the counts of 1s increase. Therefore, now we look the *unbalanced* rules. Obviously, we can not observe the CAs (designed by the same rule) for all possible sizes, so, an experimentation is performed for a few consecutive CA sizes; here it is 4 to 10. Next, we present the results. The *first* column of Table 2 refers to the count of 1s in the unbalanced rules and it gradually increasing (please look at the consecutive rows). The *second* column represents the count of 0s which decreases row wise. Total possible 2-state 5-neighborhood rules with x number of 0s and y number of 1s depicted in *third* column of the table i.e. $\binom{32}{y}$. The final column refers to the count of strictly single length cycle rules containing y 1s and generates at least *two* cycles in the respective CAs. When y increases, the number of possible rules also increase; therefore, in that case we perform the experimentation not on the entire space of rules, rather some consecutive rules are taken with the objective to extract the strictly single length cycles which are capable of generating MACAs with at least two attractors. For example, out of 906192 rules where $y = 6$ (that is number of 1's in the rule), only 12117 are holding the properties of our requirement. Analogous to the previous strategy, now we proceed with increasing the number of 0s in a rule with majority 1s. See Table 3. This table maintains the same columns description with only variation is that here, the count of 0 is less in the rules. Now, increase of y (in consideration with either 0 or 1) may lead to generate *balanced* rules where the count of 0s and 1s are same. In that also, we perform the experimentation. As the possible count is too high i.e., $\binom{32}{16}$, so we select some balanced rules and search for our *desired*

Table 3. Number of strictly single length cycle rules with *majority* 1s where the count of cycles are at least *two*.

# of 0s in rule	# of 1s in rule	Total possible rules	# of strictly single length cycle rules
1	31	32	6
2	30	496	109
3	29	4960	1123
4	28	35960	7282

CAs. The result says that out of 601080390, only 2130 (0.24%) are strictly single length cycle rules which can generate CAs with at least 2 cycles.

The above mentioned two tables in this section (referring to unbalanced rules) and the experimentation on balanced rules bring the following observation.

Observation 1. *2-state 5-neighborhood balanced and unbalanced rules (up to 25% 0 s or 12.5% 1 s) can be chosen as the candidate rules for generating strictly single length cycle CAs with at least two cycles.*

The last two tables demonstrate another interesting observation that the number of strictly single length cycles with at least two cycles are inversely proportional to y. The reason behind is that when number of y increases in rule it may increase the number of non self replicating RMTs and a collection of such non self replicating RMTs are found which may create multi length cycle which is not our desired CA. Among those, a set of unbalanced strictly single length cycle rules with combinations of y 0s or y 1s (in this case, y is at most 7) and even balanced strictly single length cycle rules generate CAs with count of single length cycles almost constant or very gradually increasing (i.e. increasing linearly and not exponentially) with increase in CA size. The CAs designed by these set of rules are actually maintaining approximately same cycle structure and they follow *stable* cyclic behavior. Next we look at Table 4. The first column represents the rule. Next two columns refer to the count 0 s and 1 s in the rules respectively. Columns *four* to *ten* refer to the cycle structures of strictly single length cycle rules for CA sizes 4 to 10. CAs designed by rule 4277534718 maintains same cycle structure [3(1)] for all possible investigated CA sizes. Every row of the column shows that the cycle structures are stable. This table also shows that 2-state 5-neighborhood CA can generate MACAs with *moderate* number of attractors (single length cycles) which is instrumental for any *good* CA classifier. Next, we use a small pool of strictly single length cycle rules with moderate number of cycles for the pattern classification.

3.2 5-Neighborhood CA as Pattern Classifier

In this section, we term single length cycles as "attractor". We choose the rules that have limited number of attractors so that they can be used for determining the class variable for data. Suppose a chosen rule $\mathcal{R}$ has 2 attractors - 'A' and

Table 4. A very few balanced and unbalanced 5-neighborhood rules with *stable* cyclic behavior

Rule	# of 0s in rule	# of 1s in rule	$n = 4$	$n = 5$	$n = 6$	$n = 7$	$n = 8$	$n = 9$	$n = 10$
84	29	3	3	3	3	4	4	4	5
1244158	16	16	3	3	3	4	4	4	4
1118	26	6	3	3	3	4	4	4	5
4277534718	4	28	3	3	3	3	3	3	3

'B' (here, 'A' and 'B' are decimal representation of the attractors). Let a dataset with 2 class information (class 0 and class 1) is taken and all of the instances are encoded to binary configurations (as 2-state CA is used here). We call these configurations as *useful* configurations. We then use the rule $\mathcal{R}$ to generate the configuration space of all useful configurations and observe which of the two attractors attract the particular instance. If attractor 'A' attracts more instances from class 0 than from class 1, it is given the class as 0. If the attractor attracts equal number of instances from both the classes, it can be given the either class. The predicted class for each attractor is determined by the majority voting of the actual classes associated with that attractor. Let us consider a scenario. An 4-cell CA designed by rule 1114110 generates 3 attractors (see Fig. 1). Let 0011, 1110, 0000, 1111, 0101 and 1010 are from class 0. The useful configurations from class 1 are 1000, 1011 and 1101. Therefore attractor 1100 represents more configurations from class 0, attractor 0000 represents only one configuration from class 0; whereas attractor 1101 identifies more configurations from class 1. Thus this CA mark attractors 0000 and 1100 as the *representative* of class 0 and 1101 as the representative of class 1. Let us explain the different steps of the process of classification.

1. **Encoding:** The datasets contain continuous as well as categorical attributes along with a class defined for each tuples. Therefore, we need to *map* real data to CA configurations. For that we use an *encoding* technique similar to the one mentioned in [9]. The categorical attribute is encoded by *one-hot* encoding and for continuous attribute, we use *frequency* encoding.
   - *Frequency encoding*: For each continuous attribute, the attribute's value range needs to be determined. Next, the range is divided into a specified number of intervals (bins). Every bin contains *close* data which are in ascending order. Next, if there are k bins, then map the data of the each bin to a binary string of length $k-1$. For example, let there be 3 bins. If the first one is represented as 00, then the second and third bins are represented as 01 and 11 respectively.
   - *One-hot encoding*: For each categorical attribute, we need to identify the count of *unique* data values and represent each data value using m-bit binary data where m is the count of the unique data values and any two different data values maintain distance as *two* in their representation. For example, if color is the categorical attribute and it has *three* unique data values red, green and blue. Then they are represented using one-hot coding encoding as follows respectively: 001, 010 and 100.
   - Let a dataset contain p continuous attributes and q categorical attributes where the first categorical attribute contains a_1 unique data elements, the second categorical attribute presents a_2 unique values and in analogous manner, we get a_q unique data elements for q^{th} categorical attribute. Now, let the number of bins be k for every continuous attribute. Thus the length of the encoded binary string can be computed as $n = (p * k) + (a_1 + a_2 \cdots + a_q)$. This n is actually the size of the CA which is to be used for the training and testing phases of the dataset.

2. **Training Procedure:** For training of our classifier model, the CA rules are used to define the class of their attractors based on which class instances the attractors attract the most.
   - As the inputs of training phase, we consider training dataset consisting of useful configurations and associated class labels; a set of 2-state 5-neighborhood strictly single length cycles with moderate number of attractors; length of useful configurations (fixed for all patterns) i.e., size of CAs.
   - For every useful configuration, we need to figure out the corresponding *attractor* where the configuration converges. This attractor is called as *useful* attractor as it used by the data object of the given dataset. For every useful attractor, we track how many objects are attracted from class 0 and class 1.
   - For each useful attractor, the counts of objects are compared with respect to class 0 and class 1. The class with the higher count is assigned to the attractor. In the event of a tie, class 0 and class 1 are assigned alternatively.
   - Let define the following variables: X_1 represents the total number of patterns in the training set; k represents the number of distinct and useful attractors found for a particular rule; A_i denotes the maximum class counts for each useful attractor. Let us calculate the training efficiency as:

$$\textit{Training Efficiency} = \left(\frac{\sum_{i=1}^{k} \max(\text{class 0 count}, \text{class 1 count})}{n}\right) \times 100$$

$$= \left(\frac{\sum_{i=1}^{k} Ai}{n}\right) \times 100$$

3. **Testing procedure and useful metrics:** After the training phase is done, the model is tested over the same set of rules used in training phase to observe the testing metrics. The attractors that have been labeled as class have been used to detect the class of original instances of the datasets. The testing accuracy and other metrics are calculated based on whether the attractor detected correct class or not.
   - The test dataset is imported in terms of useful configurations; the same set of rules of training phase is also imported for evaluation.
   - For each CA rule, the predicted class for each useful attractor obtained from the training phase is to be calculated.
   - Next, the initialization of true positives (TP), false positives (FP), and false negatives (FN) for each class of the dataset needs to be done.
   - Now, for every useful configuration in test dataset, the corresponding attractor needs to be identified for a given rule which has been used during training. Suppose a test configuration is from class 1 and its corresponding attractor in test phase represents class 1 in training phase.

Then the corresponding configuration is *correctly* predicted; otherwise the prediction is wrong. Thus a comparison needs to be done between the predicted class and the actual class of the test pattern. If the prediction matches the actual class, increment the corresponding true positive (TP) counter. If the prediction is incorrect, increment the false positive (FP) counter for the predicted class and the false negative (FN) counter for the actual class. In this way, *Accuracy*, *Precision*, *Recall* and *F-score* are to be calculated using the following formulas.
The accuracy is calculated as the percentage of correctly classified test patterns:

$$Accuracy = \left(\frac{\text{Correct Predictions}}{\text{Total Test Patterns}}\right) \times 100$$

The precision for class C (target class) is defined as the percentage of correct positive predictions for that class:

$$Precision_C = \frac{\text{TP}_C}{\text{TP}_C + \text{FP}_C} \times 100$$

where TP_C is the number of true positives and FP_C is the number of false positives for class C.

The recall for class C (target class) is defined as the percentage of actual positives correctly predicted:

$$Recall_C = \frac{\text{TP}_C}{\text{TP}_C + \text{FN}_C} \times 100$$

where FN_C is the number of false negatives for class C.

The F-score for class C (target class) is the harmonic mean of precision and recall:

$$F\text{-}score_C = \frac{2 \times \text{Precision}_C \times \text{Recall}_C}{\text{Precision}_C + \text{Recall}_C}$$

A CA classifier is said to be *good* for a dataset if all these four testing metrics evaluating higher values.

4 Results

This section reports the performance of our proposed 2-state 5-neighborhood CA based classifier on some real datasets retrieved from UCI ML repository (https://archive.ics.uci.edu/). Our model uses a collection of 150 CA rules (comprising

Table 5. Description of the Dataset

Name of the dataset	# of attributes in the dataset	# of objects in the dataset	Maximum bin size for continuous attributes	Useful configurations	CA size
Raisin	7	900	3	117	21
Rice	7	3810	3	252	21
Breast Cancer	9	116	2	71	18

of balanced and unbalanced) for training and testing. A brief description of the used datasets are as follows: In Table 5, *first* column represents the names of datasets used, from UCI ML repository. The data sets have few attributes whose count is represented in the *second* column. Here, all are continuous attributes. For that *fourth* column refers to the maximum bin size. It means that different continuous attributes can use different bin sizes. In the current discussion, all continuous attributes have the same bin size for a given dataset. Here, bin size refers to the length of the encoded binary string for the attributes. For example, in Raisin dataset, all attributes are continuous and we assign 3-bit for the bin size. As there are 7 attributes, therefore, each object of Raisin dataset is mapped to 21-bit $(7 * 3)$ binary string. The number of objects (instances) present in each of the datasets is represented by the *third* column. CA size (n) is referred by *fifth* column based on the bin size and the number of continuous attributes. For example, Raisin dataset uses 21-bit CA as the number of continuous attributes is 7 and bin size is 3. Out of 150 rules that are used for training and testing of the model, a small set of rules generates optimal results. Look Table 6 for those effective rules for good classification. In this table, we get balanced (rows 2 and 7), unbalanced rules with more 1 s (8 and 9) and unbalanced rules with more 0 s (rows 1, 3–5 and 7). The following results claim the efficacy of these rules in different datasets.

Table 6. Rules Categorization

CA rules	No. of states containing '1'	No. of states containing '0'
84	3	29
1244158	16	16
304	3	29
86	4	28
1118	6	26
1114110	16	16
754	6	26
4277534718	28	4
4227792879	29	3

Table 7. Raisin Dataset: 5-neighborhood CA based classifier results

CA rules	Efficiency	Accuracy	$Precision_1$	$Recall_1$	$F\text{-}score_1$
84	70.73%	77.14%	75%	100%	85.71%
1244158	69.51%	80%	76.92%	95.24%	85.11%
304	74.393%	80%	85%	80%	82.93%
86	71.95%	82.86%	82.61%	90.48%	86.36%
1118	73.17%	85.71%	86.36%	90.48%	88.37%
1114110	**80.49%**	**85.71%**	**90.48%**	**90.48%**	**90.48%**
4227792879	69.51%	68.57%	67.86%	90.48%	77.55%

First, we discuss on *Raisin* dataset. Table 7 shows the performance of 2-state 5-neighborhood CA model on Raisin dataset. Here, we have three types of rules (balanced, unbalanced with majority 1 and unbalanced with majority 0). Here, rule 1114110 produces the best results among other 150 CAs. Therefore, CA 1114110 is a *good* classifier for raisin dataset. Next, we compare CA 1114110 with some of the benchmark classifiers. For raisin dataset, we use the following popular classifiers: k-nearest neighbors (KNN), linear discriminant analysis (LDA), support vector classifier (SVC), extreme gradient boosting (XGB) and ridge (RIDGE) classifier. See Table 8. Here, as metrics, here we use *accuracy* and F-score with respect to class 1. CA 1114110 is compared with the results reported in Table 8 for the benchmark algorithms [16].

Our second dataset is *Rice* dataset. Table 9 depicts the role CA rules in classification of Rice dataset. Here, CA 1244158 is the best among the 150 CA classifiers for Rice dataset. Again we compare the performance of CA 1244158 on

Table 8. Comparison: CA 1114110 vs benchmark algorithms for Raisin dataset

Metrics	KNN	LDA	SVC	XGB	RIDGE	CA rule (1114110)
Accuracy	83.2%	85.7%	85.44%	86.88%	85.77%	85.71%
$F\text{-}score_1$	83.6%	86.39%	83.43%	86.22%	87.05%	90.48%

Table 9. RICE: 5-neighborhood CA based classifier results

CA rules	Efficiency	Accuracy	$Precision_0$	$Recall_0$	$F\text{-}score$
84	68.18%	61.84%	67.5%	62.79%	65.06%
1244158	**84.09%**	**80.26%**	**81.82%**	**83.72%**	**82.76%**
86	67.61%	69.74%	79.41%	62.79%	70.13%
1118	76.70%	65.79%	75.76%	58.14%	65.79%
1114110	82.95%	75%	73.08%	88.37%	80%
4277534718	56.82%	60.53%	59.428%	95.35%	73.21%

Table 10. Comparison with existing benchmark algorithms for Rice dataset

Metrics	KNN	SVC	RF	NB	CA rule (1244158)
Accuracy	83.2%	72.6%	91.8%	90.6%	80.26%
Precision	88.3%	73.3%	91.8%	90.6%	81.82%
Recall	88.3%	72.6%	91.8%	90.6%	83.72%
F-score	88.3%	71.4%	91.8%	90.6%	82.76%

Table 11. Breast Cancer Dataset: 5-neighborhood CA based classifier result

CA rules	Efficiency	Accuracy	$Precision_1$	$Recall_1$	$F\text{-}score_1$
84	**81.63%**	**63.64%**	**58.82%**	**90.91%**	**71.43%**
1244158	85.71%	50%	50%	100%	66.67%
86	75.51%	50%	50%	90.91%	64.52%
754	79.59%	63.64%	83.33%	45.45%	58.82%
1114110	85.71%	59.09%	55%	100%	70.97%
4277534718	75.5%	54.55%	52.63%	90.91%	66.7%

Table 12. Comparison with existing benchmark algorithms for Breast Cancer Dataset

Metrics	KNN	NB	DT	RF	CA rule (84)
Precision	48%	66%	70%	71%	58.82%
Recall	48%	60%	70%	71%	90.91%
$F\text{-}score_1$	48%	59%	70%	70%	71.436%

the same dataset with respect to some existing algorithms. We use the following algorithms for the comparison: KNN, SVC, random forest (RF) and naive bayes (NB) classifier. Table 10 shows the comparative analysis based on all four metrics between CA 1244158 and the respective four benchmark algorithms [5].

Our final dataset is *Breast Cancer* dataset. Table 11 represents the performance of our proposed classifiers on Breast cancer dataset. The best classifier is 84. Next, we use the following algorithms for comparison with our proposed classifier model: KNN, Decision Tree (DT), Random Forest (RF) and Gaussian Naive Bayes (NB) classifier. See Table 12 for the analysis report. Here, we use three metrics: precision, recall and F-score. From the experimentation, one interesting observation is that CA rule 1114110 acts as *good* classifier for all three different datasets.

5 Conclusion

It is observed that the attractor dynamics of 2-state 5-neighborhood balanced and unbalanced rules produce optimal classification results. This work considers only strictly single length cycles CAs. Here, those rules have been chosen for the CA based classifier which produce relatively less number of cycles irrespective of the CA sizes. Another interesting feature of the candidate rule is that such rule can design CAs of different sizes either with same cycle structure or with minimal variation in the cycle structures. By maintaining this stable cyclic behavior property, the CA attractors can be identified which can then be labeled as distinct classes that correlate with the classes of the used datasets. The optimal training and testing metrics demonstrate the potential of 5-neighborhood binary CA as a powerful tool for classification purpose.

References

1. Adak, S., Naskar, N., Maji, P., Das, S.: On synthesis of non-uniform cellular automata having only point attractors. J. Cell. Autom. **12**(1–2), 81–100 (2016)
2. Chaudhuri, P.P., Chowdhury, D.R., Nandi, S., Chatterjee, S.: Additive cellular automata – theory and applications, vol. 1. IEEE Computer Society Press, USA, ISBN 0-8186-7717-1 (1997)
3. Das, S., Mukherjee, S., Naskar, N., Sikdar, B.K.: Characterization of single cycle ca and its application in pattern classification. Electron. Notes Theor. Comput. Sci. **252**, 181–203 (2009)
4. Ganguly, N.: Cellular Automata Evolution : Theory and Applications in Pattern Recognition and Classification. Ph.D. thesis, Bengal Engineering College (a Deemed University), India (2004)
5. KIRBAŞ, İ., ÇİFCİ, A.: Machine learning-based rice grain classification through numerical feature extraction from rice image data
6. Maji, P., Shaw, C., Ganguly, N., Sikdar, B.K., Chaudhuri, P.P.: Theory and application of cellular automata for pattern classification. Fund. Inform. **58**, 321–354 (2003)
7. Maji, P.: Cellular Automata Evolution for Pattern Recognition. Ph.D. thesis, Jadavpur University, Kolkata, India (2005)
8. Maji, P., Chaudhuri, P.P.: RBFFCA: a hybrid pattern classifier using radial basis function and fuzzy cellular automata. Fund. Inform. **78**, 369–396 (2007)
9. Mukherjee, S., Bhattacharjee, K., Das, S.: Cycle based clustering using reversible cellular automata. In: International Workshop on Cellular Automata and Discrete Complex Systems, pp. 29–42. Springer (2020)
10. Mukherjee, S., Sindhuja, J.: Characterization of 3-state cellular automata with single length cycles. In: Cellular Automata Technology, pp. 137–152. Springer, Cham (2024)
11. Naskar, M.B.: Characterization and Synthesis of Non-Uniform Cellular Automata with Point State Attractors. Ph.D. thesis, Indian Institute of Engineering Science and Technology, Shibpur, India (2015)
12. Naskar, N., Adak, S.: Designing multi-class pattern classifier using ca under periodic boundary condition. In: 2018 Fifth International Conference on Emerging Applications of Information Technology (EAIT), pp. 1–4 (2018)
13. Naskar, N., Das, S., Sikdar, B.K.: Characterization of nonlinear cellular automata having only single length cycle attractors. J. Cell. Autom. **7**(5–6), 431–453 (2012)

14. Sethi, B., Roy, S., Das, S.: Asynchronous cellular automata and pattern classification. Complexity **21**(S1), 370–386 (2016)
15. Wolfram, S.: A New Kind of Science. Inc, Wolfram Media (2002)
16. Yılmaz, E.K., Oğuz, T., Adem, K.: A cnn-based hybrid approach to classification of raisin grains. In: International Conference on Frontiers in Academic Research, vol. 1, pp. 460–467 (2023)

A Study on Role of Distant Neighbor over Elementary Cellular Automata

Kamalika Bhattacharjee[1(✉)], Bhumika Sikdar[2], and Swasti Jain[3]

[1] Department of Computer Science and Engineering, National Institute of Technology, Tiruchirappalli 620015, Tamilnadu, India
kamalika.it@gmail.com
[2] Kolkata, West Bengal, India
[3] Jabalpur Engineering College, Jabalpur 482011, Madhya Pradesh, India

Abstract. This research investigates the role of distant neighbor in Elementary Cellular Automaton (ECA). Findings reveal distinct patterns and behaviors across different ECA rules when a CA cell accepts distant neighbours' interaction. The study employs spacetime diagrams, density plots, and heatmaps to visualize and quantify the interactions, comprehensively analyzing the emergent patterns. Results demonstrate that the distant neighbor interaction can significantly alter the behavior of a CA, leading to new insights into the mechanisms driving complexity and chaos. A classification of the ECAs, based on the robustness of the ECA rules on distant neighbor interaction, is reported. Such classification enables better modeling of economic and biological systems.

Keywords: Elementary cellular automata · Distant neighbor · Wolfram's classes · Robustness · Classification

1 Introduction

Elementary Cellular Automata (ECAs) symbolize the simplest class of one-dimensional CAs, where each cell in a lattice can take a binary state (0 or 1) and transmute based on its current state and the states of its immediate neighbors. In the early 1980s, Wolfram's seminal work has emphasized the potential of cellular automaton (CA) to comprehend the complexity and emergent behaviors in discrete systems and stipulated the classification of the ECAs into four distinct classes: Class I, II, III, and IV [11] based on their dynamical behaviors. Since then, many works have been recorded exploring the dynamics of CA [3,6,8–10].

Traditionally, evolution of a CA cell involves local interactions between the immediate neighbors. Despite extensive research on CAs, the influence of distant

B. Sikdar—Independent Researcher

This work is done as a project in the Indian Summer School on Cellular Automata 2024.

S. Kamilya et al. (Eds.): ASCAT 2025, CCIS 2499, pp. 101–114, 2025.
https://doi.org/10.1007/978-3-031-94121-4_8

neighbor interaction on the evolution remains an arena that has not been fully explored [1,4,7]. In [7], 2-state 3-input non-local CAs are classified based on the dynamical behaviors and mean-field parametrization. The concept of robust universal computation has been introduced for evolution of such a CA, where a cell considers randomly chosen 3 neighbors. It is argued that the non-local CA, performing universal computation, can be easily found. The authors of [1] consider some fixed long-distance neighbors in addition to the local neighborhood and propose new rules for DCT. In [4], on the other hand, long-distance CAs (LDCAs) are explored. It has been found that rule 73 can be Turing universal. However, the properties of LDCAs are not governed by the ECAs.

The additional neighbor can be considered as an external element that alters the behavior of a system. The study of distant-neighbor interaction, therefore, can provide deeper insights into the system's dynamics pointing to the mechanisms driving complexity and chaos. In this background, this research aims to analyze the resilience of a system to this external intruder under minimal conditions. We have studied the behavior of CAs considering that only one cell has a distant neighbor (one-way interaction) in addition to the local neighbors. The effect on the evolution of ECAs is explored. It is observed that some ECAs are robust as the distant neighbor interaction has zero or minimal impact on its dynamics.

Section 2 of the paper provides the background of this research work. The desired integration of distant neighbor is formalized in Sect. 3. Section 4 reports the classification of ECAs based on robustness. Detailed experimental results and analysis are in Sect. 5. Section 6 concludes the work.

2 Background

This work considers the CA of 1-dimensional with 3-neighborhood. Each cell is updated based on its current state and the states of its two immediate (left and right) neighbors. These CAs are referred to as ECAs. If S_i^t be the present state (at time t) of the ith cell and the next state function (local rule or simply, *rule*) R_i, then its next state (at $t+1$) is $S_i^{t+1} = R_i\left(S_{i-1}^t, S_i^t, S_{i+1}^t\right)$, where S_{i-1}^t, S_{i+1}^t are the present states of left neighbor (cell $i-1$), and right neighbor (cell $i+1$). So, there are $2^{2^3} = 256$ rules [11] and among them 88 are called *minimal* as they form equivalence classes under simple transformations [12]. Wolfram [11] classified all the ECA rules into four categories.

Class I: Almost all initial configurations converge to a fixed homogeneous state.
Class II: The initial configurations converge to more than one fixed state. Separated periodic regions are produced.
Class III: Such CAs produce chaotic patterns.
Class IV: Exhibits behavior intermediate between periodic and chaotic. Such a CA produce complex patterns.

For Class I rules, the information regarding the changes in initial configuration may die out. For Class II, it may remain localized. In the case of Class III, the changes may propagate throughout the automaton. For Class IV rules, the information is carried by some localized structures. However, all these are true for classical CAs defined over infinite lattice. To imitate the behavior of classical CAs, we take periodic boundary conditions over lattice sizes n, ($n = \mathbb{Z}/n\mathbb{Z}$). Each cell of the lattice updates its state based on the ECA rule except an arbitrarily chosen cell $Cell_A$, that has an additional neighbor ($CellB$) at a distance Δ. However, $Cell_B$ is not dependent on $Cell_A$.

3 Integration of Distant Neighbor on Evolution of ECAs

The immediate neighbors often influence the evolution of a CA. Similarly, the introduction of long-range interaction may cause the system to exhibit class transitions. If such interaction becomes more significant, then the system may show signs of criticality.[1] The current investigation is done on the role of distant-neighbormin ECA by incorporating all possible logical operations between $Cell_A$ and its distant neighbor $Cell_B$. For example, logical **OR** (R_α) is defined as

$$R_\alpha\text{: } S^{t+1}_{Cell_A} = R(S^t_{Cell_A-1}, S^t_{Cell_A}, S^t_{Cell_A+1}) \vee S_{Cell_B} \text{ .}$$

Others are R_β (**AND**)/R_γ (**XOR**)/R_δ (**NOR**)/R_ϵ (**NAND**)/R_ζ (**XNOR**).

The study prioritises Class II/III/IV rules [11]. Class I rules are excluded because such rules exhibit minimal variability in dynamics. The methodology involves simulation of all minimal ECAs (with/without distant neighbor) using Python leveraging libraries NumPy (for numerical computations) and Matplotlib (for plotting). The *space-time diagrams* are generated. The active cell densities are computed via *density plots*, and the differences are elucidated through *heatmaps*.

Two lattice sizes are considered, ***small*** ($n = 101$) and ***large*** ($n = 1001$). The chosen initial configurations are **(i)** $Cell_A$ is active (state 1) and all other cells are set to 0, and **(ii)** *Arbitrary initial configuration* (each cell having equal probability of occupying either 0 or 1). However, the analysis shown here is mainly for the initial configuration **(i)**, for lattice size 101 and 1001. Arbitrary initial configurations and different lattice sizes are considered to verify the results.

To observe whether the distance Δ has any impact on the dynamics, the Δ is varied as (1) $5, 10, 15, 20, \cdots, n-1$ and (2) $2, 4, 8, 16, \cdots, n-1$. The system is evolved for a single short-term time duration of 100 steps in order to capture the early-stage dynamics. If cannot be decided, then it is evolved for longer duration.

4 Classification of ECAs Based on Robustness

To quantify and categorize the impact of interaction of the distant neighbor, the concept of robustness is introduced as a measure of the extent to which a rule's dynamics is perturbed-that is, the space-time diagram of ECA_d (with distant neighbor) undergoes alterations in comparison to the ECA. Based on the stability and resilience portrayed, the rules are classified into *most robust*, *partially robust*, *weakly robust*, and *least robust*.

[1] Criticality in CA often refers to the balance between order and chaos [2].

1. Most Robust: A rule is *most robust* if its evolution is not much affected by the distant-neighbor, regardless of the kind of logical operation. There can be changes in heatmap but the resulting space-time diagram mostly retained, exemplifying a high degree of stability. There exists a subset of such rules, where the ECA remains completely unchanged under a logical operation. Such a rule is *completely robust* under that operation (Fig. 1). Here, heatmap does not show density differences.

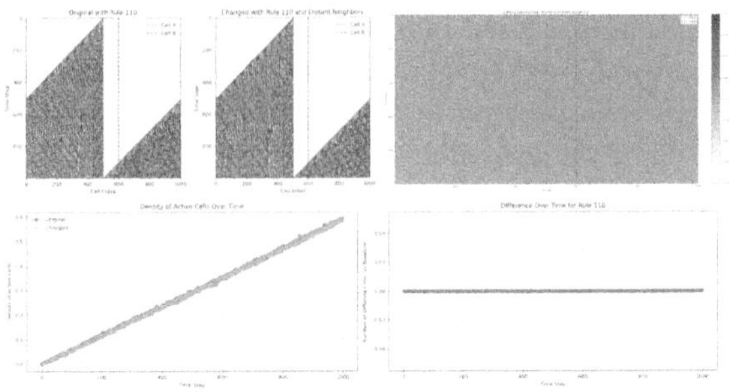

Fig. 1. ECA 110 showing complete robustness under R_α

2. Partially Robust: ECAs in this category show moderate changes. There may be localized changes or variations in density, but the fundamental nature of the system's evolution remains largely preserved. Such rules exhibit a moderate level of resilience, adapting to perturbations.

3. Weakly Robust: *Weakly robust* ECAs display noticeable changes in their spacetime diagrams when interacting with the distant neighbor. The ECA_d differ significantly, potentially leading to transformations in dynamics, such as changes from order to more complex or even chaotic, but not entirely erratic.

4. Least Robust: A rule is *least robust* if the introduction of distant-neighbor interaction leads to a complete alteration in its evolution. The changes are often dramatic, chaotic, diffused, or complex in dynamics. Such rules are most susceptible to external perturbation, indicating a system prone to instability.

Miscellaneous: A rule of miscellaneous group might be mostly robust for one specific cell distance but becomes least robust when the distance is altered.

5 Results and Analysis

This section reports the impact of distant neighbor on ECAs of Class II/III/IV.

5.1 Class II

The sample evolutions of Class II ECAs and ECA_ds are shown in Fig. 2 and Fig. 3. Table 1 reports the robustness classification of Class II rules.

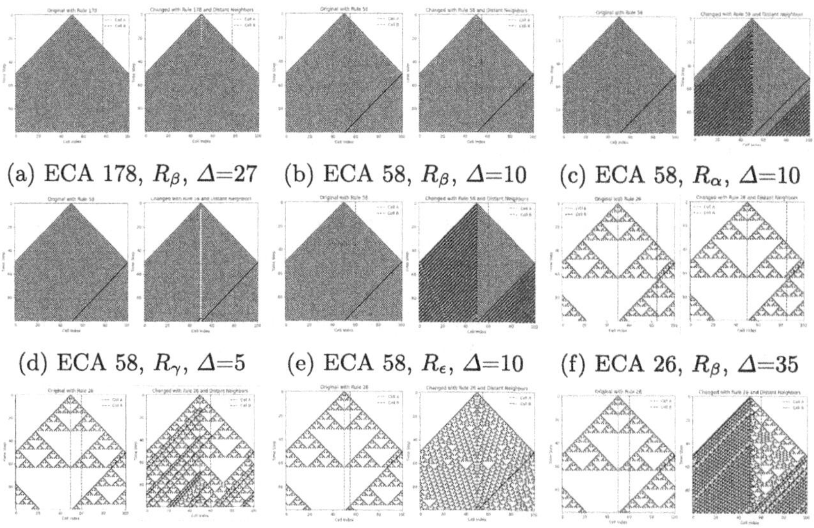

(a) ECA 178, R_β, Δ=27 (b) ECA 58, R_β, Δ=10 (c) ECA 58, R_α, Δ=10

(d) ECA 58, R_γ, Δ=5 (e) ECA 58, R_ϵ, Δ=10 (f) ECA 26, R_β, Δ=35

(g) ECA 26, R_α, Δ=10 (h) ECA 26, R_γ, Δ=5 (i) ECA26, R_ϵ, Δ=10

Fig. 2. Sample evolutions of Class II ECAs for n = 101 and time-steps = 100

1. Operation R$_\beta$: The ECAs **1, 2, 10, 24, 26, 33, 34, 42, 51, 56, 72, 74, 104, 130, 138, 152, 154, 162, 170, 184, 200,** and **232** are *completely robust* under R_β. Whereas, the ECAs **46, 76, 78, 94, 108, 132, 134, 140, 142, 156, 164,** and **172** exhibit *consistent patterns.* The Class II rules are *mostly robust* under R_β (Table 1). The ECAs **58** (Fig. 2b) and **178** (Fig. 2a) have very similar initial space-time patterns except for a diagonal line in ECA 58 (Fig. 2b). This phenomenon contains a resemblance to some changed space-time diagrams in ECA **26** (Fig. 2i) and ECA **154**. However, R_β has no effect on ECA **26** (Fig. 2f) and ECA **154** and minimally alters ECA **58** (Fig. 2b).

2. Operations R_α and R_γ: The ECAs **4, 12, 36, 44, 72, 76, 104, 108, 132, 140, 164, 172, 200, 204,** and **232** show *complete robustness* under R_γ. The ECA **51** showcases *consistent patterns* under R_γ. The ECAs **1, 4, 5, 12, 13, 28, 29, 33, 36, 37, 44, 51, 72, 76, 77, 78, 104, 108, 132, 140, 156, 164, 172, 178, 200, 204,** and **232** are *completely robust* under R_α. However, the ECA **50**, for alternate Δ, and the ECA **62**, for a few random distant neighbors, are *completely robust* under R_α. The appearance of diagonals is very similar for every distant neighbor for the ECAs **58**, **26**, and **154** under R_α (Fig. 2g and

2c) and R_γ (Fig. 2h and 2d). The Class II rules are *most robust* under R_α and R_γ.

3. Operation R_ϵ: The ECAs **4, 12, 36, 44, 76, 108, 132, 140, 164, 172,** and **204** show *complete robustness* and the ECAs **51, 57, 72, 73, 104, 130, 138, 142, 162,** and **170** show *consistent patterns* under R_ϵ. The operation R_ϵ, has unique effects on the ECAs **57** and **73**, and that is different from other five operations. The R_ϵ changes the dynamics of rule **73** (Fig. 3a/3b). With arbitrary initial configuration, it does not exhibit *consistent patterns*. The change in the dynamics (Fig. 3b), due to distant neighbor, is biased to the place of the neighbor. In ECA **57**, among the 6 operations, only R_ϵ changes the pattern (Fig. 3c and 3d). ECA **26** (Fig. 2i) and **58** (Fig. 2e) shows very similar patterns under R_ϵ,.

4. Operations R_δ and R_ζ: The ECAs **51, 72, 76, 77, 104, 108, 132, 140, 164, 172** and **51, 72, 76, 77, 104, 108, 132, 140, 164, 172** yield *consistent patterns* under R_δ and R_ζ respectively. Each negation operation also has quite similar effects on the ECAs **26, 154** and **58**. The comparable pattern generation is shown in Fig. 2i and 2e for ECA **26** and **58** respectively, under R_ϵ.

(a) ECA 73, R_ϵ, Δ=35, n=1001, t=1000

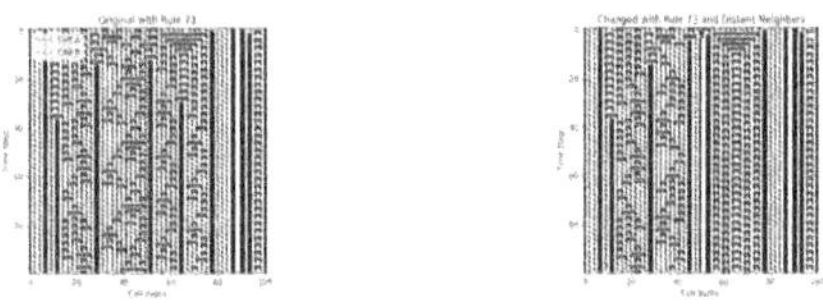

(b) ECA 73, R_ϵ, Δ=30, n=101, t=100

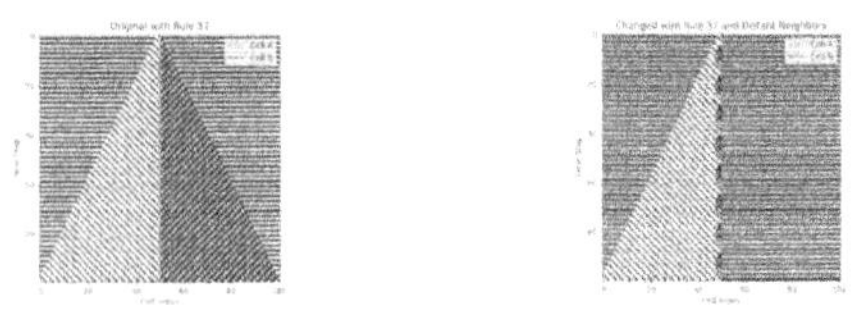

(c) ECA 57, R_ϵ, Δ=5, n=101, t=100

(d) ECA 57, R_ϵ, Δ=25, n=101, t=100

Fig. 3. Dynamics of ECA 73 and ECA 57 for R_ϵ (t = time-steps)

Table 1. Robustness classification of Class II ECAs

n = 101 and 1001; Time steps = 100 and 1000				
ECAs	Most robust	Partially robust	Weakly robust	Least robust
1	R_γ, R_β, R_α	R_ζ, R_δ		
2	R_γ, R_β, R_α			R_ϵ, R_δ, R_ζ
3	R_ϵ, R_δ, R_ζ	R_γ, R_α, R_β		
4	R_γ, R_α, R_β, R_ϵ, R_δ, R_ζ			
5	R_γ, R_α, R_β, R_ϵ, R_δ, R_ζ			

(*continued*)

Table 1. (*continued*)

n = 101 and 1001; Time steps = 100 and 1000				
ECAs	Most robust	Partially robust	Weakly robust	Least robust
6	R_γ, R_α, R_β			R_ϵ, R_δ, R_ζ
7	R_γ, R_α, R_β, R_ϵ, R_δ, R_ζ			
9				R_γ, R_α, R_β, R_ϵ, R_δ, R_ζ
10	R_γ, R_α, R_β			R_ϵ, R_δ, R_ζ
11				R_γ, R_α, R_β, R_ϵ, R_δ, R_ζ
12	R_γ, R_α, R_β, R_ϵ, R_δ, R_ζ			
13	R_α, R_γ, R_β, R_ϵ, R_δ, R_ζ			
14	R_γ, R_α, R_β			R_ϵ, R_δ, R_ζ
15		R_ϵ, R_δ		R_γ, R_α, R_β, R_ζ
19	R_γ, R_α, R_β, R_ϵ, R_δ, R_ζ			
23	R_γ, R_α, R_β, R_ϵ, R_δ, R_ζ			
24	R_β	R_γ, R_α		R_ϵ, R_δ, R_ζ
25	R_ϵ		R_γ, R_α, R_β, R_δ, R_ζ	
26	R_β		R_γ, R_α	R_ϵ, R_δ, R_ζ
27	R_ϵ, R_ζ		R_γ, R_α, R_β, R_δ	
28	R_γ, R_α, R_β, R_ϵ	R_ζ	R_δ	
29	R_γ, R_α, R_ϵ, R_β, R_δ, R_ζ			
33	R_γ, R_α, R_β, R_ϵ, R_δ, R_ζ			
34	R_γ, R_α, R_β			R_ϵ, R_δ, R_ζ
35		R_γ, R_δ, R_ζ	R_α, R_β, R_ϵ	
36	R_γ, R_α, R_β R_ϵ, R_δ, R_ζ			
37	R_α, R_γ, R_ϵ, R_β, R_δ, R_ζ			
38	R_γ, R_α, R_β			R_ϵ, R_δ, R_ζ
42	R_γ, R_α, R_β			R_ϵ, R_δ, R_ζ
43				R_γ, R_α, R_β , R_ϵ, R_δ, R_ζ
44	R_γ, R_α, R_β , R_ϵ, R_δ, R_ζ			
46	R_γ, R_α, R_β			R_ϵ, R_δ, R_ζ
50	R_γ, R_α, R_β, R_ϵ, R_δ, R_ζ			
51	R_γ, R_α, R_β , R_ϵ, R_δ, R_ζ			
56	R_β	R_γ, R_α		R_ϵ, R_δ, R_ζ
57	R_γ, R_α, R_β, R_δ, R_ζ		R_ϵ	
58	R_β, R_δ		R_ϵ	
62	R_γ, R_α, R_ϵ	R_β, R_δ, R_ζ		
72	R_γ, R_α, R_β, R_ϵ, R_δ, R_ζ			
73		R_δ		R_ϵ
74	R_γ, R_α, R_β			R_ϵ, R_δ, R_ζ
76	R_γ, R_α, R_β, R_ϵ, R_δ, R_ζ			
77	R_γ, R_α, R_β, R_ϵ, R_δ, R_ζ			
78	R_γ, R_α, R_β, R_ϵ, R_δ, R_ζ			
94	R_γ, R_α, R_β, R_ϵ, R_δ, R_ζ			
104	R_γ, R_α, R_β, R_ϵ, R_δ, R_ζ			
108	R_γ, R_α, R_β, R_ϵ, R_δ, R_ζ			
130	R_γ, R_α, R_β			R_ϵ, R_δ, R_ζ
132	R_γ, R_α, R_β, R_ϵ, R_δ, R_ζ			

(*continued*)

Table 1. (*continued*)

n = 101 and 1001; Time steps = 100 and 1000				
ECAs	Most robust	Partially robust	Weakly robust	Least robust
134	R_γ, R_α, R_β			R_ϵ, R_δ, R_ζ
138	R_γ, R_α, R_β		R_δ, R_ζ	R_ϵ
140	R_γ, R_α, R_β, R_ϵ, R_δ, R_ζ			
142	R_γ, R_α, R_β		R_δ, R_ζ	R_ϵ
152	R_β	R_α, R_γ		R_ϵ, R_δ, R_ζ
154	R_β		R_γ, R_α	R_ϵ, R_δ, R_ζ
156	R_γ, R_α, R_ϵ			*R_β, R_δ, R_ζ
162	R_γ, R_α, R_β			R_ϵ, R_δ, R_ζ
164	R_γ, R_α, R_β, R_ϵ, R_δ, R_ζ			
170	R_γ, R_α, R_β		R_δ, R_ζ	R_ϵ
172	R_γ, R_α, R_β, R_ϵ, R_δ, R_ζ			
178	R_γ, R_α, R_β, R_ϵ, R_δ, R_ζ			
184	R_β	R_γ, R_α		R_ϵ, R_δ, R_ζ
200	R_γ, R_α, R_β, R_ϵ, R_δ, R_ζ			
204	R_γ, R_α, R_β, R_ϵ, R_δ, R_ζ			
232	R_γ, R_α, R_β, R_ϵ, R_δ, R_ζ			

5.2 Class III

The behavior of Class III ECAs, under the different logical operations, is summarized in Table 2.

1. Operation R_β: The ECAs that have the original pattern as an *intricate* Pascal's triangle, such as, **18**, **90** (Fig. 4a), **146**, **154** and **26** (Fig. 2f), R_β cannot change or diffuse the pattern (*completely robust*). The ECAs **22** (Fig. 4b) and **60** (Fig. 7e and 7f), that have a *little less intricate* Pascal's triangle, the R_β can change them slightly, and yields *consistent patterns.* The ECA **150** (Fig. 4f) building blocks are similar to those of ECA **126** (Fig. 4e). Therefore, R_β yields a *consistent pattern* for all distant neighbors. For the ECA **126**, which has an *extremely less intricate* Pascal's triangle, the R_β diffuses it more (Fig. 4e). The ECA **122** is mostly robust for R_β (Fig. 4c).

Table 2. Robustness classification of Class III ECAs

n = 101 and 1001 ; Time steps = 100 and 1000				
ECAs	Most robust	Partially robust	Weakly robust	Least robust
18	R_β		R_γ, R_α	R_ϵ, R_δ, R_ζ
22	R_β		R_γ, R_α	R_ϵ, R_δ, R_ζ
30	R_γ, R_α, R_δ, R_ζ			
45	R_γ, R_α, R_β, R_ζ	R_ϵ	R_δ	
60	R_α, R_β, R_δ		R_γ, R_ϵ, R_ζ	
90	R_β		R_γ, R_α	R_ϵ, R_δ, R_ζ
105		R_ϵ, R_δ	R_γ, R_α, R_β	R_ζ
122	R_β		R_α, R_γ	R_ϵ, R_δ, R_ζ
126			R_γ, R_α	R_β, R_ϵ, R_δ, R_ζ
146	R_β		R_γ, R_α	R_ϵ, R_δ, R_ζ
150	R_α	R_β	R_δ, R_ϵ, R_γ	R_ζ

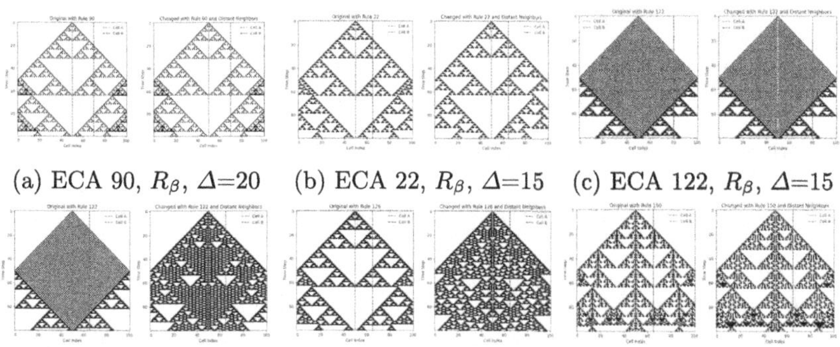

(a) ECA 90, R_β, Δ=20 (b) ECA 22, R_β, Δ=15 (c) ECA 122, R_β, Δ=15

(d) ECA 122, R_ϵ, Δ=15 (e) ECA 126, R_β, Δ=15 (f) ECA 150, R_β, Δ=15

Fig. 4. Sample Class III evolutions for n = 101 and time-steps = 100

2. Operation R_ϵ: For the ECA **90** (Fig. 5c and 5d), **18** (Fig. 5a and 5b), **122** (Fig. 5k and 5l), and **105**, the R_ϵ creates *consistent patterns* for every distant neighbor. The changed diagrams for the ECAs **18**, **90** and **122** are of a particular pattern. The ECA **122** shows diffused but *consistent patterns* for R_ϵ (Fig. 4d). The ECA **50** (Class II) (Fig. 5i and 5j), and the ECA **122** (Fig. 5k and 5l) have similar original space-time patterns and also show *consistent patterns* for R_ϵ. Usually R_β and R_ϵ exhibit quite opposite effects on Class III rules, except for ECA **126**, where both the operations heavily diffuse the pattern.

3. Operations R_α and R_γ: The ECAs **60** (Fig. 7c) and **150** are *completely robust* under R_α (Table 2). For the ECAs **90, 18, 122** and **146,** the patterns

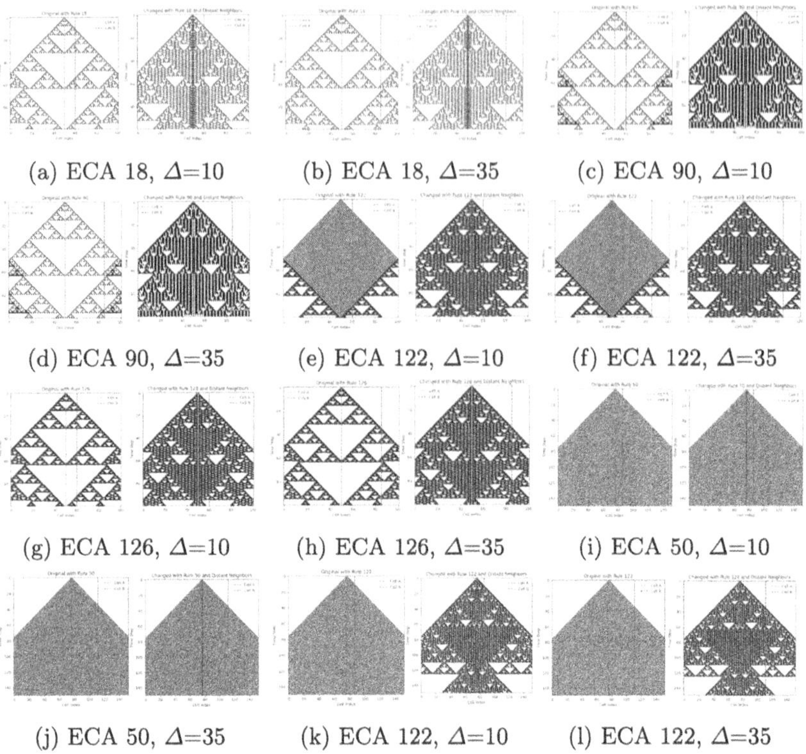

(a) ECA 18, Δ=10 (b) ECA 18, Δ=35 (c) ECA 90, Δ=10

(d) ECA 90, Δ=35 (e) ECA 122, Δ=10 (f) ECA 122, Δ=35

(g) ECA 126, Δ=10 (h) ECA 126, Δ=35 (i) ECA 50, Δ=10

(j) ECA 50, Δ=35 (k) ECA 122, Δ=10 (l) ECA 122, Δ=35

Fig. 5. ECA evolutions under R_ϵ

under R_γ and R_α are almost same for every equal position of the distant neighbor. The majority of Class III ECAs are *weakly robust* under R_α and R_γ.

4. Operations R_δ and R_ζ: For the ECAs **90** and **18**, the R_ζ and R_δ yield almost the same patterns for every equal position of the distant neighbor. The ECAs **105** and **126** show *consistent patterns* under R_δ. Most of the Class III rules are *least robust* under R_δ and R_ζ.

5. Attractor Configuration: For a few arbitrary initial configurations and Δ, the ECA **30** (Fig. 6c, 6a and 6e) and the ECA **45** (Fig. 6b, 6f and 6d) follow transition from chaotic to attractor dynamics under R_β, R_ϵ and R_δ. Therefore, for R_β, R_ϵ, and R_δ, the ECA **30** cannot be categorized in terms of robustness. The ECAs **30** and **45** have strong intrinsic criticality/power-law behavior [5].

5.3 Class IV

The behavior of Class IV ECA rules is described in Table 3.

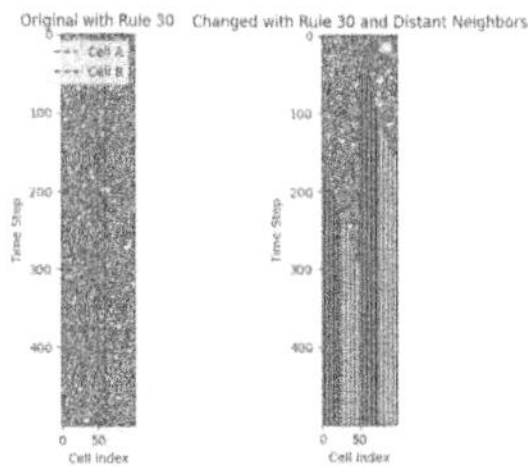

(a) ECA 30, R_ϵ, Δ=9

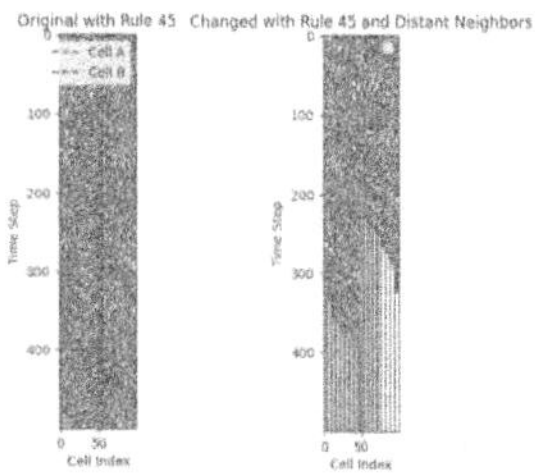

(b) ECA 45, R_β, Δ=5

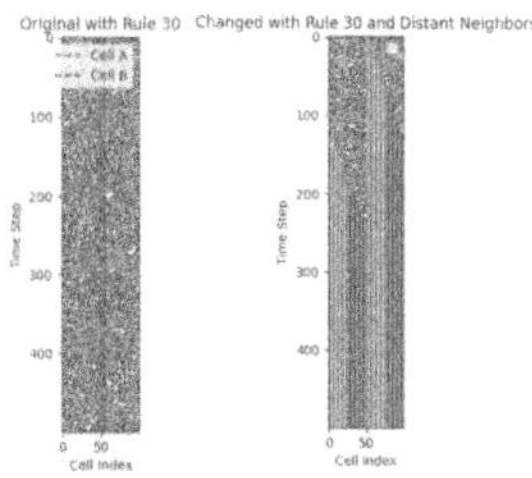

(c) ECA 30, R_β, Δ=4

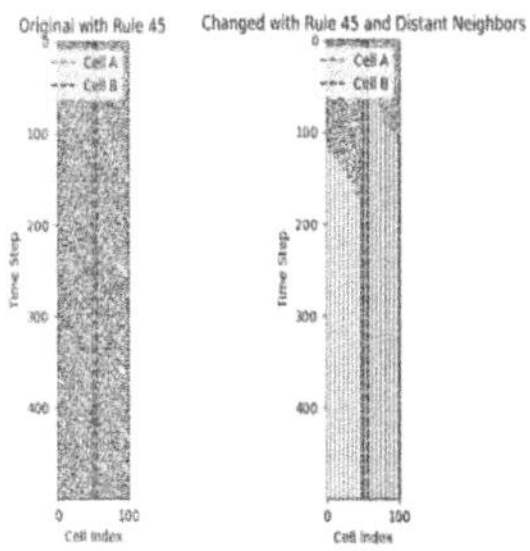

(d) ECA 45, R_δ, Δ=5

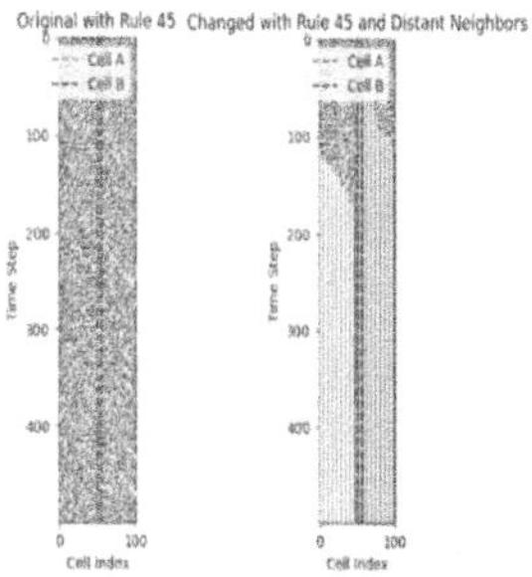

(e) ECA 30, R_δ, Δ=9

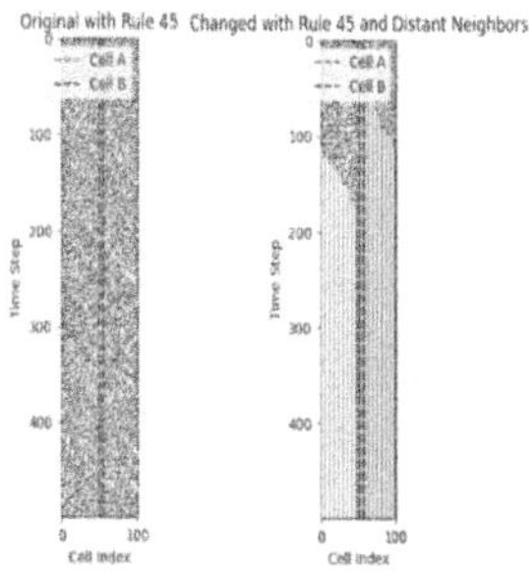

(f) ECA 45, R_ϵ, Δ=5

Fig. 6. ECAs with chaotic-to-attractor transitions at $n = 100$, time-steps = 500

Table 3. Robustness classification of Class IV ECAs

n = 101 and 1001 ; Timesteps = 100 and 1000				
ECA	Most robust	Partially robust	Weakly robust	Least robust
41				R_γ, R_β, R_α, R_ϵ, R_δ, R_ζ
54	R_δ	R_γ, R_α, R_β, R_ϵ, R_ζ		
106	R_γ, R_α, R_β			R_ϵ, R_δ, R_ζ
110	R_α	R_β	R_γ, R_ϵ	R_δ, R_ζ

1. Operations R_α, R_β and R_ϵ: The complex ECA **110** is *completely robust* under R_α (Fig. 7a and 7b). It yields *consistent patterns* under R_β and is categorized as *partially robust* (Fig. 7d and 7g). In spite of several changes in the heatmap, the space-time diagram is visually not much altered. However, for arbitrary initial configurations, the ECA **110** under R_α yields abrupt transitions between two different complex patterns for a few Δs (Fig. 7i and 7j). Similar abrupt changes have been observed for R_β and R_γ as well. The ECA **110** under R_α and R_β are quite similar to ECA **60** and **150** (Class III). For these ECAs, R_α yields no change (Fig. 7c, 7h, 7a and 7b), however for R_β, the changes are consistent for all Δs (Fig. 7e, 7f, 7d and 7g). The ECA **41** exhibits consistent patterns under R_β and the rule **106** is *completely robust* under the same. Just like R_β, the R_ϵ also produces *consistent patterns* in ECA **41**. These two logi-

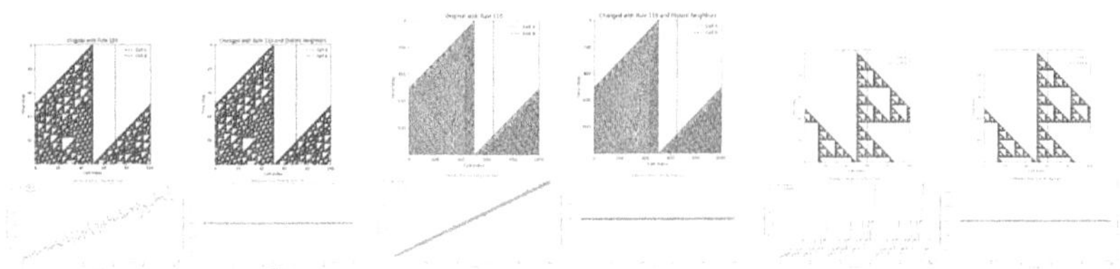

(a) ECA 110, R_α, Δ=20, n=101, t=100 (b) ECA 110, R_α, Δ=150, n=1001, t=1000 (c) ECA 60, R_α, Δ=10, n=101, t=100

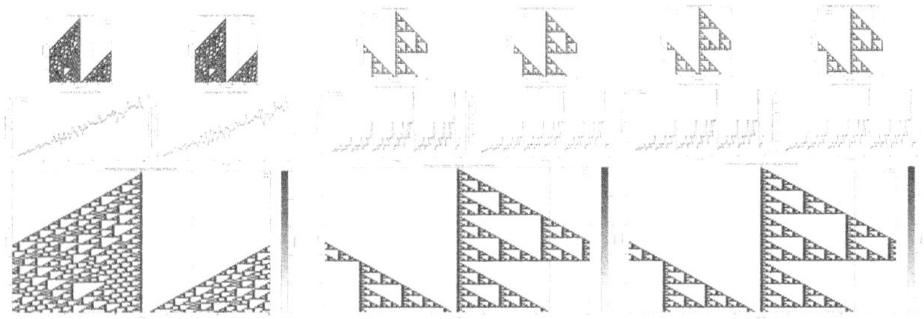

(d) ECA 110, R_β, Δ=20, n=101, t=100 (e) ECA 60, R_β, Δ=10, n=101, t=100 (f) ECA 60, R_β, Δ=35, n=101, t=100

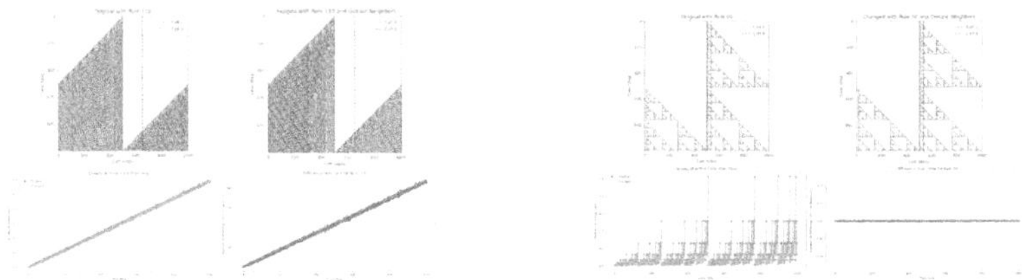

(g) ECA 110, R_β, Δ=150, n=1001, t=1000 (h) ECA 60, R_α, Δ=35, n=1001, t=1000

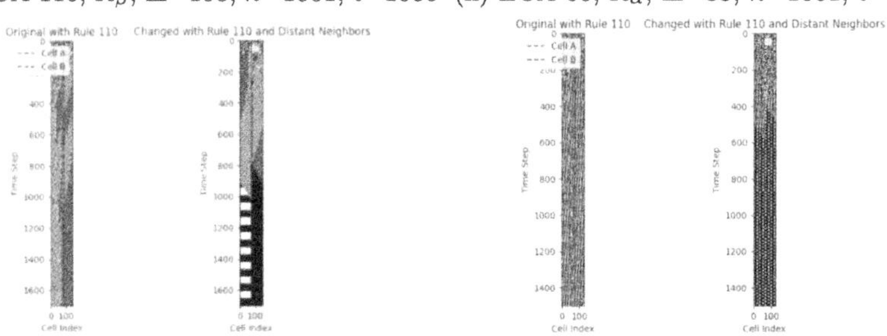

(i) ECA 110, R_α, Δ=10, n=150, t=1700 (j) ECA 110, R_α, Δ=10, n=128, t=1500

Fig. 7. ECAs 110 and 60 over R_α and R_β

cal operations, which are expected to behave in contrast, produce very similar space-time patterns for a Δ.

2. Operations R_γ, R_δ and R_ζ: These operations do not have any noteworthy effect on any Class IV ECAs. They neither produce *consistent patterns*, nor *com-*

plete robustness. However, ECA **110** under R_γ yields almost the same patterns as in R_ϵ for every equal Δ (Table 3).

5.4 Complete Robustness and Consistent Pattern Emergence

Completely robust ECAs are tabulated in Table 4 for *single active cell* (*CellA*) initial configuration. They maintain original patterns for all Δ, under different operations. However, the ECA **72** and **200** for R_β, and ECA **232** for R_α, show *complete robustness* for all configurations.

Table 4. ECAs showcasing complete robustness

n = 101, and 1001 ; Timesteps = 100 and 1000				
Logical operation	R_γ	R_α	R_β	R_ϵ
ECAs	Class II - 4, 12, 36, 44, 72, 76, 104, 108, 132, 140, 164, 172, 200, 204, 232	Class III - 150, 60 , Class II - 1, 4, 5, 12, 13, 28, 29, 33, 36, 37, 44, 51, 72, 76, 77, 78, 104, 108, 132, 140, 156, 164, 172, 178, 200, 204, **232** , Class IV - 110	Class III - 90, 18, 146, Class II - 1, 2, 10, 24, 26, 33, 34, 42, 51, 56, **72**, 74, 104, 130, 138, 152, 154, 162, 170, 184, **200**, 232, Class IV - 106	Class II - 4, 12, 36, 44, 76, 108, 132, 140, 164, 172, 204

In Table 5, *consistent patterns* refer to similar alterations appearing for every Δ, for *single active cell* initial configuration. However, ECA **18** for $R\epsilon$ and ECAs **51** and **108** for $R\delta$ showcase *consistent patterns* for all configurations.

Table 5. ECAs showcasing consistent patterns

n = 101 and 1001 ; Timesteps = 100 and 1000						
Logical Operation	R_γ	R_α	R_β	R_ϵ	R_δ	R_ζ
ECAs	Class II - 51	-	Class II - 46, 76, 78, 94, 108, 132, 134, 140, 142, 156, 164, 172, Class III - 22, 150, 60, Class IV - 41, 110	Class II - 50, 51, 57, 72, 73, 104, 130, 138, 142, 162, 170 , Class III - 90, **18**, 122, 105, Class IV - 41	Class II - **51**, 72, 76, 77, **104**, 108, 132, 140, 164, 172 , Class III - 105, 126	Class II - 51, 72, 76, 77, 104, 108, 132, 140, 164, 172

6 Conclusion

This research investigates the influence of distant neighbor on the evolution of cellular automata through various logical operations to provide valuable insights into the sensitivity and robustness of ECA rules. Among the six logical operations (AND, OR, XOR, NAND, NOR, XNOR), **OR** and **AND** show maximum robustness while the negations show the least, and **NAND** being the most sensitive among them. The **OR** operation is more likely to exhibit *complete robustness* while **NAND** is more likely to exhibit *consistent patterns*. Operation **AND** is equally likely to show both. The ECAs from different classes have shown similarities in robustness to some operations. Such rules are **110, 60** and **150**. Rule **18** shows *consistent patterns* across all configurations for **NAND**. Although the transition into an attractor configuration has been recorded for certain rules and operations and rule **110** shows abrupt transitions under the same, no class transition have been observed. The work follows experimental approach to investigate the role of distant neighbor in ECAs. The future research can explore theoretical basis of the purterbation in ECAs due to interaction of distant neighbor.

Acknowledgement. We would like to express our sincere gratitude to Prof. Sukanta Das and Prof. Biplab K. Sikdar for their invaluable support, advice, and guidance.

References

1. Andreica, A., Chira, C.: Using a hybrid cellular automata topology and neighborhood in rule discovery. In: Pan, J.-S., Polycarpou, M.M., Woźniak, M., de Carvalho, A., Quintián, H., Corchado, E. (eds.) HAIS 2013. LNCS (LNAI), vol. 8073, pp. 669–678. Springer, Heidelberg (2013). https://doi.org/10.1007/978-3-642-40846-5_67
2. Creutz, M.: Cellular automata and self organized criticality. World Scientific, River Edge, NJ (1997)
3. Dennunzio, A., Formenti, E., Kurka, P.: Cellular automata dynamical systems (2012)
4. Kang, L.: Complex behavior in long-distance cellular automata. Complex Syst. **23**(2) (2014)
5. Laval, J.: Power law behavior of elementary cellular automata (2021)
6. Li, W., Packard, N.: The structure of the elementary cellular automata rule space. Complex Syst. **4**(3), 281–297 (1990)
7. Li, W.: Phenomenology of nonlocal cellular automata. J. Stat. Phys. **68**, 829–882 (1992)
8. Li, W., Packard, N.H., Langton, C.G.: Transition phenomena in cellular automata rule space. Physica D **45**(1–3), 77–94 (1990)
9. Martinez, G.J.: A note on elementary cellular automata classification. J. Cellular Automata **8** (2013)
10. Sutner, K.: Classification of cellular automata. Encyclopedia Complex. Syst. Sci. **3**, 755–768 (2009)
11. Wolfram, S.: Statistical mechanics of cellular automata. Rev. Mod. Phys. **55**(3), 601–644 (1983)
12. Wolfram, S.: Tables of cellular automaton properties. Theor. Appl. Cellular Automata **57** (1986)

Tracking Rule Evolution in 2D Cellular Automata: Analyzing State Changes and Image Transformation Dynamics

Sugam Shaw, Arijit Ghosh(✉), and Sudhakar Sahoo

Institute of Mathematics and Applications, Bhubaneswar 751029, Odisha, India
{arijit.maths,sudhakar.sahoo,arijit.maths}@gmail.com

Abstract. This paper examines and explores the analytical behavior of two-dimensional (2D) Cellular Automata (CA) rules based on Moore's neighborhood and different Boolean operators, to better understand rule evolution and its influence on image transformation dynamics. The study investigates both computational and geometrical properties of the patterns generated by CA, revealing that repeated application of both fundamental and derived rules leads to highly regular behavior in finite configurations. Using simple initial setups and nine movement directions of XOR CA, it is possible to understand the development mechanism of Fredkin's self-replication rules. This work also focuses on the use of 2D uniform and non-uniform periodic boundary conditions in image processing for pattern generation, exploring image formation, replication, multiplication, self-reproduction as well as predicting the number and generation of pattern replicas for other operators like OR, AND, NOR, NAND, etc. The concept of generations and rule sequence is shown to play a critical role in achieving desired image formations.

Keywords: 2D CA · Moore's Neighborhood · Image Processing · Fundamental and Derived Rules · Pattern Recognition · Pattern Formation · Uniform and Non-Uniform Linear Rules

1 Introduction

Cellular automata (CA) are influential models in image processing, capturing complexity of patterns and structures within digital imagery. Originating from computational theory and formalized by mathematician John von Neumann, CA have become a foundational concept in complex systems and pattern formation [12,24]. In 2D CA, each cell's state is updated simultaneously in discrete time steps, leading to complex and often visually intricate patterns from simple initial conditions as described in Conway's game of life [14].

The relationship between CA and image processing emerges from their shared reliance on local rule-based transformations. By designing CA with specific rules and neighborhood configurations, we can manipulate images in a controlled manner, with each cell in the automaton representing a pixel in the image [2,6].

S. Kamilya et al. (Eds.): ASCAT 2025, CCIS 2499, pp. 115–133, 2025.
https://doi.org/10.1007/978-3-031-94121-4_9

Object tracking estimates the position and properties of moving objects in video sequences, assigning unique IDs to track and count them. In CA, tracking examines the evolution of individual pixels (or cells) under static (uniform) rules or dynamically changing (non-uniform) rules. Unlike object tracking, it focuses on micro-level transformations in grid-like structures. Tracking analyzes how a pixel influences or is influenced by neighbors, identifying patterns of stability, replication or chaos. This provides insights into local rule effects and global dynamics, treating CA grids as evolving images shaped by rule-driven interactions.

The inspirational work on the self-replication incepted by the pioneer von Neumann has rapidly engrossed into the minds of other researchers and they have dived deep into the 2D CA realm, churning out further pearls like, Conway's Game of Life, Wire World, Fredkin's Replicator, Hopfield Network, Langton's Loops, Sayama's Loop and Evoloop, etc. Out of which our work has a sparse relation with the Fredkin's replicator. Gradually, many researchers dived deep and done more robust works, using the 2D CA using Moore's neighborhood like edge detection, image processing, pattern formation, etc. Our work is basically a type of image processing and not only generating multiple copies from a single seed by the nine fundamental rules or by combination of these rules for different operators but to track the evolution of the 2D CA by exclusively studying the rule behavior in the formation of the image.

The formula that have been computed for tracking of group 2 rules and the concept of pixel formation is not confined for XOR operator only. The formula for pixel formation applies for any operators like OR, AND, NAND, etc. and also for multiple states and other binary operations. It is because of the XOR operator only that few pixels becomes zero and the image is not formed there if the background image is zero.

In earlier papers where extensive work has been done on 2D CA especially the replication and multiple image formation have been churned out basically for the group 2 rules. Here in this paper a rigorous discussion has been made on the denseness, sparseness and overlapping property of images.

The use of CA in image processing provides a powerful tool for exploring transformations and tracking rule evolution, where each iteration modifies pixel states according to CA rules [10,23]. Through the analysis of these transformations, we can observe unique behaviors and structural changes in images, which are valuable in areas such as image enhancement [2], pattern recognition [22], genomics and artificial life research [4,8,20]. How a pixel is moving in the 2D grid (i.e. to find the the trajectory of a pixel)? Fixing a position, how other pixel values are traversing through this? This paper investigates the dynamics of state changes and visual transformations within 2D CA, offering insights into how rules impact image evolution and how they can be imposed for various image processing tasks.

Paper Organization: Section 2 recapitulates the intercepts from previous theories. Section 3 has the main work of the paper, where the tracking of image formation for uniform and non-uniform rules has been done for different category of 512 rules with operators and for different generations. The concept of rule sequence and its use in digital imaging has been mentioned in Section. 4. Section 5 presents the conclusion and future work of this paper.

2 Glimpse Into Ruliology

In the 2D Cellular Automata (CA) model using Moore's neighborhood, each cell's state is influenced by its current state and the states of the eight surrounding cells (as illustrated in Table 1). Analyzing the tracking of a CA rule means observing how cell state changes over time. Moore's neighborhood allows for a broad range of rule behaviors, each leading to distinct outcomes: some produce stable structures, others create oscillations and some exhibit complex, chaotic patterns reminiscent of ruliology concepts.

Table 1. (a) Fundamental Rules in Moore's Neighborhood System, (b) Cell Position in Moore's Neighborhood System.

(a)

64	128	256
32	1	2
16	8	4

(b)

$(x-1, y-1)$	$(x-1, y)$	$(x-1, y+1)$
$(x, y-1)$	(x, y)	$(x, y+1)$
$(x+1, y-1)$	$(x+1, y)$	$(x+1, y+1)$

In this 2D CA model, the central cell is surrounded by its eight nearest neighbors, each associated with a unique rule number, resulting in a total of 512 possible linear rules. Fundamental rules, which depend on only one neighbor, are simpler, while more complex rules depend on multiple neighboring cells. For example, rule 15 $(1+2+4+8)$ refers to the bottom-right 2×2 cells. Each rule is grouped by the number of '1's in its binary representation, dividing them into nine categories, as detailed in [1,18,22].

This CA model uses periodic boundary condition, where cells at the edges wrap around to the opposite edge and XOR ($\oplus$) operation, where a rule's binary pattern dictates which neighbors contribute to a cell's next state [5–7].

Over multiple generations, the rule is applied iteratively, updating each cell's state based on its own and its neighbors' states. After processing, the initial and all its subsequent configurations are visualized to track the movement and evolution of active cells over time [13,20].

2.1 Behavior of Fundamental Rules

Rule 2 & 32: Rule 2 displays translation of an image towards left i.e. left shift by one cell. For an initial 2D configuration represented by an $r \times c$ matrix, it takes a number of iterations equal to the number of columns for returning to the starting configuration. Similarly, rule 32 performs a right translation by one cell, requiring the same number of iterations (equal to the matrix's column count) for returning to the initial configuration. Thus, showing the reversible nature at the same time number conserving nature too.

Rule 8 & 128: Rule 8 shifts the configuration upward by one cell, requiring a number of iterations equal to the matrix's row count to return to the initial state. Likewise, rule 128 performs a downward shift by one cell, requiring the same number of iterations as the number of rows to restore the original configuration.

Rule 4, 16, 64 & 256: These rules exhibit distinct behaviors. With periodic boundary conditions, the 2D CA behaves like a torus. Rule 4, for example, shifts each cell one step diagonally to the northwest. This rule's path covers every cell before returning to the initial configuration, requiring rc iterations. For instance, in a 3×4 matrix with elements a_{ij}, starting from the center cell a_{11}, the sequence becomes $a_{22}, a_{33}, a_{14}, a_{21}, a_{32}, a_{13}, a_{24}, a_{31}, a_{12}, a_{23}, a_{34}, a_{11}$. Similar traversal occurs for rules 16, 64 and 256. However, in an $r \times r$ square matrix, the iterations reduce to n, corresponding to the number of elements along the diagonal.

2.2 Behavior of Derived Rules

The derived rules are formed by the combination (taking XOR operation) of two or more of the fundamental rules [1,16,17]. In this present paper, we arbitrarily take up a rule say rule 6, apply on an initial image and study the dynamics of the formation of image.

2.2.1 Rule-6

Rule 6 = Rule (2 + 4). Starting with a binary grid where only the center pixel is set to 1 (the 'seed') and all others are 0, this initial configuration provides the initiation from which CA will evolve over generations. By applying Rule 6 over several generations (e.g. 8 generations), the CA updates each cell based on its neighbors, following the rule's binary pattern as shown in Fig. 1. As the generations progress, cells may activate (switch to 1) or deactivate (switch to 0) in response to the evolving state. The resulting image typically shows a complex pattern that spreads or transforms the original active state, illustrating the effect of the rule across the grid.

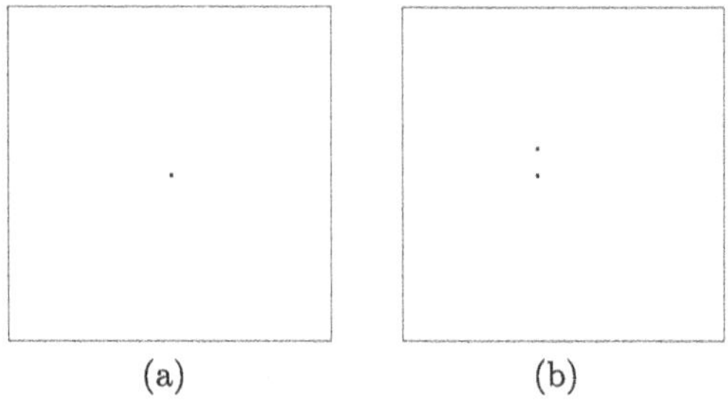

Fig. 1. (a) Initial image considered at the center of a 100×100 matrix (generation 0), (b) Image for rule 6 at generation 8.

In Fig. 1(b) the final image formed at generation 8 is shown. However, during the transition there are other images also being formed at various other generations. The progression from the initial to the final image is tracked across iterations through a rooted tree like structure named as traversal tree, which hierarchically represents the positions of active cells over time. Each level of the tree corresponds to a specific generation and each node indicates the coordinates of an active cell within that generation. Figure 2 illustrates how active cells shift and propagate across generations, visually mapping their movement within the grid.

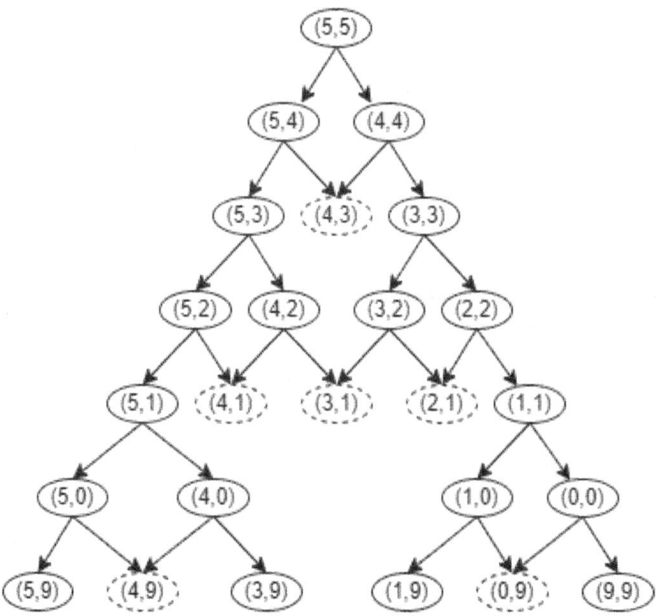

Fig. 2. Traversal Tree of Rule 6 for Generation 0 - 6.

The traversal tree depicts the relationship, with nodes in each generation connected to their respective parent nodes from the previous generation. The movement and evolution of this active cell across generations is as follows:

1. Generation 0 (Root Node): Active cell at (5, 5).
2. Generation 1: Based on the applied rule, the active cell might produce new active cells at positions (4, 4) and (5, 4) each being child of (5, 5).
3. Generation 2: Reapplying the rule to the cells from Generation 1 to determine their offspring: $(4, 4) \rightarrow (4, 3)$ & $(3, 3)$ and $(5, 4) \rightarrow (5, 3)$ & $(4, 3)$. Due to XOR operation $(4, 3)$ stays as the leaf node shown as dotted node, while $(5, 3)$ & $(3, 3)$ propagates further and the process continues upto desired generation.

The traversal tree visually captures the spread of active cells across generations, clarifying interaction dynamics under specific rules. This structure allows tracking of each cell's movement and activation lineage, revealing patterns like clustering, spreading and complex behaviors. It also facilitates analysis of metrics such as active cell count per generation, average cell movement and overall growth of active regions in the grid, offering insights into emergent CA behaviors.

3 Analysing Images from Different Generations

The fundamental rules exhibit a vivid elegant types of behavior like translation, reversibility and number conserving nature [19].

3.1 Tracking of Individual Pixel After n^{th} Generation

The fundamental rules are most obvious it is basically the directional flow of the values of the pixel as shown in Table 2. Since, we are considering the periodic boundary condition at every pixel we operate by modulus of r and by modulus of c, where r and c are respectively the number of rows and columns in the input image.

Table 2. State of the cell (x, y) after n^{th} generation of group 1 rules over an initial matrix $A_{r \times c}$

Fundamental Rules (R_{G1})	Positional Values	Direction of Translation
1	(x, y)	·
2	$(x, y+n)$	⟵
4	$(x+n, y+n)$	↖
8	$(x+n, y)$	↑
16	$(x+n, y-n)$	↗
32	$(x, y-n)$	⟶
64	$(x-n, y-n)$	↘
128	$(x-n, y)$	↓
256	$(x-n, y+n)$	↙

The group 2 rules are a combination of two fundamental rules i.e. two group 1 rules. The general propagation of the pixels is represented in a tabular form as shown in Table 3. On applying a rule from group 2 say, R_{G2_i}, it decomposes into two fundamental rules R_{G1_j} and R_{G1_k}. Now, from Table 2 (group 1) we obtain the corresponding position values $(x \pm n_1, y \pm n_1)$ for R_{G1_j} and $(x \pm n_2, y \pm n_2)$ for R_{G1_k}, where n_1 & $n_2 = 0, 1, 2, \ldots$ are variables according to Table 2 (group 1). We will collect the respective $x-$coordinate values from the two rules say, $(x \pm n_1, x \pm n_2)$ and obtain the positional values x_1 according to Table 3 (group 2). In a similar manner we will obtain y_1. Then we will apply the formula at equation. 1 for group 2. The coefficients that appear follows the pascal triangle rule. The formula by which pixel at a given position (x, y) generated at n^{th} generation can be predicted for group 2 rules is:

$$(x, y)^n = \sum_{g=0}^{n} {}^n c_g (x_1 \% r, y_1 \% c) \tag{1}$$

Table 3. Generalization of pixel formation for group 2 rules

'x' coordinates of group 1 rules $\longrightarrow$ Positional Values (x_1)	'y' coordinates of group 1 rules$\longrightarrow$ Positional Values (y_1)
$(x, x+n) \longrightarrow (x+g)$	$(y, y+n) \longrightarrow (y+g)$
$(x, x-n) \longrightarrow (x-g)$	$(y, y-n) \longrightarrow (y-g)$
$(x+n, x) \longrightarrow (x+n-g)$	$(y+n, y) \longrightarrow (y+n-g)$
$(x+n, x-n) \longrightarrow (x+g)-(n-g)$	$(y+n, y-n) \longrightarrow (y+g)-(n-g)$
$(x+n, x+n) \longrightarrow (x+n)$	$(y+n, y+n) \longrightarrow (y+n)$
$(x-n, x-n) \longrightarrow (x-n)$	$(y-n, y-n) \longrightarrow (y-n)$
$(x+n, x) \longrightarrow (x+n-g)$	$(y+n, y) \longrightarrow (y+n-g)$
$(x-n, x) \longrightarrow (x-n+g)$	$(y-n, y) \longrightarrow (y-n+g)$

where n is the number of required generation, % denotes the modulus operation and $\sum$ denotes XOR operations of the binary values at the pixels.

Consider, rule 12 = rule (8 + 4). Rule 8 $\rightarrow (x+n, y)$ and rule 4 $\rightarrow (x+n, y+n)$. Now, from Table 3 (group 2), we have $(x+n, x+n) \rightarrow x+n$ and $(y, y+n) \rightarrow y+g$.

$$(x, y)^n = \sum_{g=0}^{n} {}^n c_g ((x+n) \% r, (y+g) \% c) \tag{2}$$

where $g = 0, 1, 2, \ldots, n$ and n = total number of generations.

Example: In a 20×20 matrix, let $(x, y) = (10, 10)$, represents a pixel in active state while all other pixels remain in non-active state (initial image). The XOR operation is applied to the pixel values. Arbitrarily, at a generation (say generation 4) and at any position (say (10, 10)) there may occur an active state. If it does, we determine the positions (base case) due to which an active pixel has occurred.

$$\begin{aligned}
\text{At generation 0: } (10,10)^0 &= (10,10)(\text{Initial configuration})\\
\text{At generation 1: } (10,10)^1 &= (11,10) \oplus (11,11)\\
\text{At generation 2: } (10,10)^2 &= (12,10) \oplus 2(12,11) \oplus (12,12)\\
\text{At generation 3: } (10,10)^3 &= (13,10) \oplus 3(13,11) \oplus 3(13,12) \oplus (13,13)\\
\text{At generation 4: } (10,10)^4 &= (14,10) \oplus 4(14,11) \oplus 6(14,12) \oplus 4(14,13) \oplus (14,14)
\end{aligned}$$

Corresponding to rule 12 (group 2), the binary traversal tree in Fig. 3, illustrates the pixels of specific generations that contribute to the image formation

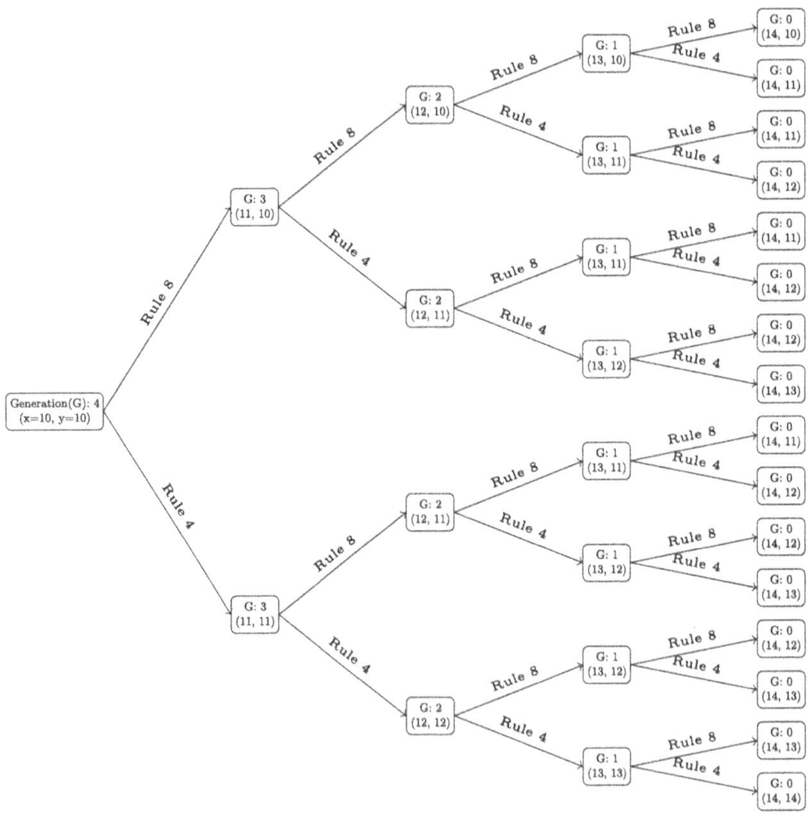

Fig. 3. Level-based recursion tree for backtracking generation for rule 12 (group 2), the labels (height) show the number of generations.

at generation 4 (identifiable by the tree's height). Similarly, selecting a rule from group 3 results in a ternary tree. Proceeding likewise, similar patterns can be observed for other groups by constructing an $n - ary$ tree using the backtracking method. There are a total of 9 groups, and an $n - ary$ tree for each group serves as a useful tool for tracking pixel formation from an initial image. Beyond the binary tree, for $n - ary$ trees where $n = 3, 4, 5, 6, 7, 8, 9$, the backtracking method can be applied to determine the coefficients. However, as n increases, the complexity grows significantly, particularly from the ternary tree onward.

The XOR operation causes pixels with even coefficient terms to disappear. Consequently, only the pixels corresponding to odd coefficient terms remain, which is reflected in the final output image. The final output image is observed at generations 2^k for $(k = 0, 1, 2, \ldots)$ and at these generations the output image will contain only two pixels. The Table 4 for group 2 rules not only provide the positions of the pixels but also the coefficients, which denotes the number of times a pixel repeats itself of the given tree in Fig. 3. In a similar manner, the number of pixels formed for other groups can be tracked.

Table 4. Counts of generation 0 nodes for rule 12

(a) Generations

Position	Count
(11, 10)	1
(11, 11)	1

(b) Generations

Position	Count
(12, 10)	1
(12, 11)	2
(12, 12)	1

(c) Generations

Position	Count
(13, 10)	1
(13, 11)	3
(13, 12)	3
(13, 13)	1

(d) Generations

Position	Count
(14, 10)	1
(14, 11)	4
(14, 12)	6
(14, 13)	4
(14, 14)	1

3.2 Density of Binary Image at Different Generation

3.2.1 Image Analysis for Uniform Rules After 2^k Generations

From an initial configuration, applying the rule vector iteratively, a well-defined predictable pattern emerges after 2^k generations [3,18,21]. At 2^k generations, by the application of transformation rules, each pixel undergoes a multiplication process. However, when multiple pixels transform to the same position, overlapping occurs, leading to the combination of their values. Consequently, although theoretically the output is expected to yield distinct pixel values, overlapping sometimes results in pixel distributions where the cardinality matches that of the initial image, as illustrated in Fig. 6(b).

Our analysis and proposed formula are independent of specific states and operations. The formula (equation. 1) is not limited to states in $\mathbb{Z}_2$ or XOR operations rather, it provides a generalized framework for precisely tracking the positions of pixels and the number of active cells, regardless of the state space or transformation rules applied. Figure 4 illustrates how the number and positions of pixels vary when applying rule 6 with different operators such as OR, AND, NAND and NOR.

(a) (b) (c) (d)

Fig. 4. Rule 6 applied on an image matrix with a central active pixel for generation 8 using operators (a) OR, (b) AND, (c) NAND & (d) NOR.

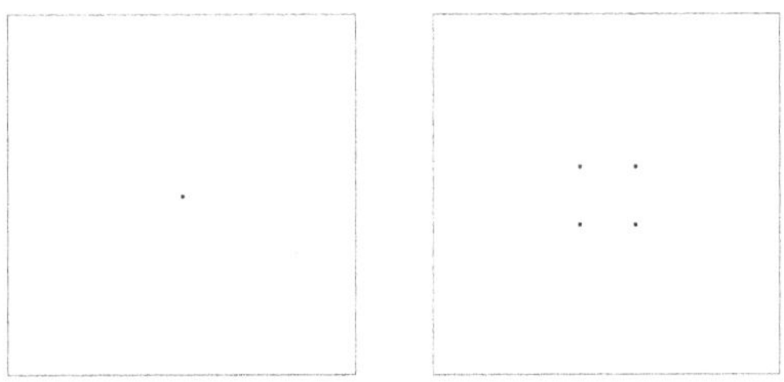

(a) (b)

Fig. 5. (a) Initial Image with 1 at center (Left), (b) Output using Rule 340 at generation 8.

Example: Consider a 100×100 pixel matrix as shown in Fig. 5(a) as initial configuration. Applying rule $340 = \text{rule } (4 + 16 + 64 + 256)$ (group 4), we obtain the output in Fig. 5(b) [1]. Since four fundamental rules are involved, four distinct pixels emerge after 2^k generations. Using the output in Fig. 5(b) as the input image, each transformation generates $4 \times 4 = 16$ pixels over successive iterations. This is observed at 2^k generations, specifically at 4, 8, 16 and 32 generations, as shown in Fig. 6 (a), (b), (c) & (d), respectively for XOR operation and also shown in Fig. 7 for OR operation.

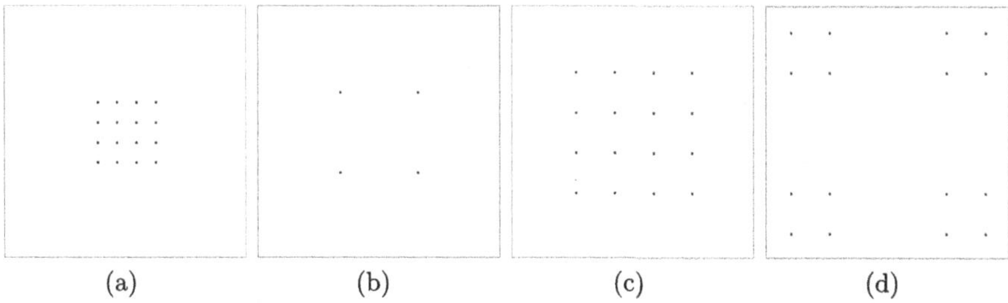

Fig. 6. Image patterns formed by rule 340 using XOR operator at Generations (a) 4, (b) 8, (c) 16 & (d) 32 taking initial image as Fig. 5(b).

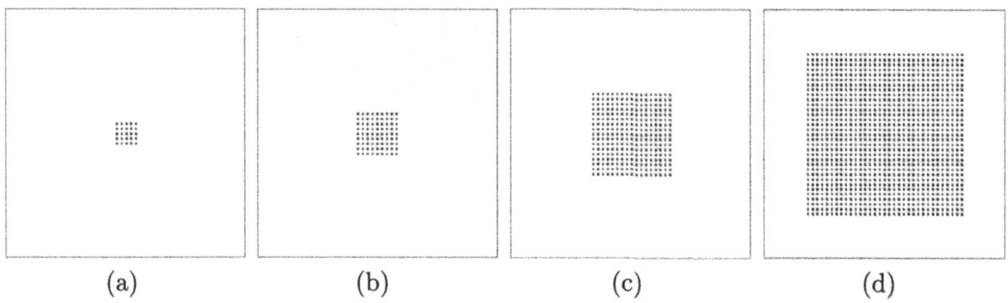

Fig. 7. Image patterns formed by rule 340 using OR operator applied to initial image as Fig. 5(b) at Generations (a) 4, (b) 8, (c) 16 & (d) 32.

The cardinality of active pixels produced is determined by the number of fundamental rules involved in a derived rule, i.e. it corresponds to the group number to which the rule belongs. The group number of a rule indicates the number of copies formed for a given pixel. For example, a rule from group 1 conserves the number of pixels (as in number-conserving cellular automata), with the pixel images either remaining fixed or translating laterally or diagonally, depending on the specific rule within the group. Rules from group 2 create two copies of the image, group 3 rules create three copies and so forth. The number of copies formed can be predicted by identifying the rule's group.

The number of pixels formed can be predicted by the the formula: (number of active pixels in the initial matrix)×(number of pixels formed by a group's rule). Though the cardinality of the pixels were determined but the positions of the pixels will differ. In one of the generations, the number of pixels formed is same as the number of pixels in the initial matrix, this happens due to the super impose of pixels. Such superimpose behavior can happen in any one of the 2^k generation thus an unpredictable behavior. This is a kind of unusual behavior rather disadvantage that at all the 2^k generations we are not sure to get the distinct number of pixels.

At generations of 2^k, where $k = 0, 1, 2, \ldots$ multiple copies of the initial pattern appear for XOR operation [1]. As generation progresses, the pixels become more dispersed, forming clusters or groups. Each cluster contains as many pixels as there are fundamental rules within the applied rule. The theory suggests that as the number of generations increases, the distance between these clusters also grows. The applied rule determines the number of copies formed and the generation count influences the distance between pixels. Similar analysis can also be done for other 2D CA rules and for other operators. Figure 8 shows the dynamics of some of the CA rules, other operators and at different generations.

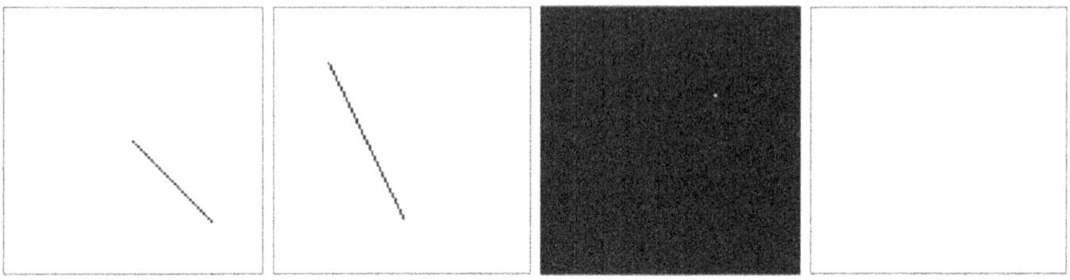

(a) (b) (c) (d)

Fig. 8. Images for different rules (R), for different operators (O) and at different generations (G) denoted by <R, O, G> for an initial matrix with a central active pixel: (a) <65, *OR*, 30>, (b) <399, *NOR*, 29>, (c) <17, NAND, 17> & (d) <9, AND, 1>.

3.2.2 Image Analysis for Non-Uniform Rules After 2^k Generations

So far, uniform rules have been applied across all generations. However, incorporating non-uniform rules can explore new patterns and behaviors. Non-uniformity can be introduced in four ways: (1) By applying rules from different groups within the same generation, (2) Using different rules from the same group within the same generation, (3) Switching to different group rules across generations or (4) Following a completely random sequence of rules for each generation.

Example: Consider the image as a single pixel at the center of a 100×100 pixel matrix. Applying sequentially, rule 6 (group 2) for generation 4, rule 54 (group 4) for generation 8, rule 3 (group 2) for generation 8, rule 32 (group 1) for generation 8. Total number of pixels formed in the final image is $2 \times 4 \times 2 \times 1 = 16$ pixels. By applying 2D CA rules from different groups, we can determine the number of pixels generated.

Generalizing, suppose we apply a rule from group R_1 at generation 2^{k_1}, followed by group R_2 at generation 2^{k_2} and so on, up to group R_p at generation 2^{k_p}, the total number of pixels generated (determined by the group numbers, not the specific generation numbers) is $R_1 \times R_2 \times \ldots \times R_p$. This would result in the following predicted sequence: a single pixel transforms to 3 pixels under rule 7 (group 3), which then expands to 12 pixels under rule 170 (group 4). In this way, we can predict the number of copies formed and design a rule sequence to achieve a specific output (Fig. 9).

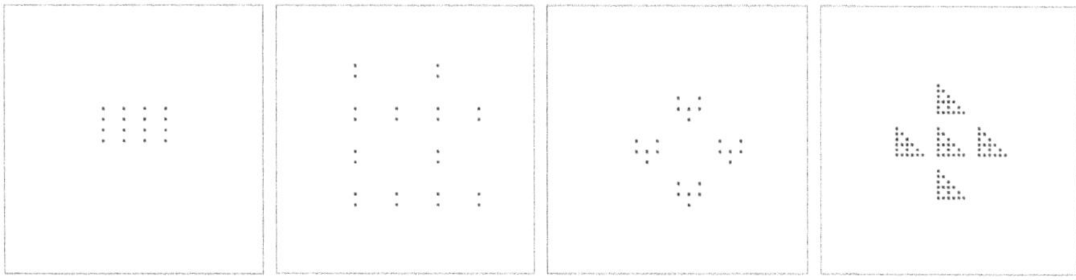

(a) (b) (c) (d)

Fig. 9. Non-Uniform Rules for Various Rule sequences: (a) 6(g=4), 54(g=8), 3(g=8), 32(g=8), (b) 48(g=4), 340(g=16), 7(g=16), 32(g=8), 128(g=8), (c) 96(g=4), 162(g=4), 170(g=16), (d) 324(g=1), 455(g=4), 171(g=16) (g=generations).

In exploring image formation under uniform rules, we previously obtained a maximum of 16 copies at specific positions. Now, the goal is to generate more copies in varied pixel positions by carefully selecting sequences of 2D CA rules. This approach allows us to monitor creation of number of pixels and their locations.

For instance, if we require 50 pixels in a single image matrix as output, with input if only one pixel, we can do so by choosing a rule from group 2 offering a selection of ${}^9C_2 = 36$ possible rules and two rules (may or may not be different) from group 5 offering a selection of $({}^9C_5)^2 = 63504$ possible rules. Hence we are left with a choice of 2286144 rules from group 2 & 5. This method enables us to identify the specific rule sequence responsible for generating the target output, illustrating a case of reversible 2D CA. Statistically, this can be described as the distribution of pixels within the matrix.

3.2.3 Tracking Formation of Pixels at Exact Location for Different Generations

In Sect. 3.2.1, we examined a case where, starting with an initial configuration of one pixel and applying rule 340, we obtained four pixels. Using these four pixels as input in the next step produced 16 pixels. This behavior, identified through empirical observation, holds true for generations $n = 2^k$, where $k = 0, 1, 2, \ldots$ and the number of output pixels is determined by the number of fundamental rules involved in the given rule. However, this concept has not been extended to a general $n \in \mathbb{N}$. In this section we put forth the number of pixels for any $n \in \mathbb{N}$.

We propose a systematic approach to determine the possible number of pixels at generation n, when n is not limited to powers of two. For example, take $n = 30$, which can be expressed in exponential form of 2, as $2^a + 2^b + 2^c + 2^d$, with $a = 4, b = 3, c = 2, d = 1$. This section presents a general formula for calculating the pixel value at a given generation based on the rule number and the binary representation of the generation number for predicting image generation:

$$n = 2^a + 2^b + 2^c + 2^d + \ldots, \text{ where } a, b, c, d \ldots \in \mathbb{N} \tag{3}$$

The number of terms in the above expression allows us to predict the number of pixels in the resulting image. This formula not only estimates the pixel count but also indicates the positions where these pixels will appear in the output image.

Consider, rule 7 (111) applied to a single pixel at generation 30. Number of ones in Rule 7 = 3 and the generation number 30 has 4 terms, resulting in $3^4 = 81$ pixels. Therefore, starting with a single pixel, at generation 30 and applying rule 7, 81 pixels are obtained.

Let, p = Number of ones in the rule's binary representation and q = Number of terms in the generation sum. The general formula for the cardinality of active pixels in the resulting image for a rule and at n generation is stated as:

$$\text{Number of active pixels in output image} = p^q \tag{4}$$

However, for overlapping positions due to specific operator, the number of active pixels in the output image will be maximum of p^q.

Hence, we can predict that by the application of rule 7 on a single pixel input image, there will be 3, 9, 27 & 81 pixels in the output image at generations 2, 6, 14, 30 respectively, as shown in Fig. 10.

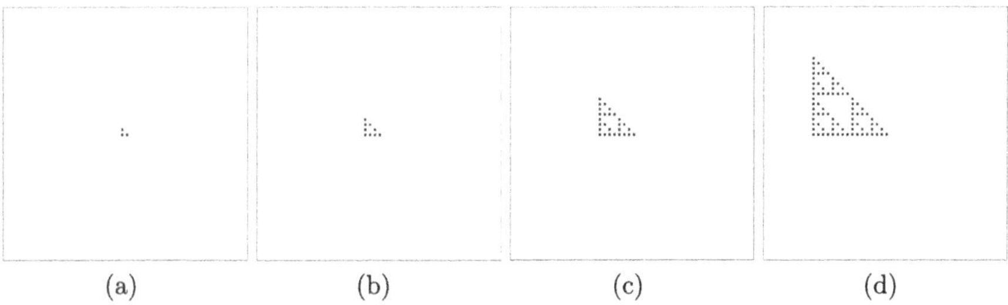

Fig. 10. Rule 7 applied on an image matrix with a central pixel producing images for generation 2, 6, 14, 30 respectively.

3.3 Permutation of Fundamental Rule Positions and Their Behavior

The rule matrix shown in Table 1 outlines properties of 2D CA, such as translation and generates diverse images from a given initial configuration [1,15]. A natural question arises: what happens if we alter the positions of the rules within Table 1? With these changes, there are "9! possible conventions" for the rule. Although each altered 3×3 neighborhood matrix maintains the same cardinality of images produced from a given input, their spatial positions may vary.

The naming convention in Table 1 shows how each cell's state is influenced by its neighbors and different rule arrangements alter cell interactions. By rearranging the rule in the matrix, we modify how neighboring cells influence each other's states without affecting the density of resulting pixels (images). This reshuffling rearranges the images at different positions while preserving their total number.

Final states represent the configurations after all CA generations. Despite rearranging the rule matrix, the consistency in copy counts in these final states indicates that the distribution of configurations remains stable.

The rule structure of 2D CA confirms the definiteness of pattern evolution, however some emergent patterns are resilient to changes this property suggests the robustness of certain patterns within CA.

The simulation result shows, with various rule arrangements in 2D CA how these changes impact pattern formation, without disrupting the core dynamics that produce consistent types of patterns. This property suggests that certain characteristics of complex systems may remain stable despite modifications in local interactions, hinting at underlying principles that govern the emergence of complexity and offering valuable insights into the resilience of complex systems. This could lead to a deeper understanding of the relationship between rules and the resulting structures in the automata.

In Fig. 11 , we present images generated by applying rule 30 at generation 32 to an initial configuration centered in a 100×100 pixel matrix, using the rule setup shown in Table 1.

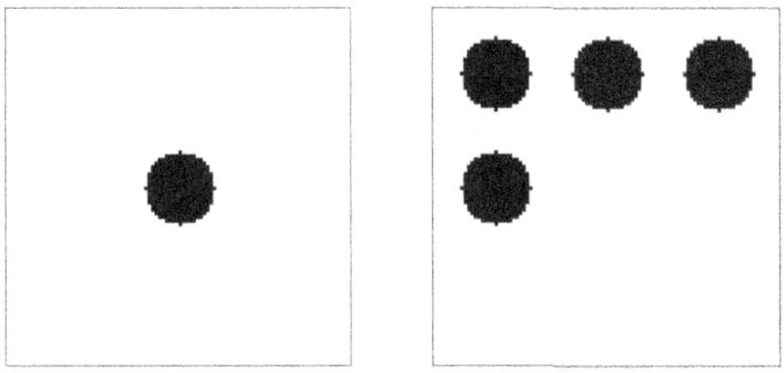

(a) (b)

Fig. 11. Normal application of rule 30 at generation 32 for rule position as in Table 1: (a) Initial Image, (b) Final Image.

In the image generation process, each fundamental rule produces images at positions opposite to its placement in the permuted rule matrix. For example, consider a permutation shown in Fig. 12(a) which gives an output image 12(b). Rule 2 is positioned at $(x-1, y)$, so its image is vertically opposite at $(x+1, y)$, where Rule 16 is located. Likewise, rule 4's image forms diagonally opposite at $(x+1, y-1)$, where rule 128 is positioned, rule 8's image appears horizontally opposite at the location of rule 64 and rule 16's image forms vertically opposite at the position of rule 2. Similarly, from 12(c) we obtained 12(d). This pattern holds for any permutation of fundamental rules, any generation and for any rule from 1 to 512. Thus, by analyzing this setup, one can predict not only the number of images formed due to the system's number-conserving nature but also the precise positions where images will appear. These images demonstrate the consistent behavior of the rules applied in various permutations according to the rule setup shown in Table 1.

1	**2**	4
8	32	64
128	**16**	256

(a)

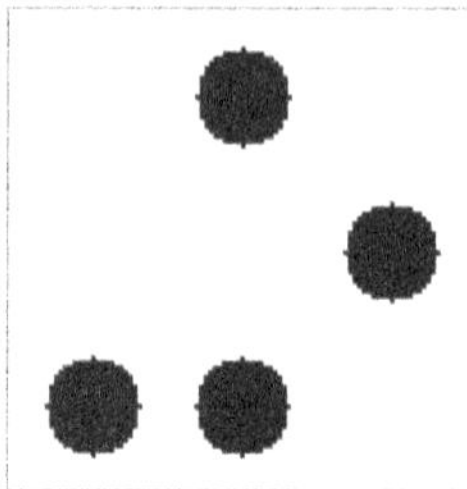

(b)

256	**2**	1
8	32	**16**
128	4	64

(c)

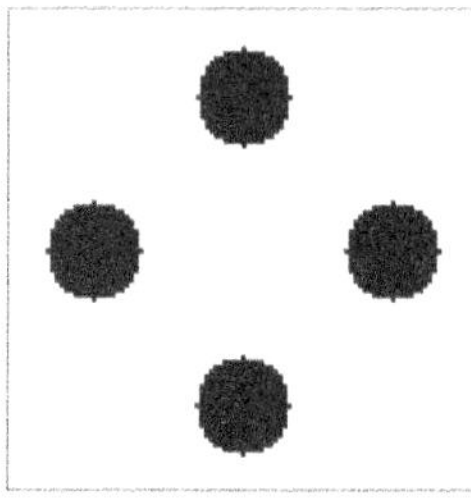

(d)

Fig. 12. Rule 30 generation 32: (a) & (c) Various alterations of rule matrix from Table 1, (b) & (d) Respective Image Matrices showing invariance in cardinality though differing in positions.

4 Rule Sequence with Asynchronous CA

In this section we have used the concept of rule sequence that can produce interesting patterns. Starting from an active pixel in a 2D grid, the given CA rule sequence when applied asynchronously in different time steps sometimes can efficiently produce the same image as produced by a synchronous CA. It is also possible to control the trajectory of the pixel through these linear rules. To study the patterns formed by a rule sequence, we have used three parameters namely the rule number, the operator used and the position where the CA rule is applied at each time step t for $t = 1, 2, \ldots, n$.

Consider an $r \times c$ matrix with an active cell at the position (x, y) as in Fig. 1(a). Figure 13 shows a spiral pattern using a sequence of rules 8, 2, 128 & 32. The rule sequence being applied from pixel position $(x-1, y)$ is given as $< 8^{2i}, 2^{2i}, 128^{2i+2}, 32^{2i+2} >$, where $i = 2p + 1$, $p = 0, 1, 2, \ldots$

Fig. 13. Starting from a center pixel the rule sequence $<8^2, 2^2, 128^4, 32^4, 8^6, 2^6, 128^8, 32^8, \ldots>$ iteratively generates a spiral pattern in counter clockwise direction.

The rule sequence is not unique, for if 2 is being replaced by 3, 7, etc. i.e. a rule obtained by adding 2 with any other fundamental rule produce the same spiral pattern as in Fig. 13.

Given a rule sequence it will generate patterns. Some patterns are chaotic, simple or periodic. The rules being applied are asynchronous in nature. The rule sequence is flexible. The chosen sequence determines the resulting image. Conversely, the desired image dictates the selection of appropriate rules. The inspiration for this type of behavioral pattern is similar to the way it emerges in Langton's ant [9,11]. Thus, exploring the space of rule sequences holds potential for digital image creation, analyzing the evolution of nonlinear rules and approximating an image by piecewise line segments using nine fundamental linear rules.

5 Conclusion and Future Work

This paper has explored the analytical behavior of two-dimensional cellular automata for image transformation, demonstrating how specific nine fundamental CA rules can lead to periodic and complex pattern formations. By examining rule evolution and boundary conditions, the study reveals insights into how CA dynamics can be applied to achieve controlled transformations in images. Tracking cell states updates in 2D CA evolutions can be done by using traversal tree diagram for different neighborhood positions as defined in the paper. Although this deterministic mathematical engineering approach focused on fundamental rules, the results provide a foundation for future exploration of CA in image processing and related fields. This work opens pathways for further research into the versatility of CA-based approaches in areas such as pattern recognition, reversible system and natural process simulations.

References

1. Choudhury, P., Nayak, B., Sahoo, S., Rath, S.: Theory and applications of two-dimensional, null-boundary, nine-neighborhood, cellular automata linear rules (2008)

2. Diosan, L., Andreica, A., Enescu, A.: The use of simple cellular automata in image processing. Studia Universitatis Babe-Bolyai Informatica **62**, 5–14 (2017). https://doi.org/10.24193/subbi.2017.1.01
3. Ekhad, S.B., Sloane, N.J.A., Zeilberger, D.: Odd-rule cellular automata on the square grid (2015). https://arxiv.org/abs/1503.04249
4. Ghosh, A., Behera, S., Padhi, S., Sahoo, S.: Analysis of one-dimensional 4-state cellular automata rules and DNA evolutions. In: Das, S., Martinez, G.J. (eds.) Proceedings of Second Asian Symposium on Cellular Automata Technology, pp. 143–160. Springer Nature Singapore, Singapore (2023)
5. Khan, A.R.: On two dimensional cellular automata and its VLSI applications. Int. J. Electr. Comput. Sci. IJECS-IJENS **10**, 124–129 (2010)
6. Khan, A., Choudhury, P., Dihidar, K., Mitra, S., Sarkar, P.: VLSI architecture of a cellular automata machine. Comput. Math. Appl. **33**(5), 79–94 (1997)
7. Khan, A., Choudhury, P., Dihidar, K., Verma, R.: Text compression using two-dimensional cellular automata. Comput. Math. Appl. **37**(6), 115–127 (1999). https://doi.org/10.1016/S0898-1221(99)00080-2
8. Langton, C.G.: Self-reproduction in cellular automata. Physica D **10**(1), 135–144 (1984). https://doi.org/10.1016/0167-2789(84)90256-2
9. Langton, C.G.: Studying artificial life with cellular automata. Physica D **22**(1), 120–149 (1986). https://doi.org/10.1016/0167-2789(86)90237-X. proceedings of the Fifth Annual International Conference
10. McCaskill, J.S., Packard, N.H.: Analysing emergent dynamics of evolving computation in 2D cellular automata. In: Martín-Vide, C., Pond, G., Vega-Rodríguez, M.A. (eds.) Theory and Practice of Natural Computing, pp. 3–40. Springer International Publishing, Cham (2019)
11. Moreira, A., Gajardo, A., Goles, E.: Dynamical behavior and complexity of Langton's ant. Complexity **6**(4), 46–52 (2001). https://doi.org/10.1002/CPLX.1042
12. von Neumann, J., Burks, A.W.: Theory of self reproducing automata (1966). https://api.semanticscholar.org/CorpusID:62696120
13. Packard, N.H., Wolfram, S.: Two-dimensional cellular automata. J. Stat. Phys. **38**, 901–946 (1985). https://api.semanticscholar.org/CorpusID:14972808
14. Peña, E., Sayama, H.: Life worth mentioning: complexity in life-like cellular automata. Artificial. Life **27**(2), 105–112 (2021). https://doi.org/10.1162/artl_a_00348
15. Qadir, F., Peer, M.A., Khan, K.A.: A novel method for generating self replicate patterns based on two dimensional cellular automata, twenty five neighborhood model. Int. J. Comput. Appl. **47**(2), 43–48 (2012)
16. Rao, U., Lakshmanan, J.: Replication of a binary image on a one-dimensional cellular automaton with linear rules. Complex Systems. **27**, 415–430 (2018). https://doi.org/10.25088/ComplexSystems.27.4.415
17. Rao, U., Lakshmanan, J., Mathew, T.: Replication in two-dimensional cellular automaton with linear rules. Journal of Cellular Automata **15**, 377–405 (2021)
18. Sahin, U., Uğuz, S., Akin, H.: The transition rules of 2d linear cellular automata over ternary field and self-replicating patterns. International Journal of Bifurcation and Chaos **25**, 1–24 (01 2015). https://doi.org/10.1142/S021812741550011X
19. Sahoo, S., Choudhury, P., Pal, A., Nayak, B.: Solutions on 1D and 2D density classification problem using programmable cellular automata. J. Cell. Automata **9** (2009)
20. Sayama, H., Nehaniv, C.L.: Self-reproduction and evolution in cellular automata: 25 years after evoloops (2024)

21. Sloane, N.J.A.: On the Number of ON Cells in Cellular Automata, p. 13-38. Cambridge University Press (2018). https://doi.org/10.1017/9781316650295.003
22. Uğuz, S., Sahin, U., Akin, H., Siap, I.: Self-replicating patterns in 2D linear cellular automata. Int. J. Bifurcat. Chaos **24**, 1430002 (2014) (2014). https://doi.org/10.1142/S021812741430002X
23. Wolfram, S.: Cellular Automata and Complexity. In: Collected Papers, pp. 5–7. Addison-Wesley Publishing Company, Reading, Massachusetts (1994)
24. Wolfram, S.: A New Kind of Science, vol. 1. Wolfram Media Inc, Champaign, IL, USA (2002)

Exploring Opinion Dynamics Through Social Influence, Online Communication, and Leadership: A Fuzzy Cellular Automaton Approach

Aradhya Chakrabarti[1], Niharika Rana[2], Sanchit Baweja[3], and Rinkaj Goyal[4](✉)

[1] School of Computer Engineering, Kalinga Institute of Industrial Technology, Odisha 751024, India
[2] Bennett University, Greater Noida 201310, India
[3] Maharaja Surajmal Institute of Technology, New Delhi 110058, India
[4] University School of Information, Communication and Technology, Guru Gobind Singh (GGS) Indraprastha University, New Delhi 110078, India
rinkajgoyal@gmail.com

Abstract. This simulation study presents an opinion dynamics model derived from the Hegselmann-Krause (HK) bounded confidence model, incorporating fuzzy cellular automata and the theory of Normative Social Influence (NSI) to investigate opinion changes of population via social interactions. The heterogeneous HK model, which describes population of agents, updating their opinions based on their neighbour's average opinion within each agent's confidence threshold, is extended to include online communication (1-many), leader's influence, and dissenters, thereby modelling real-world scenarios exhaustively. Opinions and other parameters introduced in the model are fuzzified to accurately describe uncertainty in their values and tolerances towards change. It is seen that in a population, more zealots (or dissenters) cause groups to form around different opinions and reduce agreement, while higher confidence in one's views helps keep a balance of opinions. Diplomacy is asserted to be a natural state for consensus in a diverse population and the rate of convergence to consensus is observed to show polynomial growth. This study also explores the dynamics of fringe groups led by a central figure and their influence on local populations. It highlights that when over three-fourths of agents are willing zealots, the time taken for extremist opinion dissolution exhibits a cubic polynomial increase with population size.

Keywords: Opinion formation · Neutrality · Fuzzy Cellular Automata · Normative Social Influence

A. Chakrabarti, N. Rana, S. Baweja and R. Goyal—All authors contributed equally to this work.

S. Kamilya et al. (Eds.): ASCAT 2025, CCIS 2499, pp. 134–148, 2025.
https://doi.org/10.1007/978-3-031-94121-4_10

1 Introduction

In any social setting, the exchange of opinions among individuals and groups is inevitable. With the rise of online platforms like social media and forums, people now interact across the globe more easily than ever. Opinion dynamics, a subfield of social dynamics, uses mathematical and computational tools to study how opinions spread within interacting individuals and communities [6]. Various models have been developed to explain how social interactions shape opinions [25], with the Hegselmann-Krause (HK) model [19] being particularly notable. In the HK model, a population of n agents is characterized by their opinions $x_i \in [0, 1]$, and each agent updates their opinion by averaging the opinions of neighbors within its confidence set i.e. those whose opinions differ by at most d (the confidence threshold), and agents may vary in their individual confidence thresholds [10]. Normative Social Influence (NSI) [4], an external factor building on the concept of conformity [9], also plays a role in opinion evolution. To simulate this, we use a two-dimensional fuzzified cellular automaton with periodic boundary conditions, incorporating multiple social factors into understandable terms. By fuzzifying opinions and parameters, along with employing the Herfindahl-Hirschman Index and statistical tools, we verify previously proposed outcomes (See 5) and establish rules for how opinions evolve through interactions. Traditional models of opinion dynamics deal with only changes in opinions held by interacting individuals, not specifying details like their neighborhoods, external influences or even their individual preferences. The model devised herewith incorporates all such changes as shown in Table 1.

Table 1. Proposed CA Model vs. Other Opinion Dynamics Models.

Model	Interaction	Neighborhood	External Influences
Bounded Confidence (DW) [3]	1:1 (binary)	Random Selection	None
Hegselmann-Krause [19]	1:Many (confidence set)	Confidence Set	None
CA Model	1:Many	8 Nearest+Distant Neighbors	NSI

2 Terminologies

Cellular Automaton (CA): It is a model of a physical system, where space and time are discrete and interactions are local. A CA is characterized by five properties, namely the number of spatial dimensions, width of each side of the lattice, width of the neighborhood of each cell, states of the CA cells and the CA transition function TF, which computes the next state of each cell w.r.t the current state of its neighboring cells [17]. In this study, a 2D CA with periodic boundary conditions (See Fig. 1) is utilized, with a fuzzified version of the CA later replacing the states of the CA cells with the fuzzy sets corresponding to their crisp values [16].

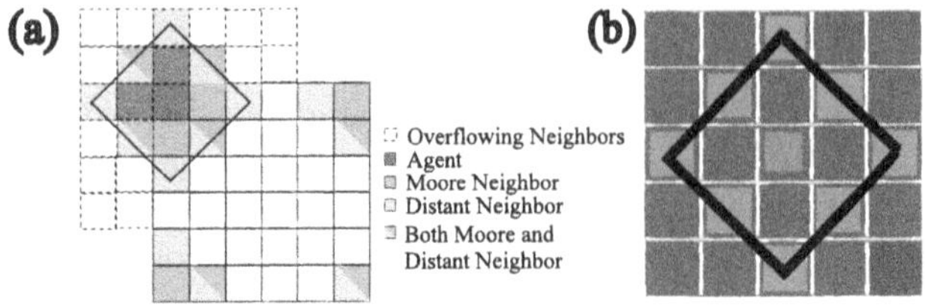

Fig. 1. (a) CA Periodic Boundary Conditions. Neighbors that extend beyond the grid boundaries are wrapped around to corresponding cells within the grid. (b) Distant neighbors of an agent (red) at radius=2 (green) (Color figure online)

Conformity: It is the act of matching attitudes, beliefs, and behaviors to group norms, politics or being like-minded. It can be mathematically described using the construct of an ideal opinion [22], i.e. at any time t, the ideal opinion of an agent or cell on a two-dimensional lattice is given by:

$$x_i^*(t) = \bar{x}(t) + \delta_i.\sigma(t) \tag{1}$$

where, $\bar{x}(t)$ = mean opinion of the neighbors of the agent; δ_i = distinctiveness factor of the agent (willingness of the agent to conform to the norms of the interacting neighbors); and, $\sigma(t)$ = standard deviation of the opinions of the neighbors of the agent. After every interaction an agent has with its interacting neighbors, its opinion changes as follows:

$$x_i(t+1) = x_i(t) + k[x_i^*(t) - x_i(t)] \tag{2}$$

where, k = adjustment rate (or coefficient) for opinion updation.

Normative Social Influence (NSI): Building upon conformity, it describes the tendency of an individual to conform to fit in with the group due to fear of rejection. It is usually associated with compliance, where a person changes their public behavior but not their private beliefs. Unlike informational social influence, where levels of involvement and complexity are greater, when normative social influence is involved in interactions, higher levels of conspicuousness, contact and advice solicitation frequency are observed [13].

Gaussian Fuzzification: It is used to convert crisp values (opinions and parameters) to fuzzy values to regulate the degree of membership in our model, to help describe their real values computationally. The Gaussian membership function [20] is described as follows (where σ = standard deviation, c = mean of Gaussian function):

$$\mu(x;\sigma,c) = e^{\frac{-(x-c)^2}{2\sigma^2}} \tag{3}$$

It is symmetric around the mean, ensuring that the degree of membership decreases equally on both sides of the mean.

Herfindahl-Hirschman Index (HHI): It is a commonly used statistical measure for market concentration which describes the degree to which opinions are integrated into a consensus or fragmented into several opinion groups. HHI = 1 indicates complete convergence of opinions. HHI = 0 indicates complete fragmentation [12]. This index is utilized exhaustively here for experimentation purposes.

$$HHI = \sum_{i=1}^{n} (\frac{s_i}{N})^2; HHI \in [0, 1] \tag{4}$$

where n = number of clusters, s_i = number of agents in the ith cluster, and N = number of all agents.

3 The Model

The devised model is derived from the heterogeneous version of the HK model and incorporates the theory of NSI as an external influence affecting opinion formation. Unlike the Bounded Confidence model of opinion dynamics [3] where interactions are binary, the updation of opinion of an agent in the HK model with opinion $x_i(t)$ occurs as follows [5]:

$$x_i(t+1) = \frac{1}{|N_i(t)|} \cdot \sum_{j \in N_i(t)} (x_j(t) - x_i(t)) \tag{5}$$

where, $|N_i(t)| = |\{j \in V || x_i(t) - x_j(t)| \le \epsilon_i\}|$ is the cardinality of neighbor set $N_i(t)$ of agent i at time t with confidence level ϵ_i varying for different agents, and V denotes set of n agents. The population in the model is represented by a finite two-dimensional lattice / grid of cells (or agents), and based upon the agent's ability to interact online, online neighbors are defined as its distant neighbors (See Fig. 1). Moreover, a fully asynchronous updating scheme for the evolution of the Cellular Automaton is used. That is, at each step, only one agent's opinion is updated [2] although some parameters of other agents may get updated.

3.1 Model Definition

The Cellular Automaton constituting the population in the model is characterized by the following four parameters:

- Lattice: $L = Z^2$ (two-dimensional Grid)
- States: S = *{Strongly Disagree, Disagree, Neutral, Agree, Strongly Agree}*
- Neighborhood vector: $M = \{(-1, 0), (1, 0), (0, -1), (0, 1), (1, 1), (1, -1), (-1, 1), (-1, -1)\}$
- Local Rule (R): Described through the CA Transition function.

An agent in this model is not only characterized by its opinion, but also by many agent and population-level parameters defined in Table 2.

Table 2. Agent and Population level parameters

Name	Description	Value	isDynamic
Agent level parameters			
Q(t)	Opinion of the agent at time step T	[0, 1]	Yes
$\delta(t)$	Distinctiveness factor of the agent at time step T	[0, 1]	Yes
k	Adjustment rate for NSI coefficient	[0, 1]	No
t	Tolerance of the agent against social interaction [8]	[0, 1]	Yes
x	Whether an agent is a Dissenter, i.e. able to get negatively influenced via the NSI coefficient	{0, 1}	Yes
y	Whether an agent is a Leader (i.e. has the ability to interact with more online neighbors)	{0, 1}	Yes
z	Whether an agent has the ability to communicate Online.	{0, 1}	No
d	Confidence threshold of the agent	[0, 1]	No
$\psi(t)$	NSI Coefficient for the agent at time step T	[0, 1]	Yes
ω	Connectivity of the agent (i.e. the radius up to which it can access online neighbors) [12]	[2, 5]	No
θ	Accessibility of the agent (i.e. the percentage of its distant neighbors that it can interact with) [12]	[0, 1]	No
Population level parameters			
α	Learning rate (a constant updating factor for $\delta_i(t)$)	[0, 1]	No
γ	Percentage of Dissenters / Leaders within the population (to be randomly interspersed)	[0, 1]	No
η	Leader weight (external influence for opinion updation)	[0, 1]	No
κ	Bias for initialization of accessibility of agents within the population, default value = 2.	[0, 1]	No

To implement NSI as an external influence affecting opinion formation, it is perceived as a cause for the updation of the distinctiveness factor (δ):

$$\delta_i(t+1) = min(max(\delta_i(t) + \alpha * (\bar{\delta}_{neighbors}(t) - \delta_i(t)), l), h) \tag{6}$$

where, $alpha \in [0, 1]$: constant updation factor for δ; $\bar{\delta}_{neighbors}(t)$ = mean distinctiveness factor of the interacting neighbors of the agent at any time step t; and, l, h = lower and upper bounds for δ. The NSI coefficient is defined as:

$$nsi = k * [x_i^*(t) - x_i(t)] \tag{7}$$

where, k = adjustment rate (or coefficient) for opinion updation; $\bar{x}(t)$ = mean opinion of the agent neighbors; and, $x^*(t)$ = ideal opinion of the population [22].

Distant Neighbors: The agents in the CA lattice which can interact with an online-connected agent are defined by its *distant neighbors*, i.e., a proper subset of the set of its neighbors at radius r determined by two parameters; accessibility and connectivity. The set of distant neighbors of an agent includes the first θ neighbors at $r = \omega$ (See Fig. 1).

***Initialization of Accessibility(θ)*:** Finally, each agent, with bias (κ) in the population can have varying values of accessibility (See Table 2), which is initialized in the following way based on whether it is a leader (y) or not:

$$\theta = random(0,1)^{(\frac{1}{\kappa})},\ y = True\ ; val = random(0,1)^{\kappa}; \theta = \frac{val}{val+1},\ y = False$$

The CA model is non-linear in nature and as such only one agent changes its opinion at a time. However, multiple parameters affect interactions among agents and the distinctiveness factor is also being updated for interacting agents, as described in Algorithm 1. For each parameter, including the opinion, leadership, distinctiveness factor, Algorithm 2 describes subroutines for evolution. Certain augmentations are also described used to study social constructs later.

Algorithm 1 Transition Function (Rules of interaction)

```
cell ←Random agent in population
neighbors ←local neighbors
if cell.online_access = True then
    neighbors += Set of distant neighbors
confidence_set ← Initialized set of neighbors falling under confidence threshold
foreach agent ∈ neighbors do
    if |agent.opinion − cell.opinion| ≤(d - 2t) then
        confidence_set += agent // d is confidence threshold and t is tolerance of cell
cell.fuzzy_opinion ← Average fuzzy opinion of agents in confidence set
if cell.is_leader = True then
    update_if_Leader()
if cell.is_dissenter = True then
    update_if_Dissenter()
elseif cell.is_dissenter = False then
    update_if_not_Dissenter()
updateDelta()
```

Algorithm 2 Supporting Update Functions

```
Function update_if_Dissenter(cell)
    foreach category ∈ opinion_categories do
        cell.fuzzy_nsi ← cell.k · (IdealOpinion − cell.fuzzy_opinion[category])
        cell.fuzzy_opinion− = cell.fuzzy_nsi
Function update_if_Leader(cell)
    foreach category ∈ opinion_categories do
        cell.fuzzy_nsi ← cell.k · (IdealOpinion − cell.fuzzy_opinion[category])
```

```
    cell.fuzzy_opinion ← cell.fuzzy_opinion.(1 − leader_weight) + cell.fuzzy_opinion.leader_weight) + cell.fuzzy_nsi
Function update_if_not_Dissenter(cell)
    foreach category ∈ opinion_categories do
        cell.fuzzy_nsi ← cell.k · (IdealOpinion − cell.fuzzy_opinion[category])
        cell.fuzzy_opinion+ = cell.fuzzy_nsi
Function updateDelta(cell) //Update distinctiveness factor of the current cell
    newDelta ← cell.delta + learning_rate ∗ (AvgDelta(cell) − cell.delta)
// Rules for Parameter Updates (Augmentation):
if neighbors.are_dissenters or {cell.is_dissenter and cell.delta > threshold } then
    cell.is_dissenter ←True else False
end if
if cell.is_leader then
    if |avg_neighbor_opinion - cell.opinion| > threshold then
        cell.leader ←False else {SetOfNeighborsWithLowTolerance} ←cell.opinion
end if
if |neighbors.mean_tolerance - cell.tolerance| < threshold then
    cell.tolerance ←neighbors.mean_tolerance
end if
```

Figure 2 picturizes evolution of opinions of an agent in the CA. In case (a), due to the opposing average opinions of its neighbors, it takes a neutral opinion. In case (b), the dissenting agent retains its opinion, even with opposing opinion of its neighbors.

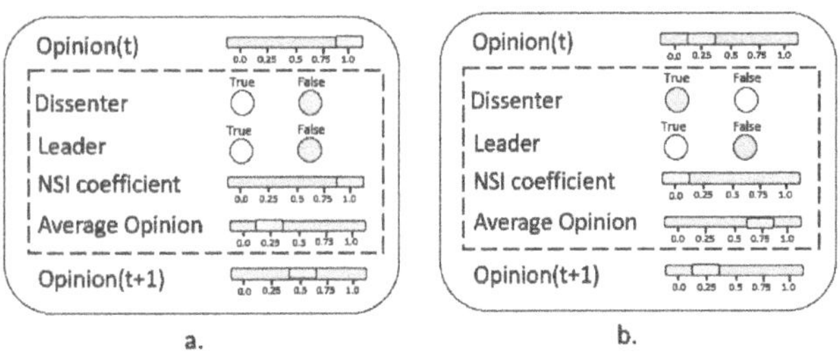

Fig. 2. Example Evolution of opinion of agent in two scenarios (a), (b). Here each numeric value refers to a fuzzy category, see Sect. 3.2

3.2 Fuzzification

In order to directly translate real opinions and inclinations, in the form of categories of values, fuzzification is employed. Every parameter, having a crisp numerical value is converted to a Gaussian fuzzy set, the mean value representing the parameter itself and a standard deviation characterizing an uncertainty about the opinion [1]. In classical models of opinion dynamics, it is assumed that interactions occur using crisp values of parameters, which is not always true since agents can also be influenced by sentiments, politics, products, and events. Thus, fuzzification helps in modeling uncertain opinions and uncertainty tolerances, arising due to the differences in culture backgrounds and characters of agents [11]. The fuzzy categories for opinions and parameters are:

1. 'opinion' and 'avg_opinion': {StronglyDisagree: 0.0, Disagree: 0.25, Neutral: 0.5, Agree: 0.75, StronglyAgree: 1.0}
2. 'nsi_coeff': {NonConforming: 0.0, SlightlyNonConforming: 0.25, Neutral: 0.5, SlightlyConforming: 0.75, Conforming: 1.0}

Fuzzy Rules of Interaction (See Fig. 3): Upon analyzing all the possible combinations of the fuzzy parameters considered, and the corresponding evolved opinions (i.e. 500 possibilities), two general broad categories of rules of interaction were observed:

1. First, for a cell which is a dissenter, its next state is the same as the current fuzzy state of the cell. (The crisp value may differ slightly)
2. (a) Second, for any opinion of a cell which is not a dissenter, and the current combination of values of its parameters belong to the ones specified in Table 3, its next state is Neutral.
   (b) Otherwise, the next state of the cell is the same as the current fuzzy state of the cell. Note, the augmentation to the local rule is not considered.

Table 3. Rules of Interaction for Non-Dissenters.

Q(t)	x	y	NSI Coeff.	Avg. Op(t)	Q(t+1)
Agree	False	True	Neutral	Agree	Neutral
Agree	False	True	S. Conforming	Disagree/S. Agree	Neutral
Agree	False	False	Non-conforming	Disagree	Neutral
Agree	False	False	S. Conforming	Agree	Neutral
Agree	False	False	Conforming	S. Disagree	Neutral
S. Agree	False	False	Neutral	S. Disagree	Neutral
S. Agree	False	True	S. (Non-)Conforming	Disagree	Neutral
S. Agree	False	False	S. Non-conforming	Agree	Neutral
S. Agree	False	False	S. Conforming	S. Agree	Neutral
S. Agree	False	False	Conforming	Disagree	Neutral
S. Disagree	False	True	Neutral	S. Agree	Neutral
S. Disagree	False	True	S. Conforming	Disagree Agree	Neutral
S. Disagree	False	False	S. Non-conforming	Disagree	Neutral
Disagree	False	True	Neutral/(S.) Non-conforming	Agree	Neutral
Disagree	False	True	(S.) Conforming	Disagree	Neutral
Disagree	False	False	Conforming	S. Agree	Neutral

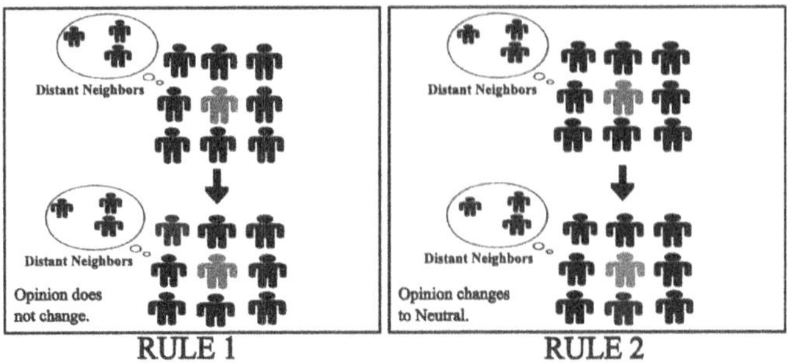

Fig. 3. Fuzzy Rules of Interaction. In RULE 1, central agent is a dissenter and opinion does not change. In RULE 2, central agent is not a dissenter and the next state is neutral if its parameters align with any of the combinations in Table 3. **Note** - Blue: Any opinion; Red: Neutral opinion.

4 Results

The model was simulated using Python 3 on a system equipped with an Intel i7 quad-core CPU clocked at 2.80 GHz, running the Windows operating system, with 16 Gigabytes of RAM, without the use of any major external Python packages. Upon analysis of various scenarios outlined hereafter, diplomacy was seen to be the steady state for consensus, for which the time taken to converge was studied. Further, the effects of presence of zealots in populations were also observed, specifically in the case of fringe groups, i.e. clustered populations.

4.1 Ideal Population

An ideal population is a CA configuration with only non-dissenting, offline, non-leader agents having the same opinion who reach a steady state of neutral opinions. The rate of convergence towards neutrality, analyzed through a chi-squared goodness-of-fit test showed polynomial growth (See Fig. 4).

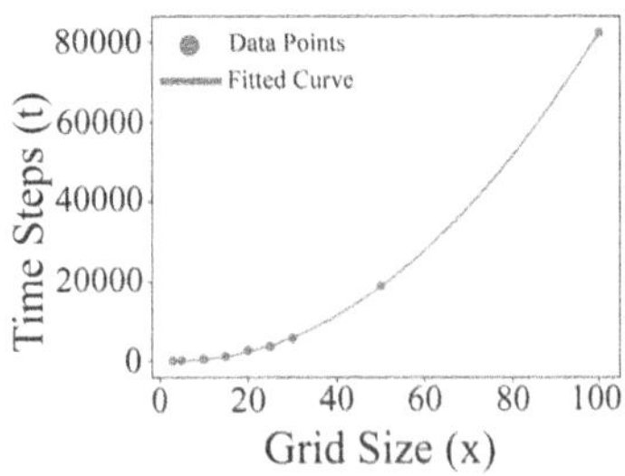

Fig. 4. Trend in convergence to steady state.

4.2 Mixed Population

A mixed population is a CA configuration where agents have varying opinions and parameters. **Note:** It was assumed that $\alpha = 0.5$; percentage of dissenters, leaders & online agents $= 0.25$; $\eta = 0.1$; d and $t = (0.1, 0.3)$ and population size $= 10000$ (see Fig. 5).

Initialization of Opinions through Uniform Distribution: Larger populations converge after a long time (See 4.1). With higher concentration of dissenters in the grid, such dissenters tend to have similarly opinionated nodes close by. The number of diplomats increases, relative to the number of other opinion-holders. (Standard Deviation of Opinions: 0.2769; Median Opinion: 0.5)

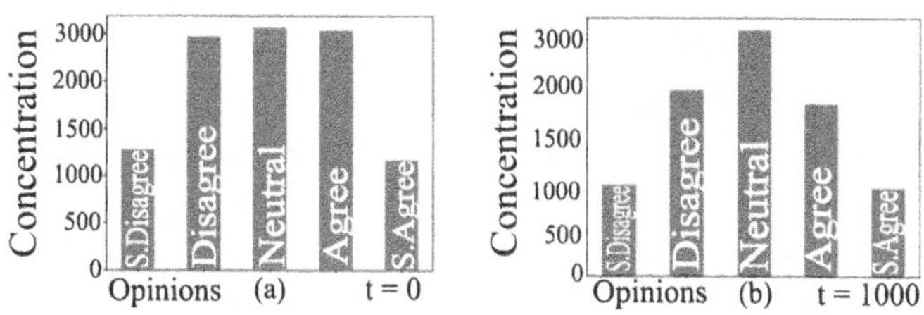

Fig. 5. Distribution of opinions from initial state (a) to final (b) with Uniform distribution. Many agents converge to neutrality.

4.3 Analysis Using Herfindahl-Hirschman Index (HHI)

HHI is used as a metric to analyse the extent of convergence in a population and formation of cliques, by varying population-level with agent-level parameters. **Note:** It is assumed that a group of >5 nearby agents having similar opinions is considered to be a clique. A uniform population of size 2500, over 1000 evolutions was considered (See Fig. 6).

HHI vs. Confidence Threshold (d): HHI fluctuates more with increase in d (w.r.t to percentage of online agents). HHI also converges to the middle (0.6–0.65) from the extremes (w.r.t percentage of dissenters). HHI converges and fluctuates more (inversely proportional to percentage of leaders) with higher d.

HHI vs. Learning Rate (α): HHI fluctuates lesser with increase in learning rate with varying percentage of dissenters. Higher the dissenters, higher the extremity of the initial HHI. HHI remains in the median range (0.55–0.65) for all percentage of online connected cells respectively, with increase in learning rate.

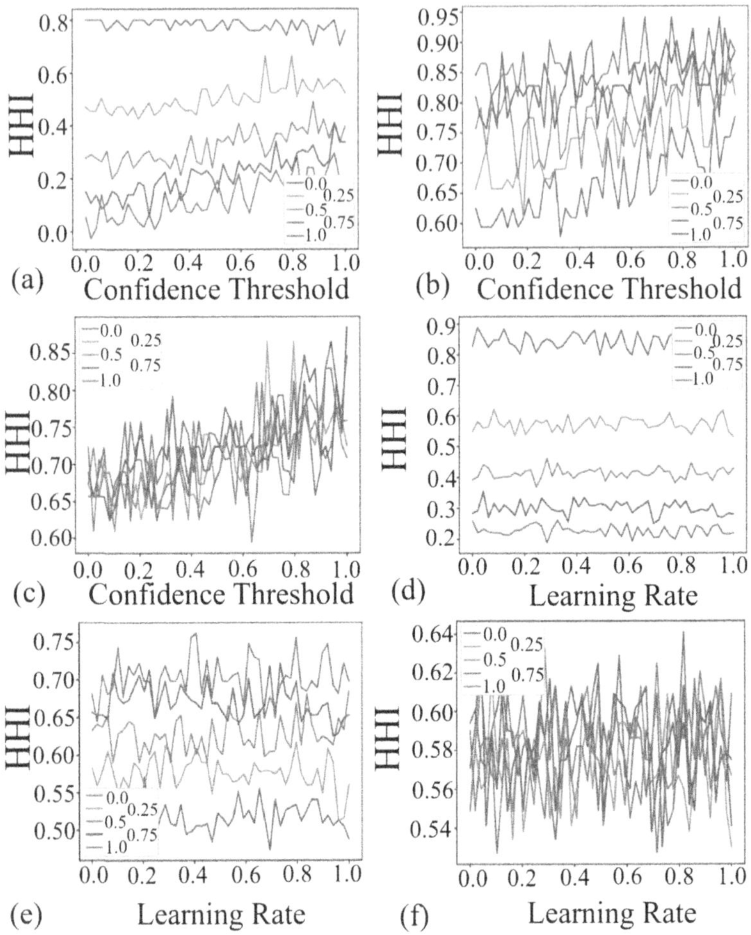

Fig. 6. HHI vs. Confidence Threshold (a, b, c) and vs, Learning Rate (d, e, f) w.r.t varying percentages of (a, d) Dissenters (b, e) Leaders (c, f) Online.

4.4 Clustered Populations

It has been observed that non-conformist behavior is more prevalent among people with higher general intelligence (IQ) [15]. Considering such individuals as outliers, who can be dissenters and leaders and have an online presence simultaneously, we consider the social structure of fringe groups that hold extremist opinions and their dissolution over time in an overwhelmingly conformist population. The previously defined extension to the local rule is employed here, enabling the dynamic variation of parameters and the opinion to get normally updated through the CA fuzzy rules. This leads to the rigid belief barriers produced by clusters (fringe) groups consisting of a central leader and zealots/extremists to break over time and converge towards consensus (Fig. 7).

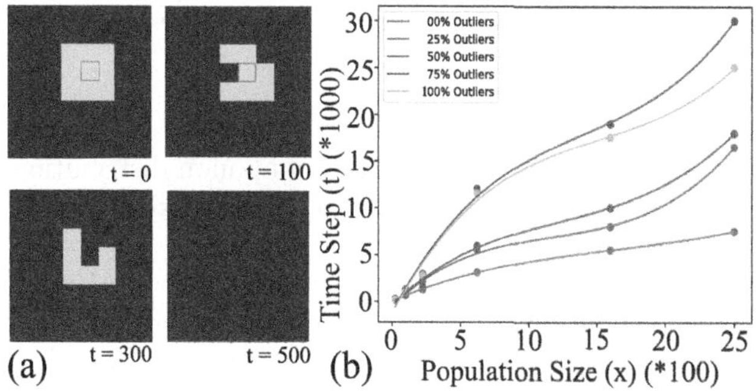

Fig. 7. (a) Dissolution of a fringe group (Opinion = S.Disagree) with no outliers (Population opinion = Neutral) over time. Central cell acts as a leader. (b) Time Steps taken for a fringe group to dissolve with varying number of outliers in the population.

The time taken for dissolution depends upon the number of outliers in the population. A general increase in the rate of dissolution is seen since eccentric opinions are more likely to spread in a population, especially to those who are willing to dissent [18]. It was observed that when the population has close to 75% outliers, the increase in time taken shows cubic polynomial behavior. However, with even higher concentrations, the increase is not as high.
Moreover, it was observed from the results of the simulation study that:

1. With increase in percentage of dissenters in a population with uniformly distributed initial opinions, clustering increases and convergence decreases. So, more number nodes of non-conformists or zealots emerge.
2. If the percentage of dissenters within the population is more, the distribution of opinions in the final configuration is more chaotic (without pattern / order). If it is less, the number of agents holding neutral opinions (diplomats) are more. Individuals who are not dissenters tend to move towards the ideal opinion (neutrality) (See Table 2).
3. Fringe groups professing extreme opinions headed by a leader dissolve towards the majority opinion over time, taking more time with more outliers in a population since they ensure the spread of the eccentric opinion, not letting them die down. It is considerably more challenging to reach a complete consensus when the population surrounding the fringe group is not at a consensus.

5 Discussion and Conclusion

The simulation study conducted herewith builds upon previous works [19] [21] [23] in the field, albeit through the perspective of a deterministic fuzzy CA model. Unlike real-world situations, this implementation allows for the exact prediction of opinions in a population when the initial configuration is known. This may prove to be helpful while trying to mitigate bias by backtracking the evolution of the model, especially in the study of fringe groups. Upon comparison of results from previous studies on opinion dynamics with the results of this simulation study, certain observations can be made:

1. The steady-state emergence of neutrality arises from inflexible non-conformists who resist persuasion, dictating opinion distribution. With more than 50% dissenters, standard deviation of opinions exceeded 0.250, signifying comparable opinions.
2. It has been previously observed that agents with a local majority opinion in a given group could be persuasive only if their average convincing power (CP) [14] was larger than the minority's conviction, making it possible for a minority to dominate the population in the long time limit and reaching full-consensus more difficult. This can be considered analogous to fringe groups within a population at a consensus. Whenever an extremist idea is professed within even a small group in a population which has varied kinds of agents, three-fourths of whom (75%) are susceptible and willing to spread them, such polarized views never die down. They are carried forward by propagation through an ever-growing network of emerging fringe groups.
3. More conformists lead to evolution to an equilibrium state. However, with increase in the number of zealots, the steady state of neutrality in opinions can rarely be reached completely [7]. Unlike previous theories about the significance of the concentration of initially neutral agents in a population [24], through this model, diplomacy can be considered a more natural state for consensus. It is also seen that full consensus is easy for the competition of up to three opinions but difficult for four and more opinions, leading to partial consensus, polarization and fragmentation.

6 Limitations of the Study

This exposition is a simulation study, which while insightful into opinion dynamics, has yet to be validated with the real-world data. The 2D Cellular Automaton framework captures local interactions but simplifies the complex networked structures of real social dynamics, providing scope for future empirical validation and extended modeling.

Disclosure of Interests. The authors have no competing interests to declare that are relevant to the content of this article.

References

1. Amirjanov, A.: Consensus in fuzzy opinion networks. Int. J. Mod. Phys. C (2023). https://doi.org/10.1142/S0129183123501358
2. Bhattacharjee, K., Naskar, N., Roy, S., Das, S.: A survey of cellular automata: types, dynamics, non-uniformity and applications. Nat. Comput. **19**(2), 433–461 (2018). https://doi.org/10.1007/s11047-018-9696-8

3. Deffuant, G., Neau, D., Amblard, F., Weisbuch, G.: Mixing beliefs among interacting agents. Adv. Complex Syst. (2000). https://doi.org/10.1142/S0219525900000078
4. Deutsch, M., Gerard, H.B.: A study of normative and informational social influences upon individual judgment. Psychol. Sci. Public Interest (1955). https://doi.org/10.1037/h0046408
5. Fu, G., Zhang, W.: Opinion dynamics of modified Hegselmann-Krause model with group-based bounded confidence. IFAC Proc. Volumes **47**(3) (2014). https://doi.org/10.3182/20140824-6-ZA-1003.02770
6. Hołyst, J., Kacperski, K.: Social impact models of opinion dynamics. Ann. Rev. Comp. Phys. (2001). https://doi.org/10.1142/9789812811578_0005
7. Javarone, M.A.: Social influences in opinion dynamics: the role of conformity. Phys. A (2014). https://doi.org/10.1016/j.physa.2014.07.018
8. Kan, U., Feng, M., Porter, M.A.: An adaptive bounded-confidence model of opinion dynamics on networks. J. Complex Netw. (2023). https://doi.org/10.1093/comnet/cnac055
9. Larsen, K.S.: Conformity in the Asch experiment. J. Soc. Psychol. (1974). https://doi.org/10.1080/00224545.1974.9923224
10. Li, H.L.: Mixed Hegselmann-Krause dynamics. Discr. Contin. Dynam. Syst. - B (2022). https://doi.org/10.3934/dcdsb.2021084
11. Liang, H., Dong, Y., Li, C.: Dynamics of uncertain opinion formation: an agent-based simulation. J. Artif. Soc. Soc. Simul. (2016). https://doi.org/10.18564/jasss.3111
12. Lim, D., Lee, H., Zo, H., Ciganek, A.: Opinion formation in the digital divide. J. Artif. Soc. Soc. Simul. (2014). https://doi.org/10.18564/jasss.2366
13. Lord, K.R., Lee, M.S., Choong, P.: Differences in normative and informational social influence. Adv. Consum. Res. (2001). https://research.ebsco.com/linkprocessor/plink?id=f2c4bd6c-a7cf-3333-aa9b-13247dec36f1
14. Lu, Y., Zhao, Y., Zhang, J., Hu, J., Hu, X.: Fuzzy Hegselmann-Krause opinion dynamics with opinion leaders. In: 2019 Chinese Control Conference (CCC), Guangzhou, China (2019). https://doi.org/10.23919/ChiCC.2019.8865519
15. Millet, K., Dewitte, S.: Non-conformity may be hidden driver behind relation. BMJ (2007). https://doi.org/10.1136/bmj.39121.870139.1F
16. Mraz, M., Zimic, N., Lapanja, I., Lebar Bajec, I.: Fuzzy cellular automata: from theory to applications. IEEE Trans. Appl. Indus. (2000). https://doi.org/10.1109/TAI.2000.889889
17. Ntinas, V.G., Moutafis, B.E., Trunfio, G.A., Ch, G.: Sirakoulis, parallel fuzzy cellular automata for data-driven simulation of wildfire spreading. J. Comput. Sci. (2017). https://doi.org/10.1016/j.jocs.2016.08.003
18. Pandey, S., et al.: Generation and influence of eccentric ideas on social networks. Sci. Rep. (2023). https://doi.org/10.1038/s41598-023-47823-0
19. Rainer, H., Krause, U.: Opinion dynamics and bounded confidence: models, analysis and simulation. J. Artif. Soc. Soc. Simul. (2002). https://www.jasss.org/5/3/2.html
20. Ross, T.J.: Automated Methods for Fuzzy Systems, chap. 7, pp. 211–244. John Wiley & Sons, Ltd (2010). https://doi.org/10.1002/9781119994374.ch7
21. Sharma, I., Chourasia, B., Bhatia, A., Goyal, R.: On the role of evangelism in consensus formation: a simulation approach. Complex Adapt. Syst. Model. **4**(1), 1–21 (2016). https://doi.org/10.1186/s40294-016-0029-4

22. Smaldino Paul, E., Epstein Joshua, M.: Social conformity despite individual preferences for distinctiveness R. Soc. Open Sci. (2015). https://doi.org/10.1098/rsos.140437
23. Weinans, E., van Voorn, G., Steinmann, P., Perrone, E., Marandi, A.: An exploration of drivers of opinion dynamics. J. Artif. Soc. Soc. Simul. (2024). https://doi.org/10.18564/jasss.5212
24. Wu, Y., Xiong, X., Zhang, Y.: Effects of convincing power and neutrality on minority opinion spreading. Mod. Phys. Lett. B (2017). https://doi.org/10.1142/S0217984917500580
25. Xia, H., Wang, H., Xuan, Z.: Opinion dynamics: a multidisciplinary review and perspective on future research. Int. J. Knowl. Syst. Sci. (IJKSS) (2011). https://doi.org/10.4018/jkss.2011100106

Cellular Automata-based Toxicity Analysis for Social Media Comments

M. J. Elizabeth[2(✉)] and Raju Hazari[1]

[1] Department of Computer Science and Engineering, National Institute of Technology, Calicut, Kerala, India
[2] Department of Computer Science, Mary Matha Arts and Science College, Mananthavady, Wayanad, Kerala, India
elizabethmj@marymathacollege.ac.in

Abstract. Toxic analysis of social media data aims to identify harmful language, including hate speech, bullying, and abusive comments, to promote healthier online environments. This paper presents a cellular automata-based approach for toxic text classification. The results show that it has the ability to capture text context and patterns through simple, localized interactions. Our proposed model encodes text data using Unicode formats, which are used as rules for the CA evolution process. The CA signals are then used as input for machine-learning models. This hybrid approach enables the classification model to detect nuanced toxic expressions. Experimental results on benchmark datasets demonstrate that the CA-based toxic analysis model achieves comparatively good accuracy in identifying class labels of toxic and non-toxic comments.

Keywords: Cellular Automata (CA) · Toxic Comments · Hate Speech · Machine Learning · Text Classification

1 Introduction

Toxic text classification has improved greatly in recent years, driven by increased user-generated data on social media platforms. However, this large amount of online content also brings new challenges. Hateful, harassing, and harmful language in texts can seriously harm online communities. Toxic text must be automatically identified and categorized to provide a more secure and welcoming online community. People and communities are seriously threatened by harmful material, which includes hate speech, cyberbullying, and other types of online harassment. It is important to recognize and minimize toxic remarks since they can have serious effects in real life and make it harder to have constructive conversations. Frequently, these remarks are impolite, hostile, nasty, or cruel. Toxic analysis helps us to reduce such hateful words from the text as well as it has potential applications across multiple industries for brand monitoring, market research, and social media analysis, and is also used to analyze employee feedback. It can be used in healthcare sectors for psychological analysis and patient

S. Kamilya et al. (Eds.): ASCAT 2025, CCIS 2499, pp. 149–164, 2025.
https://doi.org/10.1007/978-3-031-94121-4_11

feedback monitoring. Threat detection is another application area by analyzing the toxicity of the text and behaviors in security footage or online communications. Additionally, text toxicity analysis is essential for creating emotion-aware interactions that require understanding the user's emotions. However, toxic text analysis presents several challenges, such as modeling the context of conversations, addressing the dynamic shifts in offensive language among participants, and amplifying the complexity of the task.

Numerous studies have already generated favorable results in this domain. In 2009, supervised learning methods were used for detecting harassment [36]. A well-structured survey [31] was done in 2017 in which multiple application areas of hate speech classification, different classification techniques, challenges, and an overview of data considerations were explained in detail. Another survey was done by Fortuna et.al [16] in 2018. Various machine-learning (ML) methods were introduced in that paper. Logistic Regression [8], Random Forest, and SVM [19], and many other ML models [3,5,6] were analyzed for toxicity analysis. A variety of prominent deep learning models [3] are also applied in the field of text classification tasks such as Convolutional Neural Networks (CNNs) [3], Recurrent Neural Networks (RNNs) [27], Deep Averaging Networks (DANs), Denoising Autoencoders (DAEs), Long Short-term Memory (LSTM) [10] networks, Bi-directional Long Short-term Memory (Bi-LSTM) networks, Gated Recurrent Units (GRUs), and attention mechanisms. In recent years, pre-trained language models [37] such as BERT, RoBERTa, XLM, and Glove. have come into play for text classification. Their ability to capture bidirectional context and nuances in language has led to promising results. Many researchers combine multiple of these methods within a single model to enhance the performance of toxic text analysis. All these approaches completely rely on conventional language-related features, such as TF-IDF, POS, BOW, N-grams, sentiment, hashtags, mentions, retweets, URLs, character or word counts, syllables, dictionaries, syntactic and semantic features, and more. Our primary goal is to develop a machine-learning model without relying on such language-specific features for toxicity analysis.

The proposed method entails bringing the concept of cellular automata (CA) into the field of natural language processing. The CA is used to evaluate the complexities of every sentence through a weight-based approach in which word weight is calculated based on the occurrence of the words in the text along with deep learning models. We have used publicly available English, Hindi, and Bengali hate speech datasets for this analysis. The proposed method is a language-independent Deep Learning (DL) Model for toxic text or hate speech classification. In recent works, we have used CA-based machine learning models [11–15] for sentiment analysis which gives comparatively good results. This work motivates us to use the CA-based machine-learning model for the toxicity classification of texts.

The proposed approach is a DL model, implemented using the combination of cellular automata (CA) and deep-learning models. The process begins by transforming each comment in the dataset into a vectorized form using Unicode formats, which are then utilized as rules for the CA model. Through an iterative evolution process, the Cycle Length (CL) values of each CA cell are determined. Those CL values are further used for calculating the weights of each word in the text. Finally, we will get a weight matrix that is used as an input for DL models. Ultimately, the model predicts toxicity, categorizing them into two classes such as positive and negative classes.

The paper is organized as follows. The basics of cellular automata are described in Section 2. Section 3 describes the design techniques of the CA model. In Section 4, the results and its evaluation details are explained. Finally, Section 5 concludes the work.

2 Basics of Cellular Automata

The history of Cellular Automata traces back to Von Neuman's work [26], where they initially developed a CA-based model to simulate a biological self-reproduction system. Cellular Automata (CA) are discrete, spatially extended dynamical systems that have been extensively explored as models for various physical systems. They evolve within discrete space and time, featuring a lattice of cells. Each cell stores a discrete variable representing its current state at time 't'. The state of a cell at '$t+1$' is determined by its current state and the states of its neighbouring cells at a time 't'. The simplest form of CA is Elementary Cellular Automata (ECA), introduced by Wolfram [35]. ECA are fundamental and straightforward models for understanding complex behaviors that can emerge from simple local rules. In simpler terms, ECA are cellular automata where the future states of each cell depend on the current state of that cell and its two closest neighbours. These cells can be thought of as binary (typically 0 or 1), having two possible values, and the interactions between neighbouring cells lead to the evolution of the system over discrete time steps. The state of a cell at time '$t + 1$' is determined as:

$$S_i^{t+1} = f(S_{i-1}^t, S_i^t, S_{i+1}^t) \tag{1}$$

where 'f' represents the next state function, and the variables S_{i-1}^t, S_i^t and S_{i+1}^t correspond to the present states of the left, self, and right neighbour of the ith cell at time t. The function $f : \{0,1\}^3 \mapsto \{0,1\}$ can be expressed as a look-up table (see Table 1) with 8 possible combinations of inputs, each resulting in an output of either 0 or 1. There are a total of 256 possible elementary cellular automata rules, and Table 1 illustrates two examples: Rule 120 and Rule 240.

In the traditional setup of a Cellular Automata (CA), all the cells within the automata follow the same next state function, making it a "uniform" CA. However, in some instances, a more flexible approach is adopted where individual cells can follow different next-state functions or rules. This variant of CA is referred to as a "non-uniform" or "hybrid" CA. In this study, we chose to use a non-uniform elementary cellular automata configuration under periodic boundary conditions where the first and last cells are considered neighbours of each other. This non-uniform approach provides greater complexity and flexibility in modeling and simulating the system, allowing for a broader range of behaviors and patterns to emerge within the CA that helps to learn the context of text.

Table 1. Look-up table for rules 120 and 240

Present State :	111	110	101	100	011	010	001	000	*Rule*
	(7)	(6)	(5)	(4)	(3)	(2)	(1)	(0)	
(i) Next State :	0	1	1	1	1	0	0	0	120
(ii) Next State :	1	1	1	1	0	0	0	0	240

3 Proposed CA Model Implementation

In the proposed classification framework, we've created a weight-based classifier with a deep-learning model designed to assess whether the text is toxic or not. In this context, the input text is transformed into a vector format with the help of Unicode standards. Our approach involves building a language-independent machine-learning classifier based on CA. The design of our model can be categorized into two modules, which are listed below:

- Module-I: Weight Matrix Generation
- Module-II: Toxicity Prediction Process

3.1 Module-I: Weight Matrix Generation

We need to perform the data preprocessing process, text vectorization process, and the CA evolution process before creating the weight matrices.

3.1.1 Data Preprocessing

Each comment in the dataset is preprocessed using the following steps:

1. Remove extra spaces, punctuation, and new lines.
2. Remove repeated patterns, URLs, and user tags.
3. Remove single characters, emojis, and stop words from the comments.
4. Remove non-English characters from English comments, Non-Hindi characters from Hindi comments, and Non-Bengali characters from Bengali comments.
5. Remove the start and end white spaces.
6. Convert each comment into lowercase letters for English comments.

3.1.2 Text Vectorization

The preprocessed data is converted into a vectorized form using Unicode values of every character found in the text. The process for creating CA rule vectors involves the following steps:

- Identify the Unicode position for each character from the Unicode charts.

- If the Unicode position is represented in the format: 0x (p1 p2 p3 p4), where p1, p2, p3, p4 are hexadecimal characters, then the decimal value derived for the character is calculated with the expression: $256^1(16p1 + p2) + 256^0(16p3 + p4)$. Subsequently, the corresponding CA rule can be calculated using the expression: $<(16p1 + p2), (16p3 + p4)>$.
- If p1 and p2 both are '0' then the corresponding CA rule calculated using the expression is: $<(16p3 + p4)>$

A few examples are given below:

- The Unicode of the English character 'f' is 0x(0066). Here p1=0, p2=0, p3=6 and p4=6. The CA Rule is determined using the formula:

$$CA\ Rule = <(16p1 + p2), (16p3 + p4)> \tag{2}$$

  Substituting the values: CA Rule $=<(16*0 + 0, 16*6 + 6)> =<0, 102>$.
  Thus, the CA Rule for the character 'f' is represented as [102].
- The Unicode of the Hindi character 'फ' is 0x(092B). Here p1=0, p2=9, p3=2 and p4=B. The CA Rule is determined using the Eq. 2. Substituting the values: CA Rule $=<(16*0 + 9, 16*2 + B)> =<9, 43>$.
  Thus, the CA Rule for the character 'फ' is represented as [9, 43].
- The Unicode of the Bengali character 'ক' is 0x(0995). Here the CA Rules will be [9,149].

If we apply it to words then,

- Let's consider a word 'love' which has 4 letters: ['l', 'o', 'v', 'e'] and their Unicode characters are [0x(006C), 0x(006F), 0x(0076), 0x(0065)]. The corresponding CA rules are [108, 111, 118, 101]

Examples of different texts and their corresponding rule vectors are given below:
English Text- "many unsecured wi fi hot spots so many idiot wikipedia pages"
Rule Vectors - [109, 97, 110, 121, 32, 117, 110, 115, 101, 99, 117, 114, 101, 100, 32, 119, 105, 32, 102, 105, 32, 104, 111, 116, 32, 115, 112, 111, 116, 115, 32, 115, 111, 32, 109, 97, 110, 121, 32, 105, 100, 105, 111, 116, 32, 119, 105, 107, 105, 112, 101, 100, 105, 97, 32, 112, 97, 103, 101, 115]
Hindi Text- "देखिए मेंरे साथ अभी अबकी बार भोजपुरी सितारों की धारठीक बजे ।" (English Translation: "Watch with me this time the edge of Bhojpuri stars at the right time.")
The characters are separated into individual list items, making it possible to analyze each one independently. The output of character tokenization is given below:

['द','े','ख','ि','ए', ,'म','े','र','े', ,'स','ा','थ', ,'अ','भ','ी', ,'अ','ब','क','ी', ,'ब','ा','र', ,'भ','ो','ज','प','ु','र','ी', ,'स','ि','त','ा','र','ो','ं', ,'क','ी', ,'ध','ा','र','ठ','ी','क', ,'ब','ज','े', ,'।']

Rule Vectors - [9, 38, 9, 71, 9, 22, 9, 63, 9, 15, 32, 9, 46, 9, 71, 9, 48, 9, 71, 32, 9, 56, 9, 62, 9, 37, 32, 9, 5, 9, 45, 9, 64, 32, 9, 5, 9, 44, 9, 21, 9, 64, 32, 9, 44, 9, 62, 9, 48, 32, 9, 45, 9, 75, 9, 28, 9, 42, 9, 65, 9, 48, 9, 64, 32, 9, 56, 9, 63, 9, 36, 9, 62, 9, 48, 9, 75, 9, 2, 32, 9, 21, 9, 64, 32, 9, 39, 9, 62, 9, 48, 9, 32, 9, 64, 9, 21, 32, 9, 44, 9, 28, 9, 71, 32, 9, 100]

Bengali Text- "আপনাদের ওপর শরম লাগে এখানে মারছে আর আপনি ভিডিও বানাচ্ছেন." (English Translation: "Shame on you guys here and making videos")

Rule Vectors - [9, 134, 9, 170, 9, 168, 9, 190, 9, 166, 9, 199, 9, 176, 32, 9, 147, 9, 170, 9, 176, 32, 9, 182, 9, 176, 9, 174, 32, 9, 178, 9, 190, 9, 151, 9, 199, 32, 9, 143, 9, 150, 9, 190, 9, 168, 9, 199, 32, 9, 174, 9, 190, 9, 176, 9, 155, 9, 199, 32, 9, 134, 9, 176, 32, 9, 134, 9, 170, 9, 168, 9, 191, 32, 9, 173, 9, 191, 9, 161, 9, 191, 9, 147, 32, 9, 172, 9, 190, 9, 168, 9, 190, 9, 154, 9, 205, 9, 155, 9, 199, 9, 168]

3.1.3 CA Evolution and CL-Graph Generation

The CA evolution process is a sequence of state transformations driven by rules applied iteratively to a grid of cells. Each vector in the rule vector table is used as a rule for the CA evolution process, which is carried out to generate cycle-length (CL) values. The process begins with an initial configuration, representing the starting state of all cells in the grid. This configuration can be random (e.g., cells randomly initialized as 0 or 1) or structured (e.g., alternating 0's and 1's or a specific pattern). The next state of each cell depends on the states of its neighbourhood. In ECA, the neighbourhood includes the current cell and its immediate left and right neighbours (radius 1). For a 2-D CA, the common neighbourhood structures include Moore (8 neighbours) or Von Neumann (4 neighbours) configurations. The rule vector defines how the states of a cell and its neighbours determine the next state. The rule can be encoded as a single number, e.g., Rule 30 or Rule 110 in Wolfram's notation. The CA evolves by applying the rule iteratively over a predefined number of steps or until a stopping condition is met. The local dependency and the deterministic nature of CA make them suitable for modeling complex phenomena. During the evolution, a repeating pattern or cycle may emerge. The length of the repeating pattern is called the Cycle Length (CL) value. If no repeating pattern is observed within a threshold of iterations, the CL value can be assigned as a negative value or zero. By iteratively applying simple rules to the states of cells and their neighbours, CA demonstrate rich and diverse behavior, making them a powerful computational model for exploring complex systems. The steps used to generate CL values are given below:

1. *Initialization Stage*: The initial state of the ECA alternates between 1 and 0 for each cell, creating a regular pattern such as 101010... or 010101... The length of the initial configuration matches the size of the rule vector. The same pattern is used for each text in the study.
2. *State Transition*: The current state of each cell in the ECA is updated at the time of transition by using its current state and its neighbouring states. These transitions are governed by the rule number specified in the rule vector.

3. *Iteration Process*: ECA undergoes iterative state transitions, evolving over discrete time steps. The iteration value is a critical parameter, determining how long the CA evolves before stopping. The minimum number of iterations is set to '1000' with periodic boundary conditions applied. If the length of the text goes beyond '2000' then half of the length of the text is considered as iteration step. For instance, if the maximum length of the comments in the dataset is '2400', the iteration value is set to '1200'.
4. *Emergence of Patterns*: After a specific number of iterations, a repeating pattern may emerge. The Cycle Length (CL), representing the periodicity of these patterns, is a widely used metric for characterizing CA dynamics. Identifying the CL helps quantify the complexity and structure of the text when CA is applied to text data analysis, as observed in various pattern recognition tasks.
5. *Threshold Evaluation*: A threshold of '32' repetitions is set, inspired by prior studies that suggest recurring patterns below a certain limit may represent noise or transient states rather than meaningful cycles [20]. If the recurrence does not occur or fails to reach the threshold of 32 repetitions, then the CL value is taken as '0' or any negative value to denote non-periodicity.

The visualization of the evolution is shown in Fig. 1 in which the specific patterns of variable length recur periodically, as indicated by the bold lines. To illustrate this further, we provide an example that describes the input text and the corresponding CL values. From these CL values, we can generate CL graphs, which are displayed in Figs. 2, 3 and 4.

....	1	0	1	0	1	0	1	0	....
....	0	1	0	1	0	1	1	0	....
....	1	0	1	0	1	1	1	0	....
....	1	0	0	1	0	0	1	1	....
....	0	0	1	0	0	1	1	1	....
....	1	1	0	1	1	0	1	0	....
....	1	1	0	0	1	0	0	0	....
....	0	0	0	1	1	0	0	1	....
....	1	1	0	0	0	1	1	0	....
....	1	0	0	1	1	0	0	0	....
....	0	1	1	0	0	0	0	0	....
....	1	0	0	1	0	0	0	0	....
....	1	0	0	0	0	1	0	1	....
....	⋮	⋮	⋮	⋮	⋮	⋮	⋮	⋮	....

Fig. 1. CA Evolution.

English Text: "many unsecured wi fi hot spots so many idiot wikipedia pages"
CL values of English Text: [8, 8, 8, 8, 1, 1, 1, 6, 6, 6, 2, 2, 1, 1, 1, 2, 1, 2, 1, 1, 1, 1, 1, 4, 4, 2, 2, 6, 6, 3, 1, 1, 2, 18, 18, 18, 18, 18, 18, 3, 3, 1, 1, 1, 10, 10, 10, 10, 10, 10, 10, 10, 10, 10, 10, 1, 1, 4, 4, 4, 2]

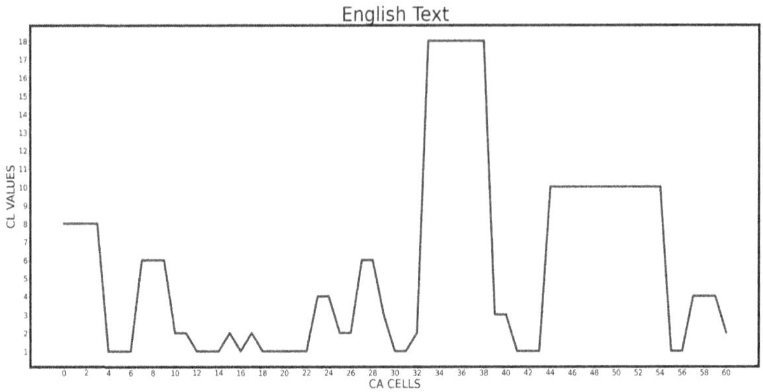

Fig. 2. CL-graph for English Text.

Hindi Text- "देखिए मेरे साथ अभी अबकी बार भोजपुरी सितारों की धारठीक बजे।" (English Translation: "Watch with me this time the edge of Bhojpuri stars at the right time.")

CL values of Hindi Text: [1, 1, 1, 1, 1, 1, 1, 1, 1, 1, 1, 1, 1, 1, 1, 1, 1, 6, 6, 6, 3, 3, 1, 1, 1, 1, 1, 1, 1, 1, 1, 1, 1, 1, 1, 1, 1, 1, 2, 2, 1, 1, 1, 1, 1, 1, 1, 1, 1, 1, 1, 1, 2, 1, 1, 1, 0, 0, 0, 0, 0, 0, 1, 1, 3, 3, 1, 1, 1, 1, 1, 1, 1, 1, 1, 1, 1, 1, 1, 4, 2, 2, 1, 1, 1, 1, 1, 1, 1, 1, 3, 3, 3, 1, 1, 1, 1, 1, 1, 1, 5, 5, 5, 5, 1]

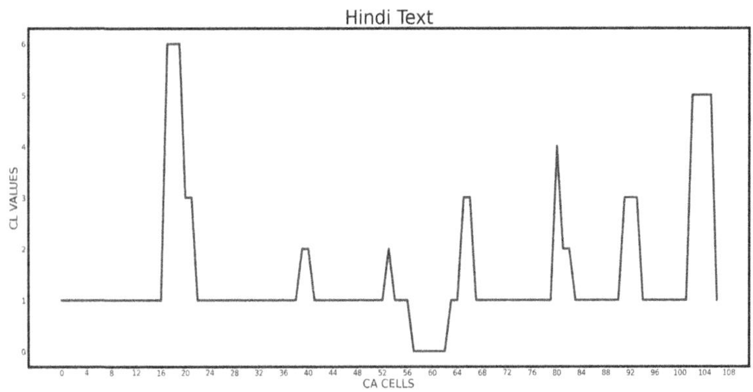

Fig. 3. CL-graph for Hindi Text.

Bengali Text- "আপনাদের ওপর শরম লাগে এখানে মারছে আর আপনি ভিডিও বানাচ্ছেন." (English Translation: "Shame on you guys here and making videos")

CL values of Bengali Text: [3, 3, 3, 3, 3, 1, 1, 1, 1, 1, 1, 1, 1, 1, 1, 9, 9, 9, 3, 3, 3, 1, 1, 1, 1, 1, 1, 1, 3, 3, 3, 1, 1, 1, 1, 1, 1, 1, 1, 1, 1, 1, 1, 1, 1, 1, 1, 1, 1, 1, 1, 1, 1, 1, 1, 6, 3, 3, 3, 3, 3, 1, 3, 3, 1, 1, 1, 1, 3, 3, 1, 1, 1, 1, 1, 1, 1, 1, 1, 2, 1, 1, 1, 2, 1, 1, 1, 1, 1, 1, 1, 1, 1, 1, 1, 1, 1, 1, 2, 1, 1, 1, 1]

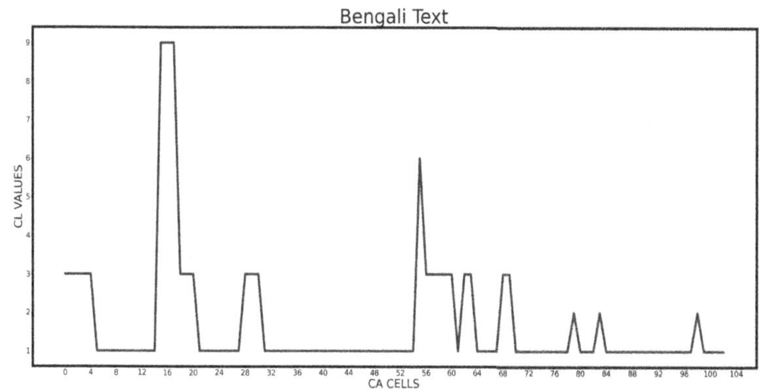

Fig. 4. CL-graph for Bengali Text.

3.1.4 Word Weight Generation Procedure

The proposed model identifies each unique word within the dataset and then we need to find out the CL sequences of each word. These CL sequences at sentence and word levels are used to determine the CL-Weight for each word. The steps for calculating CL-Weight are as follows:

- **Sentence CL-Sum (X)**: The sum of CL values of a sentence.
- **Word CL-Sum (Y)**: Sum of the CL values of the word.
- **CL-Weight**: Let α represent a word that appears in n number of sentences within the dataset. If the Word CL-Sum of α is Y_α, then the CL-Weight of α is calculated by:

$$CL - Weight_\alpha = \frac{\sum_{i=1}^{n} X_i}{Y_\alpha}$$

The weights of each word are used to generate a weight matrix for each comment in the dataset, serving as input to the deep learning models. For example, let us consider a Bengali dataset with only two sentences then, the procedure of word vectorization is as follows:
Text 1 - "দারুন মজার নাটক.." (English Translation: "Great fun drama")
CL Values - [1, 1, 1, 1, 1, 1, 1, 1, 1, 1, 1, 1, 1, 1, 1, 1, 1, 1, 1, 2, 1, 1, 1, 1, 1, 1, 1] and the sentence CL-Sum is 29.
Text 2 - "নাটক অনেক ভালো লাগছে খুবি মজার ছিলো" (English Translation: "The drama is very good and very funny")
CL Values -[4, 1, 1, 1, 1, 1, 1, 1, 1, 1, 1, 1, 1, 1, 1, 1, 1, 1, 1, 1, 1, 1, 1, 1, 1, 1, 1, 1, 9, 9, 9, 9, 3, 3, 1, 1, 1, 1, 1, 2, 1, 1, 1, 1, 1, 1, 2, 1, 1, 1, 1, 1, 1, 1, 1, 1, 1, 1, 1, 1, 1, 4, 2, 1, 1, 1, 4, 4, 4] and the sum of CL values is 123.

Each word in Text 1 and Text 2 is given in Tables 2 along with its CL values and corresponding word weights. Tables 3 shows Text 1 and Text 2 in vector form.

Table 2. CL values, Word CL-Sum and CL-Weight of words

Unique_Words	CL_Values	Word_CL-Sum	CL-Weight
দারুন	[1,1,1,1,1,1,1,1,1,1]	10	29/10=3
মজার	[1,1,1,1,1,1,1,1]	8	(29+123)/8 = 19
নাটক	[1,1,1,1,1,1,1,1,1,1]	10	(29+123)/10 = 15
অনেক	[1,1,1,1,1,1,1,1,1]	9	123/9=14
ভালো	[1,1,1,1,1,0,0,0]	5	123/5=25
লাগছে	[1,1,1,1,1,1,1,2,1,1]	12	123/12=10
খুবি	[1,1,0,0,0,1,1,1,1]	6	123/6=21
ছিলো	[1,2,1,1,1,0,0,1]	7	123/7=18

Table 3. Mappings of Reviews

Reviews	Word Vectors
" দারুন মজার নাটক "	[3,19,15]
"নাটক অনেক ভালো লাগছে খুবি মজার ছিলো "	[15,14,25,10,21,19,18]

4 Toxicity Prediction Process

In this section, we explored the experiment results, performance analysis, and comparative study with other state-of-the-art models. We have used four DL models such as LSTM, Bi-LSTM, CNN, and CNN-LSTM. We used 64 LSTM layers, 128 convolution layers, and 256 Bi-LSTM layers. The experiment is done with 'adam' as the optimizer with learning_rate=0.0005. We set both dropout and recurrent dropout rates to 0.2. The 'Relu' activation function is used in all cases. The architecture of CNN and Bi_LSTM models are shown in Figs. 5 and 6.

4.1 Datasets Description

- *Hate Speech and Offensive Language Dataset* [2]: The dataset is created using Twitter data for hate-speech detection. There are 24K comments classified as hate speech, offensive language, and neither (class labels: 0 - hate speech, 1 - offensive language, and 2 - neither offensive nor non-offensive).
- *Jigsaw Toxic Comment Classification Dataset* [7]: The dataset contains 1,59,571 comments from Wikipedia's talk pages that have been labeled by human raters as toxic or non-toxic. The types of toxicity are toxic, severe_toxic, obscene, threat, insult, and identity_hate.
- *Hindi Dataset* [24]: Hasoc task A dataset is used for the binary classification of Hate and Not Offensive Hindi texts. The dataset contains 2,963 training data and 663 texts for testing the performance measures.
- *Bengali Dataset* [30]: The dataset has 15,000 comments classified as hate speech and Non-offensive.

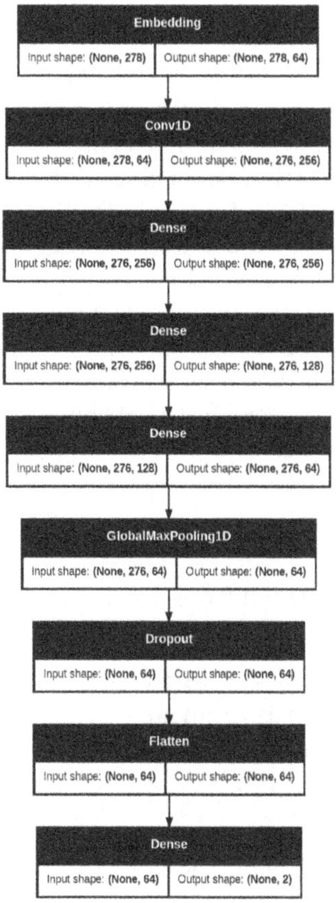

Fig. 5. CA-based CNN Model.

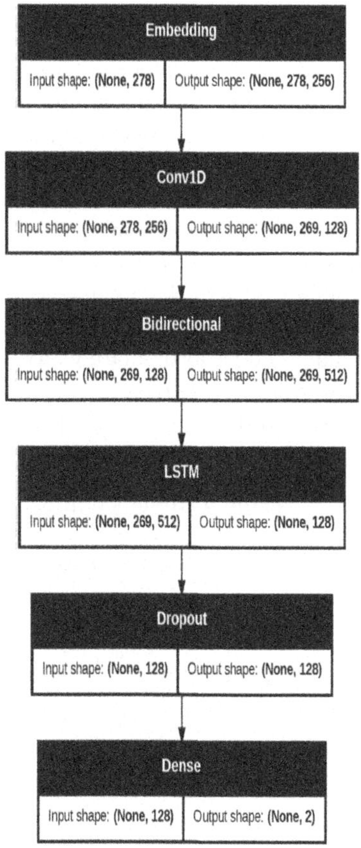

Fig. 6. CA-based Bi-LSTM Model.

4.2 Results and Performance Analysis

The results of the proposed approach are discussed with the help of accuracy, precision, recall, and F1_Score values. The F1_Score is computed using precision and recall, utilizing a confusion matrix comprising elements such as true positives, true negatives, false positives, and false negatives. This F1_Score serves as a measure to assess the classification accuracy of the model developed. CNN and LSTM achieved the overall best results but LSTM had relatively worst accuracy and F1_Score for the Hindi dataset. The overall best result is obtained with CNN for three cases except the Jigsaw dataset. LSTM produces the highest accuracy and an F1_Score of 93.73% and 92.82%, respectively, for the Jigsaw dataset. The results are given in Table 4.

4.3 Comparative Analysis

The proposed model produces approximately 92% accuracy for English hate speech datasets and it is comparable with other state of art models. The result analysis is given in Table 5 for English toxicity analysis. Our model produces better results than all the state-of-art models for the Hate and Offensive Language dataset. Table 6 shows the comparative analysis of non-English Indian languages such as Hindi and Bengali toxic text datasets. Very few works are available for Hindi and Bengali toxic text classification tasks. We have got comparatively good results for both of the languages. From Tables 5 and 6, we can conclude that the model produces more than 90% for all performance analysis metrics for English languages and approximately 85% for non-English languages. The lower accuracy of non-English languages compared to English can largely be attributed to the disparity in dataset sizes. English benefits from vast and diverse datasets, such

Table 4. Results of proposed approach for different toxic datasets

Model Used	Accuracy	Precision	Recall	F1_Score
Jigsaw Dataset [7]				
LSTM	**93.73**	**92.80**	**93.72**	**92.82**
Bi-LSTM	91.18	91.78	91.18	91.46
CNN	92.36	91.29	92.36	91.68
CNN_LSTM	93.54	92.50	93.54	92.35
Hate Speech and Offensive Language Dataset [2]				
LSTM	89.37	89.23	89.36	88.90
Bi-LSTM	86.60	87.15	86.59	85.27
CNN-LSTM	87.49	87.16	87.48	86.93
CNN	**90.62**	**90.55**	**90.61**	**90.25**
Hindi [24]				
LSTM	76.20	76.73	76.2	75.93
Bi-LSTM	78.26	78.57	78.27	78.26
CNN-LSTM	77.12	77.11	80.43	78.72
CNN	**81.01**	**81.77**	**81.00**	**80.98**
Bengali [30]				
LSTM	83.80	83.55	83.6	83.47
Bi-LSTM	83.33	83.06	83.34	82.96
CNN-LSTM	82.07	82.26	82.06	82.15
CNN	**85.20**	**85.00**	**85.20**	**85.02**

as the Jigsaw dataset, IMDB reviews, SQuAD, and Wikipedia dumps, which are well-labeled and extensively used to train deep learning models. In contrast, non-English languages often have smaller datasets, limiting the linguistic patterns that models can learn. Additionally, non-English datasets may suffer from inconsistent labeling and a lack of representation for regional dialects, further reducing model performance.

Table 5. Different models and their results in terms of F1_score for English hate speech datasets

Authors	Techniques used	Accuracy (%)	F1_score (%)
	Jigsaw Dataset[7]		
Garg et al. (2024) [17]	Explainable HSD system	93.60	98.19
Singh et al. (2022) [32]	AlexNet based on CNN	98.50	79.00
ashok et al. (2021) [21]	MCBiGRU	-	71.70
Wang et al. (2021) [34]	CNN with GloVe	-	99.50
Zhao et al. (2021) [37]	RoBERTa	-	78.22
Georgakopoulos et al. (2018) [18]	CNN	> 90.00	91.20
Aken et al. (2018) [1]	Ensemble	-	79.10
Our model	**CA based LSTM**	**93.73**	**92.82**
	Hate and Offensive Language [2]		
Mamun et al. (2024) [23]	BERT + LIME	74.92	69.83
Nandini et al. (2022) [25]	BERT + LIME	82.60	82.80
Rajput et al. (2021) [28]	Bi-LSTM	80.15	79.71
Mandl et al. (2020) [24]	LSTM	-	83.95
Davidson et al. (2017) [9]	Logistic regression with L2 regularization	82.33	90.00
Our model	**CA-based CNN Model**	**90.62**	**90.25**

Table 6. Comparative analysis of toxic datasets for Non-English Indian languages

Model Used	Accuracy	Precision	Recall	F1_Score
	Hindi [24]			
CNN (2021)[33]	78.00	85.00	76.00	76.00
RoBERTa (2021)[33]	80.00	77.80	75.40	76.30
CNN (2020) [4]	82.00	84.00	82.00	81.00
CNN (2020) [24]	–	–	–	82.02
Proposed Method(CNN)	**81.01**	**81.77**	**81.00**	**80.98**
	Bengali [30]			
SVM (2021) [30]	87.50	–	–	91.10
SVM -U (2022) [29]	–	88.9	88.7	88.7
Bengali BERT (2020) [22]	70.60	70.70	70.60	70.50
CA-based CNN	**85.20**	**85.00**	**85.20**	**85.02**

5 Conclusion and Future Work

In this work, we proposed to use cellular automata-based machine learning (CAML) to perform the toxic text classification task. The training of data relies on a weight-based approach using CA with deep-learning models. The results show that CA acquire an understanding of the context of text through its evolutionary process from the vectorized texts using Unicode standards. Our findings indicate that the local neighbourhood communication principle is preserved, enabling the ECA to be considered heterogeneous to learn the behaviour of the dataset. The proposed CAML classifier is good enough to predict the class labels of the dataset. In the future, we plan to develop a language-independent generic CAML model for analyzing the toxicity of comments of any language.

References

1. Aken, B.V., Risch, J., Krestel, R., Löser, A.: Challenges for toxic comment classification: an in-depth error analysis. In: Proceedings of the Second Workshop on Abusive Language Online (ALW2), pp. 33–42. Association for Computational Linguistics (2018)
2. Andrii, S.: Hate speech and offensive language dataset (2020). https://www.kaggle.com/datasets/mrmorj/hate-speech-and-offensive-language-dataset
3. Badjatiya, P., Gupta, S., Gupta, M., Varma, V.: Deep learning for hate speech detection in tweets. In: Proceedings of the 26th International Conference on World Wide Web Companion, pp. 759-760. International World Wide Web Conferences Steering Committee (2017). https://doi.org/10.1145/3041021.3054223
4. Bashar, M.A., Nayak, R.: Qutnocturnal@hasoc'19: CNN for hate speech and offensive content identification in Hindi language (2020). arXiv:abs/2008.12448
5. Burnap, P., Williams, M.L.: Cyber hate speech on Twitter: an application of machine classification and statistical modeling for policy and decision making. Policy Internet **7**(2), 223–242 (2015)
6. Burnap, P., Williams, M.L.: Us and them: identifying cyber hate on Twitter across multiple protected characteristics. EPJ Data Sci. **5**, 1–15 (2016)
7. Cjadams, J.S. et al.: Toxic comment classification challenge (2017). https://kaggle.com/competitions/jigsaw-toxic-comment-classification-challenge
8. Davidson, T., Warmsley, D., Macy, M., Weber, I.: Automated hate speech detection and the problem of offensive language. In: Proceedings of the International AAAI Conference on Web and Social Media, vol. 11, pp. 512–515 (2017)
9. Davidson, T., Warmsley, D., Macy, M., Weber, I.: Automated hate speech detection and the problem of offensive language. In: Proceedings of the Eleventh International AAAI Conference on Web and Social Media (ICWSM 2017). Association for the Advancement of Artificial Intelligence (2017)
10. Del Vigna12, F., Cimino23, A., Dell'Orletta, F., Petrocchi, M., Tesconi, M.: Hate me, hate me not: Hate speech detection on Facebook. In: Proceedings of the First Italian Conference on Cybersecurity (ITASEC17), pp. 86–95 (2017)
11. Elizabeth, M.J., Chaudhuri, P.P., Hazari, R.: Cellular automata-based sentiment analysis for bipolar classification of reviews. Complex Syst. (2024)
12. Elizabeth, M.J., Hazari, R.: Cellular automata-based sentiment analysis using deep learning methods. Complex Syst. (2024)

13. Elizabeth, M.J., Kommineni, A.K., Hazari, R.: Sentiment analysis for code-mixed data using cellular automata with deep learning models. In: Lecture Notes in Computer Science (including subseries Lecture Notes in Artificial Intelligence and Lecture Notes in Bioinformatics) 14978 LNCS, pp. 163–176 (2024). https://doi.org/10.1007/978-3-031-71552-5_14
14. Elizabeth, M., Panda, A.K., Chaudhuri, P.P., Hazari, R.: Cellular automata-based sentiment analysis. In: Asian Symposium on Cellular Automata Technology, pp. 53–64. Springer (2023)
15. Elizabeth, M., Parsotambhai, S.M., Hazari, R.: Cellular automata enhanced machine learning model for toxic text classification. In: International Conference on Cellular Automata for Research and Industry, pp. 346–355. Springer (2022)
16. Fortuna, P., Nunes, S.: A survey on automatic detection of hate speech in text. ACM Comput. Surv. (CSUR) **51**(4), 1–30 (2018)
17. Garg, P., Sharma, M.K., Kumar, P.: Improving hate speech classification through ensemble learning and explainable AI techniques. Arab. J. Sci. Eng. (2024). https://doi.org/10.1007/s13369-024-09540-2
18. Georgakopoulos, S.V., Tasoulis, S.K., Vrahatis, A.G., Plagianakos, V.P.: Convolutional neural networks for toxic comment classification. In: Proceedings of the 10th Hellenic Conference on Artificial Intelligence, pp. 1–6 (2018)
19. Greevy, E.: Automatic text categorisation of racist webpages. Ph.D. thesis, Dublin City University (2004)
20. Gutowitz, H.A.: Cellular Automata: Theory and Experiment. MIT Press, Cambridge, MA (1990)
21. J, A.K., S, A., Trueman, T.E., Cambria, E.: Comment toxicity detection via a multichannel convolutional bidirectional gated recurrent unit. Neurocomputing **441**, 272–278 (2021). https://doi.org/10.1016/j.neucom.2021.02.023
22. Karim, M.R., Chakravarti, B.R., P. McCrae, J., Cochez, M.: Classification benchmarks for under-resourced Bengali language based on multichannel convolutional-LSTM network. In: 7th IEEE International Conference on Data Science and Advanced Analytics (IEEE DSAA,2020). IEEE (2020)
23. Mamun, M.B., Tsunakawa, T., Nishida, M., Nishimura, M.: Hate speech detection by using rationales for judging sarcasm. Appl. Sci. (Switzerland) **14** (2024). https://doi.org/10.3390/app14114898
24. Mandl, T., Modha, S., Shahi, G.K., Jaiswal, A.K., Nandini, D., Patel, D., Schäfer, J.: Overview of the HASOC track at fire 2020: Hate Speech and Offensive Content Identification in Indo-European Languages Under Creative Commons License Attribution 4.0 International (cc by 4.0) (2020). http://ceur-ws.org
25. Nandini, D., Schmid, U.: Explaining hate speech classification with model-agnostic methods (2022)
26. Neumann, J.V.: The Theory of Self-reproducing Automata, A. W. Burks ed. University of Illinois Press, Urbana and London (1966)
27. Pavlopoulos, J., Malakasiotis, P., Androutsopoulos, I.: Deeper attention to abusive user content moderation, pp. 1125–1135 (2017). https://doi.org/10.18653/v1/D17-1117
28. Rajput, G., Punn, N.S., Sonbhadra, S.K., Agarwal, S.: Hate speech detection using static BERT embeddings. In: Srirama, S.N., Lin, J.C.-W., Bhatnagar, R., Agarwal, S., Reddy, P.K. (eds.) BDA 2021. LNCS, vol. 13147, pp. 67–77. Springer, Cham (2021). https://doi.org/10.1007/978-3-030-93620-4_6
29. Romim, N., Ahmed, M., Islam, M.S., Sen Sharma, A., Talukder, H., Amin, M.R.: BD-SHS: A benchmark dataset for learning to detect online Bangla hate speech in

different social contexts. In: Proceedings of the Thirteenth Language Resources and Evaluation Conference, pp. 5153–5162. European Language Resources Association, Marseille, France (2022)
30. Romim, N., Ahmed, M., Talukder, H., Saiful Islam, M.: Hate speech detection in the Bengali language: a dataset and its baseline evaluation. In: Proceedings of International Joint Conference on Advances in Computational Intelligence: IJCACI 2020, pp. 457–468. Springer (2021)
31. Schmidt, A., Wiegand, M.: A survey on hate speech detection using natural language processing. In: Proceedings of the Fifth International Workshop on Natural Language Processing for Social Media, pp. 1–10 (2017)
32. Singh, I., Goyal, G., Chandel, A.: AlexNet architecture based convolutional neural network for toxic comments classification. J. King Saud Univ. - Comput. Inf. Sci. **34**, 7547–7558 (2022). https://doi.org/10.1016/j.jksuci.2022.06.007
33. Velankar, A., Patil, H., Gore, A., Salunke, S., Joshi, R.: Hate and offensive speech detection in Hindi and Marathi. CoRR abs/2110.12200 (2021)
34. Wang, K., Yang, J., Wu, H.: A survey of toxic comment classification methods. CoRR (2021)
35. Wolfram, S.: Theory and Applications of Cellular Automata. World Scientific, Singapore (1986)
36. Yin, D., Xue, Z., Hong, L., Davison, B.D., Kontostathis, A., Edwards, L.: Detection of harassment on web 2.0. In: Proceedings of the Content Analysis in the WEB, vol. 2, pp. 1–7 (2009)
37. Zhao, Z., Zhang, Z., Hopfgartner, F.: A comparative study of using pre-trained language models for toxic comment classification. In: Companion Proceedings of the Web Conference 2021, pp. 500–507. Association for Computing Machinery, Inc (2021). https://doi.org/10.1145/3442442.3452313

Touching Loops and Worms Formed by Cellular Automata

Rolf Hoffmann(✉)

Technical University Darmstadt, Darmstadt, Germany
hoffmann@rgb.informatik.tu-darmstadt.de

Abstract. The objective is to form patterns with "touching" loops by Cellular Automata (CA). A loop is defined as a closed path of 1-cells in a 2D grid on a zero background and with a zero border. A path cell has two adjacent path cells, and two path cells are allowed to touch each other on a diagonal. The designed CA rule tries to build loops by overlapping tiles. It uses so-called *templates*, 5×5 matching patterns which are systematically derived from the problem specific defined set of 3×3 tiles. The rule checks the pattern being evolved against the list of templates. If the outer neighbors of a template match, then the cell's state is set to the template's center value. Noise is injected if there is no matching template or certain constraints are not fulfilled. "Worms" are special loops which are have a maximum of inner self-touching points. The designed Rule 1 variant evolves stable patterns with any kind of loop structures in a short time, whereas the Rule 2 variant evolves stable worm-only patterns in a much longer time. For different grid sizes, a selection of generated loop patterns are presented and some of their properties are discussed.

Keywords: Pattern Formation · Evolving Loop Structures · Overlapping Tiles · Matching Templates · Probabilistic Cellular Automata · Asynchronous Updating

1 Introduction

Loops are common structures that appear in a natural way or by design and construction. They can be found in various organisms (e.g. closed blood circulation system, veins in tree leaves), physical systems (e.g. topological magnetic structures like Hopfions [1] and geophysical systems (highly porous limestone, cracks on the surface of glazed ceramics), chemical systems (e.g. cyclic polymers [2], cyclic nanostructures [3]), biological systems (e.g. [4,5]), or also in the products of human activity (e.g. street networks in cities). Such reticulated patterns (as opposed to tree patterns, where loops are absent) usually arise as a result of complex interactions, and understanding the mechanisms leading to their formation are a fascinating and challenging subject of research. On the other hand, patterns containing loops can be the subject of research in the field of designing algorithms leading to their creation.

S. Kamilya et al. (Eds.): ASCAT 2025, CCIS 2499, pp. 165–178, 2025.
https://doi.org/10.1007/978-3-031-94121-4_12

The Objective. We consider 2D grid patterns of size $n \times n$ where a cell has the state value/color $s \in \{0, 1\}$. The aim is to find a CA rule that can form patterns with loops. We allow that a loop can touch itself (*self-touching)*, and that individual loops can touch each other. In addition we aim at so-called "worm patterns". Worms are special densely contracted self-touching loops. The main challenge was how to form a loop (its path and closing it) by local conditions. The solution presented is the definition of a set of tiles (local patterns acting as local conditions) that are allowed and forced to overlap (aggregate, self-assemble) taking into account certain constraints.

Related Work. Probabilistic CA rules using overlapping tiles can generate point patterns [7], domino patterns [8], and sensor point patterns [9]. In [10] loops were formed under cyclic boundary conditions, first with a Genetic Algorithm and then with a CA rule also based on overlapping tiles.

We will also follow the idea of using a probabilistic rule and overlapping tiles. In contrast to the cited work, here (i) a different and smaller set of tiles was designed in order to evolve loops that can touch, (ii) the boundaries were fixed which is more challenging than using cyclic boundary conditions, and (iii) an additional constraint was defined allowing to evolve worm-only loops.

Overlapping 1D tilings are treated in [6], in the context of recognizable languages, monoids, and two-way automata, and this paper could be a good basis for further studies and extensions for the n-dimensional case.

Our problem is related to space-filling curves and Hamiltonian cycles [11,12]. Generating such curves by a CA rule is a task that could be addressed. For space-filling curves, the field size has to be considered with respect to the possible solutions.

There are several publications on "Peterson's worms" and their derivatives understood as moving agents which may form cyclic paths under the constraint that every grid connection can be "eaten" only once, e.g. [13]. In contrast, here the loops evolve asynchronously cell by cell without using an agent and a sophisticated control algorithm. In other work (e.g. [14]) worms are complex entities composed of several cells that are able to move around. It is interesting that loop structures may appear in the *Game of Life* as "Still Life", but they appear very rarely among many other stable objects [15].

Paper Structure. In Sect. 2, the properties of loops and loop patterns are explained. The design method using defined tiles and templates is presented in Sect. 3. In Sect. 4, the designed CA rule (in two variants) is defined. The loop patterns evolved by Rule 1 are discussed in Sect. 5.1, and worm-only patterns evolved by Rule 2 are shown in Sect. 5.2. Conclusions and future work finalize the paper.

2 General Aspects of Loops and Loop Patterns

A *loop* is modeled as a *cyclically closed path* of *loop path cells* (1-cells) embedded in a grid of size $n \times n$ with a zero background and a zero boundary. A *loop pattern* is a pattern which contains at least one loop. A *loop path cell* is a 1-cell that has exactly two 1-cells in NESW, where NESW are the four orthogonal adjacent von-Neumann neighbors. Formally the following condition (so-called *loop path condition*) is true for all path cells of a loop:

$$(s = 1) \wedge ((s^N + s^E + s^S + s^W) = 2)$$

where the cell's main state is $s = 1$ for path cells and $s = 0$ for the surrounding cells. The term $s^N + s^E + s^S + s^W$ gives the number of NESW-neighbors which are in state $s = 1$. The loop path condition ensures that three path cells in sequence form a horizontal straight line segment $\begin{smallmatrix} & 0 & \\ 1 & 1 & 1 \\ & 0 & \end{smallmatrix}$ (or a vertical one, resp.), or a corner $\begin{smallmatrix} & 0 & 0 \\ 0 & 1 & 1 \\ 0 & 1 & 0 \end{smallmatrix}$ (4 variants by rotation). In order to form a loop it is required that the neighboring path cells of a path cell also fulfill the loop path condition, and in this way all path cells in continuation will form a cyclically closed path.

The loop path condition also allows corner path cells (more precisely the heads/outer corner points of corners) to touch each other on a diagonal, i.e. either the two path cells (*NorthEast, SouthWest*) or (*NorthWest, SouthEast*) can be neighbors. The outer two path cells belonging to a corner lie also on a diagonal but they are not considered as touching. So touching path cells fulfill the local patterns $\begin{smallmatrix} 1 & 0 \\ 0 & 1 \end{smallmatrix}$ or $\begin{smallmatrix} 0 & 1 \\ 1 & 0 \end{smallmatrix}$.

The loop's path shall be surrounded by 0-cells forming a *hull*. This requirement is already guaranteed by the loop path condition, with the exception that a cell diagonally located outside of a corner can be 0 (background) or 1 (another touching path cell). The hull ensures that loop paths keep a minimal distance between each other, and also within a loop itself. Hull cells are allowed to be used by several paths, from the same or other loops, i.e. hull cells may overlap.

For reasoning about loop attributes we can imagine that an agent walks clockwise in one-way around the loop until it returns back to its starting point. The agent follows its path by going stepwise straight, to the right or to the left. Attributes of loops and loop patterns are:

- $P \in \{0,1\}^{n \times n}$ – *pattern.*
- $N_{Loop}(P)$ – *number of loops in the pattern P.*
- $\lambda(i)$ – *length of loop with index i, the number of its path cells.* The loop indexes $i \geq 1$ are assigned by scanning the pattern row by row from top to bottom, and from left to right in each row.
- $\alpha(i)$ – *(enclosed area) number of cells which loop i encloses.*
- $\delta(P) = N_{ones}(P)/(n+1)^2$ – *density of 1-cells in a pattern P.* where $N_{ones}(P)$ is the number of ones in the pattern. The divisor $(n+1)^2$ is used instead of n^2 because then the maximal possible density becomes 0.5. This is the case for instance for space filling curves. So we include half of the border cells into the number of all cells being considered.
- $R(i) \geq 4$ – *number of right turns for a circuit-making agent in loop i.* A *right turn* corresponds to a (convex) *outer corner* pointing to the loop's outside.
- $L(i) \geq 0$ – *number of left turns in loop i.* A *left turn* corresponds to a (concave) *inner corner* pointing to the loop's inside. The following assertion is true: $R - L = 4$, or $L = \Delta R$ where $R = 4 + \Delta R$, and '4' corresponds to the 360° turn for making a circuit.
- $b^{NESW}(i) = (b^N, b^E, b^S, b^W)(i)$ – *number of loop i's path cells that are adjacent to the N(orth)/E(ast)/S(outh)/W(est) border.*

Let us look at the patterns depicted in Fig. 1(a–d). Pattern (a) contains two nested loops, loop 1 encloses loop 2 (the 3 × 3 so-called mini-loop), and its values are

$$\lambda(1) = 4 \times 6 = 24;\ \ \lambda(2) = 8;\ \ R(1) = R(2) = 4;\ \ L(1) = L(2) = 0;$$
$$\delta = 32/8^2 = 0.5;\ \ b^{NESW}(1) = (7,7,7,7);\ \ b^{NESW}(2) = (0,0,0,0).$$

The pattern in Fig. 1(b) has the values

$$\lambda = 24;\ \ R = 8;\ \ L = \Delta R = 4;\ \ b^{NESW} = (3,3,3,3).$$

The pattern in Fig. 1(c) has the values

$$\lambda = 26;\ \ \delta = 26/64 = 0.406;\ \ R = 7;\ \ L = 3;\ \ b^{NESW} = (6,4,3,7).$$

Figure 1(d) is a space-filling Hamiltonian circuit with the values

$$\lambda = 32, \delta = 32/64 = 0.5;\ \ \alpha = 13;\ \ R = 8;\ \ L = 4.$$

Figure 1(i) is a so-called "max-worm" with the values

$$\lambda = 24;\ \ \delta = 0.375;\ \ \alpha = 5;\ \ R = 12;\ \ L = 8;\ \ b^{NESW} = (3,3,3,3).$$

Touching Points. We can distinguish four different types of touching points:

- *self-touching point*
  - *inner:* Two cells are touching with the *inner* side of the loop path. Such a point is called "*inner self-touching point*" (Fig. 1(e)).

  - *outer:* Two cells are touching with the *outer* side of the loop path. Such a point is called "*outer self-touching point*" (Fig. 1(f)). Such points are bottlenecks within caves or bottles with an entry from outside the loop.
- *inter-touching point*
  - *external:* Two cells of different loops A and B are touching. If both outer sides of the loops' paths are touching, then the point is called "*external inter-touching point*" (Fig. 1(g)). This is the more frequent case when loop A is situated beside loop B.
  - *internal:* Two cells of different loops A and B are touching where A encloses B. Then the inner side of path A touches the outer side of path B. Such a point is called "*internal inter-touching point*", or "*enclosing inter-touching point*" (Fig. 1(h)).

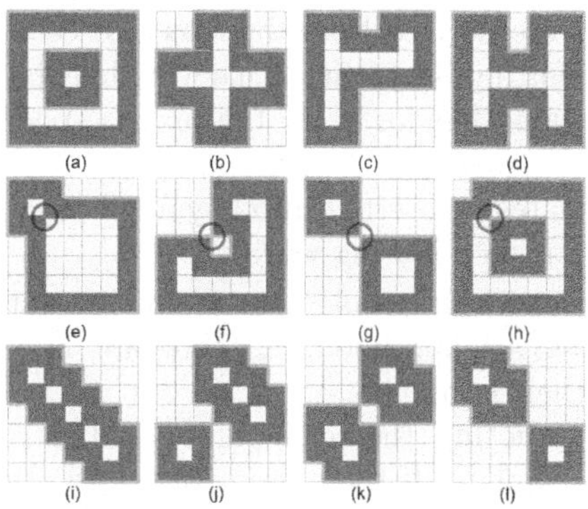

Fig. 1. Some specific 7×7 loop patterns. (a) Two nested loops. (b) Loop with 4/4 outer/inner corners, and (c) with 5/3. (d) A space-filling loop. (e) Inner self-touching, (f) outer self-touching, (g) external inter-touching, (h) internal inter-touching loop. (i) Worm of maximal size 4, (j) 2-worm inter-touching with a mini-loop of size 3×3, (k) two 1-worms, (l) 1-worm with mini-loop. Path cells in blue, inner/outer side of a loop's path in red/green. (Color figure online)

Worms. We define a "*worm*" as a loop that is strongly contracted with a maximal number of inner self-touching points, and it forms a diagonal of zeroes enclosed by ones. We do not consider the simple square 3×3 loop (mini-loop) as a worm because it has no touching point. A *k-worm* is a worm that has $k \geq 1$ inner self-touching points. The number of enclosed zeroes is $\alpha = k + 1$ which corresponds to the length of the zero-diagonal. The worm's length is $\lambda = 4(k + 2) = 4(\alpha + 1)$. A *max-worm* is the largest worm that can fit into a square grid of size n^2, which is a $(k = n - 3)$-worm. Some 7×7 patterns with worms are shown in Fig. 1(i – l), (i) shows a max-worm with the maximum of 4 inner self-touching points, (j) a 2-worm inter-touching with a mini-loop, (k) two 1-worms, and (l) a 1-worm inter-touching with a mini-loop.

3 The Design Method Using Tiles and Templates

The rule was designed on the basis of a method already applied successfully for evolving point patterns [7], domino patterns [8], sensor networks [9], and loop patterns [10]. The steps of the updated method are:

1. A *set of tiles* has to be defined that can induce the desired pattern. A *tile* is a small pattern consisting of elementary tiles (square cells of minimal size according to the grid) which are called *pixels* in order to distinguish them from the *cells* of the aimed pattern. The defined tiles (and replicates of them) are then used to fill the given space. Pixels from different tiles are allowed to overlap if there is no conflict in color. If K pixels of the same color are overlaid, then the cover level v of this cell becomes $v = K$. Depending on the set of tiles and the given space it is not always possible to cover the whole space ($v > 0$ for all cells) if that is required.
2. So-called *templates* are derived from the tiles. For each pixel of a tile a template is derived by shifting the tile in such a way that the chosen pixel becomes the center. This means that for every tile pixel a specific template is generated.
3. The set of all templates is inserted into a general probabilistic CA rule, and the rule is applied in the following way:
   - All outer templates (the templates without the center) are tested against the actual cell's neighborhood. If there are hits, they are registered together with their templates.
   - If there is only one hit, the cell's value is adjusted to the center value of the hitting template. If there are several hits, then one of the hitting templates is chosen at random for adjustment.
   - If there is no hit, then noise is injected. This means that the cell's state is changed into another state with a certain probability. In addition, noise can be injected if certain constraints are not fulfilled.

3.1 The Defined Set of Tiles

For the construction of touching loops only two basic (proto-) tiles are necessary: (a) a *straight path tile* $\begin{smallmatrix} & 0 & \\ 1 & 1 & 1 \\ & 0 & \end{smallmatrix}$ responsible for the straight continuation of a path, and (b) a *corner tile* $\begin{smallmatrix} & 0 & 0 \\ 0 & 1 & 1 \\ 0 & 1 & 0 \end{smallmatrix}$. Each tile contains three adjacent one-pixels. In order to yield the full set of tiles the symmetric ones have to be added (Fig. 2), because we aim at patterns which do not give preference to a certain direction.

3.2 The Derived Templates

What is a *template*? A template is a local matching pattern. It defines local conditions/constraints for valid patterns. For each tile T of the tile set, K templates $T_0, T_1, ...T_{K-1}$ are derived, where K is the number of valid tile pixels with the

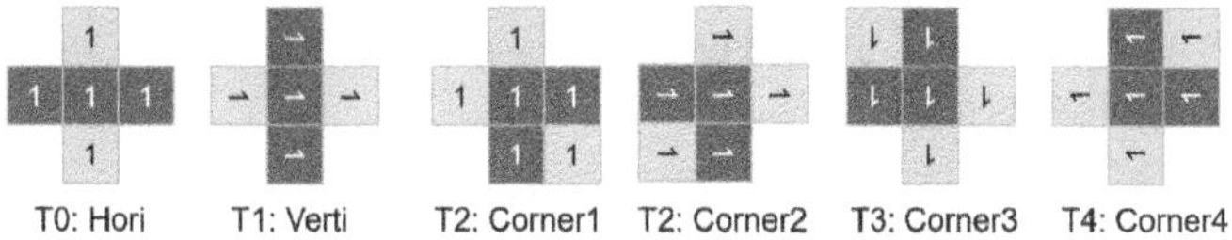

Fig. 2. The whole set of tiles, including the symmetric ones. The added numbers in the tile pixels indicate the cover/overlay level, '1' means that they are not overlaid.

```
-----  -----  -----  -----  --0--     -----  -----  -----  -----  --0--  -0---
--0--  -----  ---0-  -0---  -111-     --0--  -----  ---0-  -0---  -011-  011--
-111-  --0--  --111  111--  --0--     -011-  --0--  --011  011--  --10-  -10--
--0--  -111-  ---0-  -0---  -----     --10-  -011-  ---10  -10--  -----  -----
-----  --0--  -----  -----  -----     -----  --10-  -----  -----  -----  -----
A0     A1     A2     A3     A4        C0     C1     C2     C3     C4     C5
```

Fig. 3. The five templates *A0* – *A4* derived from tile *A0* = *Hori*, and the six templates *C0* – *C5* derived from tile *C0* = *Corner1*. The symbol '-' denotes sites which are don't cares where no valid pixel exists and where no test is performed.

value 0 or 1. The template T_i is derived by shifting the tile in way that the valid pixel i becomes the center of that template. By convention we define $T_0 = T$, i.e. the template T_0 is equal to the tile without shifting. Note that we assume a center/anchor for each tile, and also for each template. Figure 3 shows (i) the five templates *A0* – *A4* derived form tile *Hori*, where *A0* = *Hori* and *A1* = *ShiftDown(A0)*, *A2* = *ShiftRight(A0)*, *A3* = *ShiftLeft(A0)*, *A4* = *ShiftUp(A0)*, and (ii) the six templates *C0* – *C5* derived form tile *Corner1*, where *C0* = *Corner1*. Then, in then same way, the templates for the other tiles are derived. The number of all templates becomes $2 \times 5 + 4 \times 6 = 34$, a template for each tile $(2 + 4)$ and for each pixel.

4 The CA Rule

Cell State. The used CA cell state at a site (x, y) is $q = (s, h_0, h_1)$, where $s \in \{0, 1\}$ is the main pattern state (color), and $h_0 \in \{0, 1, ..., 8\}$ is a template hit with $s = 0$, and $h_1 \in \{0, 1, 2, 3\}$ is a template hit with $s = 1$, and (x, y) are the coordinates (indexes) of the cell. It can be noticed that the hit values converge to the cover levels when a pattern stabilizes.

Asynchronous Updating. We want to use *asynchronous* execution/simulation of the CA rule. For each generation (time-step $t \rightarrow t + 1$), n^2 cell operations (micro-steps $\tau \rightarrow \tau + 1$) are executed. A simple scheme is used with the following cell operation at each micro-step

1. A cell (x, y) is selected at random.
2. The new state value s' at (x, y) is computed by the CA rule (for details see next paragraph).

3. The new state value s' is immediately copied to the state variable $s(x, y)$ yielding the next state.

For output it is useful to compute also the *cover levels* for each considered configuration (displayed pattern). This can be done by testing at each site all tiles and storing the number of pixels that overlay.

Rule Details. The rule is composed of two function evaluations (steps) in sequence. In the first step a template test is performed. In the second step the state is corrected (adjusted) if there is a matching template. In the third step noise is injected if certain constraints are not satisfied. Note that the rule as a whole forms a function although it is computed stepwise.

1. TEST for template hits
   All templates are tested against the current pattern at site (x, y) where the center of each template is aligned to this site. The test is performed only on the outer neighborhood without considering the central state $s(x, y)$. So the current value of the cell is not taken into account, only the real neighbors.
   $(h_0, h_1, k) \leftarrow \mathit{ComputeHitsForAllTemplates}\ T_i \in \{T_0, T_1, ...T_{K-1}\}$
   where
   $h_0 =$ number of zero-hits (hits with template center value $= 0$)
   $h_1 =$ number of one-hits (hits with template center value $= 1$)
   $k =$ index of a matching template T_k
   $K = 34 = 4 \times 2 + 6 \times 4$, the number of templates (#valid pixels in all tiles).
2. ADJUST if hit
   The default new state is $s' = s$ (keep old state). If there is a hit, the state is adjusted to the center of a matching template.
   $s' = \mathit{center}(T_k)$ if $h = h_0 + h_1 > 0$
   where
   $\mathit{center}(T_k)$ is the value of the center pixel of a matching template T_k.
   If there is only one hit, the cell's state s is adjusted to the value of the template's center value. This means that the center color is corrected if all other pixels of the template are correct. In the case when the whole template (including the center) correctly matches, the adjustment is not necessary, it is redundant. If there are several hits, one of the hitting templates is selected at random for adjustment.
3. NOISE INJECTION
   The current new state (temporary variable) may be changed with a certain probability if the constraints are not fulfilled.

$$s' \leftarrow \begin{cases} \mathit{Random}(0/1) & \text{if } P \wedge (s' = 1) \wedge \mathit{TRUE}(\pi_1) \\ & \text{or } Q \wedge (s' = 0) \wedge \mathit{TRUE}(\pi_0) \\ & \text{or } \mathit{NOT}(W) \wedge (s' = 1) \wedge \mathit{TRUE}(\pi_2) \end{cases}$$

   where
   $\mathit{TRUE}(\pi) \Leftrightarrow \mathit{TRUE}$ with probability π else FALSE
   $P = (h_1 \neq 3)$ – *the 'loop path condition' for a closed loop*
   $Q = (h_0 = 0)$ – *this constraint forbids uncovered 0-cells*

$$W = (h_0^{North} = 6,7) \vee (h_0^{East} = 6,7) \vee (h_0^{South} = 6,7) \vee (h_0^{West} = 6,7)$$
– *the constraint for yielding worms*

Two rule variants were defined and simulated, ***Rule 1*** with the parameter settings $\pi_1 = 0.5, \pi_0 = 0, \pi_2 = 0$, and ***Rule 2*** (the *Worm-only Rule)* with the settings $\pi_1 = 0.5, \pi_0 = 0, \pi_2 = 0.5$. Rule 1 evolves loop patterns with all kinds of loops (non-touching, touching, worms), and Rule 2 patterns with worms only.

Both rules use the setting $\pi_2 = 0$ which means that empty uncovered areas are allowed in the pattern, and also the extreme case all-zero for the whole pattern. It turned out that this configuration appears very rarely (because its probability to appear becomes very low with an increasing grid size), and the evolution is faster compared to a setting $\pi_0 > 0$. Setting $\pi_0 > 0$ would exclude patterns with uncovered 0-cells (marked in red in the shown patterns). Then such interesting or demanded loop patterns like worm-only patterns with uncovered cells would be destroyed during the evolution and would appear only as transients, and most of the evolved patterns could appear very noisy. Therefore we used $\pi_0 = 0$. Nevertheless $\pi_0 > 0$ can be necessary if loop patterns without empty areas are required to be evolved by Rule 1. Further studies are required to determine the probability range of uncovered cells in the patterns for $\pi_0 = 0$.

5 Loop Patterns

5.1 Patterns Evolved with Rule 1

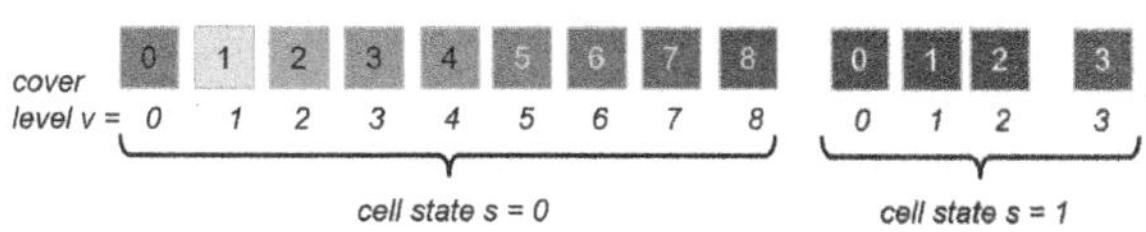

Fig. 4. The color scheme for the shown patterns, taking into account the cover level

Patterns of Size 7×7. The pattern shown in the following use the color scheme as depicted in Fig. 4. Selected evolved 7×7 patterns are shown in Fig. 5. A loop encloses another in c5 and c6, and there are four inter-touching points in c6. Loops a6 and a7 have an entry-point (bottleneck) from outside the loop to a cave, there are cells which touch each other with the outside of the loop's path. Such outer-touching points can appear for $n \geq 7$. The patterns c4 – c7 are space-filling, the number of path cells is maximal (32). The loop c7 is a non-touching space-filling curve.

Patterns of Size $n = 8, 16, 32$. A selection of evolved 8×8 loop pattern is shown in Fig. 6. (I) Row (a) shows 8 loops with *one* self-touching point, and (b1 – c2) show 10 loops with *two* self-touching points. Loops (c3 – c6) have *three*

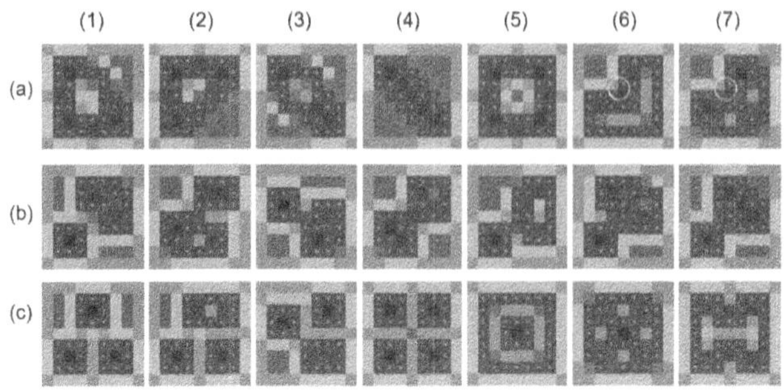

Fig. 5. A selection of 7 × 7 patterns. (a1, a2, a3) contain loops with 3 self-touching points, and (a4, a5) with 4. Loops a6 and a7 each have cells that touch each other from the outer side of the loop's path.

self-touching points, and loops (c7, c8) have *four*. Patterns (d1 – e1) have *two inter-touching* loops. Loops (e2, e3) each are loops that touch externally (with an *outer self-touching* point). (e5 – e8) are patterns with *worms*. The patterns (Fig. 6(II)) are seven 16 × 16 patterns with a different number of loops, and one 32 × 32 pattern with 5 loops.

The CA rule can evolve very many different, *stable* touching loop patterns. The rate of the evolved 8 × 8 patterns with one loop, with four (or more) was much lower than that for patterns with two or three loops.

Special regular patterns (like the ones shown in Fig. 5(c5 – c7) or Fig. 6(e1, e4 – e8) can appear, but the probability is low because the total amount of all valid loops patterns is very high. The evolution of a stable loop pattern with Rule 1 is relatively fast. After typically 10 generations partial loop structures appear in 8 × 8 patterns, but it takes more time ($t_{avrg} = 51$ on average) to change all intermediate structures into stable closed loops.

Time-Complexity. The time-complexity was estimated by executing multiple runs and averaging. The number of performed runs (each yielding a stable pattern) was 100 for $n = 3, 4, \ldots, 12$, and 50 for $n = 14, 16, 18, 20$, and 20 for $n = 24, 28$, and 10 for $n = 32$. Surprisingly, the time (time-steps t needed for a stable loop pattern) grows only approximately linear with the number of cells ($t = O(n^2)$), as we may conclude from the curve's trend (Fig. 7). This means that each cell's rule execution consumes a constant number of time-steps. (Here we got a special result where the constant is approx. 1.) In the parallel case, when a fraction α of the cells are updated in parallel, the parallel time could approximate a constant. To confirm this proposition more experiments and/or theoretical investigations are needed. One explanation for the low time-complexity could be that a large loop can evolve faster in an area that was restricted by smaller, already formed loops. The CA rule is able to evolve also larger patterns with $n > 16$ with a relatively low computational effort because of the low time complexity.

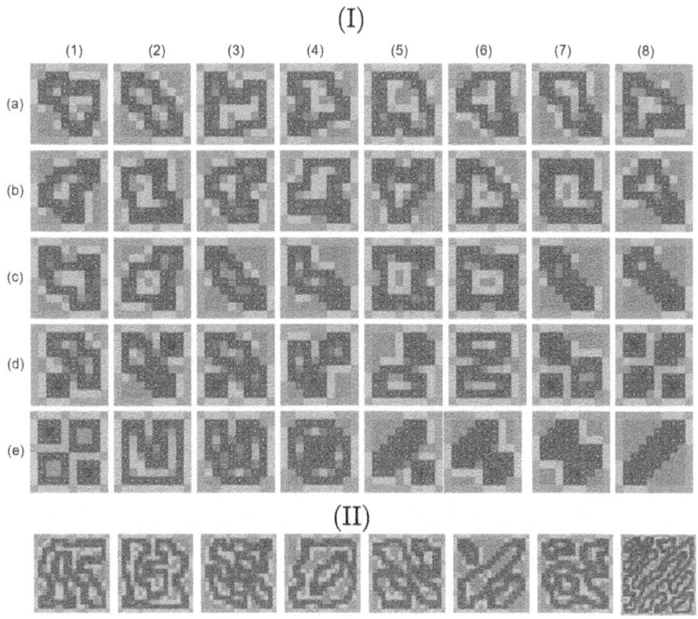

Fig. 6. (I) A selection of evolved 8×8 patterns, showing loops with (a) one self-touching point, (b1 – c2) two, (c3 – c6) three, and (c7, c8) four. Patterns (d1 – e1) contain two inter-touching loops. Loops e2 and e3 each contain an outer self-touching point, and (e5 – e8) are worm-only patterns.—(II) (1 – 7) are 16×16 patterns with $(1, 1, 2, 2, 3, 3, 4)$ loops, (8) is a 32×32 pattern with 5 loops.

Density. The maximal reachable density depends on the grid size and the loop structure(s). It is $\delta_{max} = 0.5, 0.4, 0.44, 0.41, 0.5, 0.49$ for $n = 3, 4, 5, 6, 7, 8$. The upper bound is 0.5 which can only be reached for certain n (e.g. 3, 7). The density of the evolved patterns on average was

$\delta_{avrg} = 0.381, 0.363, 0.345, 0.349, 0.363, 0.379, 0.388, 0.393, 0.398, 0.402, 0.394,$
$0.413, 0.414, 0.435, 0.435$ for
$n = 5, 6, 7, 8, 9, 10, 11, 12, 4, 16, 18, 20, 24, 28, 32.$

5.2 Worm-Only Patterns Evolved with Rule 2

The worm-only Rule 2 is more strict than Rule 1 because of the additional constraint W saying that path cells need zero-valued neighbors with cover level 6 or 7 in their Moore-neighborhood. Now worm-only patterns are the fixed points of the evolution. As the number of such patterns is much lower in comparison to the number of fixed points for Rule 1, the evolution takes much longer and is increasing super-linear in the number of cells. For example the average time to evolve always worm-only patterns for all used 60 runs was $t_{avrg} = 207, 1223, 5262$

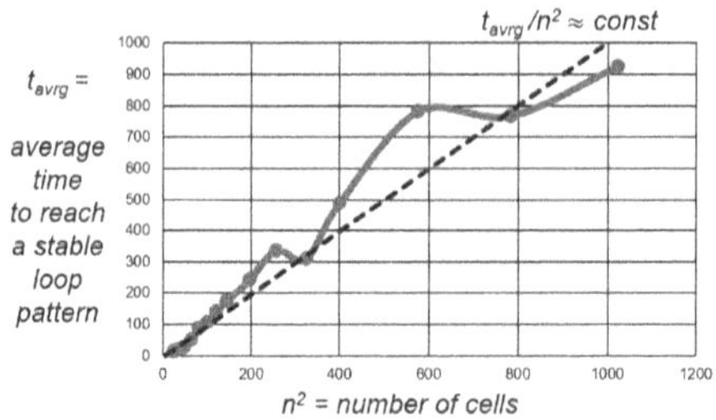

Fig. 7. In the simulation, the time-complexity is approximately $O(n^2)$. The data: t_{avrg} = 13, 19, 27.4, 51, 87.8, 106, 141.7, 172, 241, 335, 308, 487, 781, 766, 926 for n = 5, 6, 7, 8, 9, 10, 11, 12, 4, 16, 18, 20, 24, 28, 32.

for $n = 4, 5, 6$, and the time needed per cell becomes $t_{avrg}/n^2 = 13, 49, 146$ $(1 : 3.8 : 11)$. Therefore a few of the shown patterns in Fig. 8 were constructed because they did not show up under the maximal time-limit setting of 100'000 generations. The patterns (b1, b6 – b8) were designed by hand and then used as initial configurations which stayed stable, therefore they can be evolved in principal. The shown patterns have the additional property that for each of the four boundaries there exists at least one adjacent path cell.

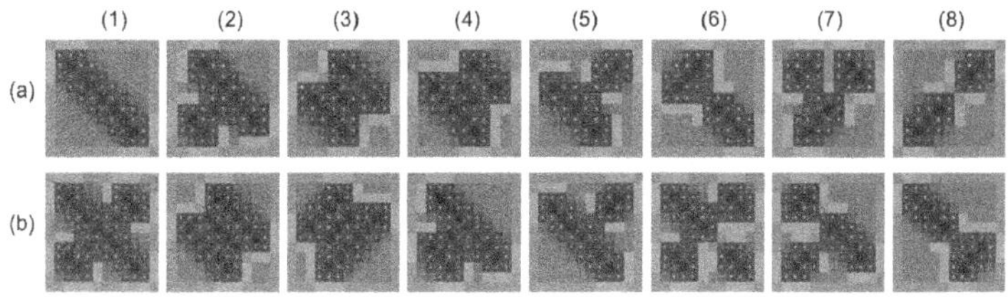

Fig. 8. A selection of worm-only patterns. (a) 9×9 patterns, and (b) 10×10 patterns.

6 Conclusions and Future Work

A CA rule was designed that can evolve stable touching loop patterns. The Rule 1 variant evolves all kind of loops, whereas the Rule 2 variant evolves worm-only patterns. First a set of 6 tiles was defined which are able to build paths in the grid. Each tile defines a 3×3 local pattern consisting of 0/1-pixels. 34 templates were derived from the tiles by shifting them in a way that each pixel becomes the center of a template. The templates were inserted into a probabilistic CA rule with asynchronous updating. All templates are tested for matching in the neighborhood of each CA cell. The rule tries to establish a template at each site by correcting the cell's value to the center value of a hitting template or by noise injection otherwise. In addition, the loop path conditions needs to be fulfilled, saying that each path cell shall have an overlap level of 3. For Rule 1 the time-complexity was estimated by simulations, it depends only linearly in the number of cells. For Rule 2 the time-complexity increases super-linear, shown for some cases.—There are several topics for future work, like

- Larger spaces should be investigated in search of specific phenomena.
- How are the loops structured under cyclic boundary conditions?
- How can certain pattern characteristics like the number of loops, the loops length, or the density be maximized or minimized in a local way?
- How can the time-complexity be computed theoretically?
- How can worm-only patterns be evolved with less time, e.g. by using other tiles and constraints?
- Loops with special connections and intersections could be considered.
- How resistant are loop patterns against perturbations?
- The generation of Hamiltonian cycles could be addressed.

References

1. Kent, N.A.: Structure and Topology in Nanoscale Magnetic Systems. Doctoral dissertation. University of California, Santa Cruz (2020)
2. Kricheldorf, H.R.: Cyclic polymers: synthetic strategies and physical properties. J. Polym. Sci. Part A: Polym. Chem. **48**(2), 251–284 (2010)
3. Olenyuk, B., Fechtenkötter, A., Stang, P.J.: Molecular architecture of cyclic nanostructures: use of co-ordination chemistry in the building of supermolecules with predefined geometric shapes. J. Chem. Soc. Dalton Trans. **11**, 1707–1728 (1998)
4. Cardote, T.A., Ciulli, A.: Cyclic and macrocyclic peptides as chemical tools to recognise protein surfaces and probe protein-protein interactions. ChemMedChem **11**(8), 787–794 (2016)
5. Wang, C.K., Kaas, Q., Chiche, L., Craik, D.J.: CyBase: a database of cyclic protein sequences and structures, with applications in protein discovery and engineering. Nucleic Acids Res. **36**(suppl 1), D206–D210 (2007)
6. Janin, D.: On languages of one-dimensional overlapping tiles. In: van Emde Boas, P., Groen, F., Italiano, G.F., Nawrocki, J., Sack, H. (eds.) SOFSEM 2013. LNCS, vol. 7741, pp. 244–256. Springer, Heidelberg (2013). https://doi.org/10.1007/978-3-642-35843-2_22
7. Hoffmann, R.: Forming Point Patterns by a Probabilistic Cellular Automata Rule. arXiv:2202.06656. Based on a *Presentation at Summer Solstice Conference on Complex Systems (June 2019)* (2022)
8. Hoffmann, R., Désérable, D., Seredyński, F.: A cellular automata rule placing a maximal number of dominoes in the square and diamond. J. Supercomput. **77**(8), 9069–9087 (2021). https://doi.org/10.1007/s11227-020-03549-8
9. Hoffmann, R., Désérable, D., Seredyński, F.: Cellular automata rules solving the wireless sensor network coverage problem. Nat. Comput. *21*, 417–447 (2022)
10. Hoffmann, R.: Generating loop patterns with a genetic algorithm and a probabilistic cellular automata rule. Algorithms **16**(7), 352 (2023)
11. Prusinkiewicz, P., Lindenmayer, A., Fracccia, D.: Synthesis of space-filling curves on the square grid. In: Peitgen, H.-O., Henrique, J.M., Pencdo, L.F. (eds.), Fractals in the Fundamental and Applied Sciences. Elsevier Science Publishers B.V., Amsterdam (1991)

12. Crossley, M.: Hamiltonian Cycles in Two Dimensional Lattices. In: Statistical Physics in Biology. Springer, Berlin/Heidelberg, Germany (2013)
13. Gardner, M.: Worm Paths. In: Knotted doughnuts and other mathematical entertainments. American Mathematical Soc. (2020)
14. Takata, D., Isokawa, T., Lee, J., Peper, F., Matsui, N.: On construction by worm-like agents on a self-timed cellular automaton. In: Cellular Automata: 10th International Conference on Cellular Automata for Research and Industry, Proceedings, vol. 10, pp. 575–84. Springer, Berlin, Heidelberg (2012)
15. Cook, M.: Still life theory. New Constr. Cell. Automata **226**, 93–118 (2003)

First Experiments on Chaotic *k* Non-Uniform Cellular Automata

Souvik Roy[1(✉)], Aryan Sukhadia[1], Harsh Modi[1], Tanay Shah[1], and Biplab K. Sikdar[2]

[1] Ahmedabad University, Gujarat, India
{souvik.roy,aryan.s5,harsh.m6}@ahduni.edu.in, svkr89@gmail.com
[2] Indian Institute of Engineering Science and Technology, Shibpur, Howrah, India
biplab@cs.iiests.ac.in

Abstract. This work reports a study on the dynamics of k non-uniform cellular automata (CA), where the cellular system is associated with two elementary CA (ECA) rules $\mathcal{R}_1$ and $\mathcal{R}_2$ (non-uniform couples). The $\mathcal{R}_1$ is the default rule, and the noise rule $\mathcal{R}_2$ is applied to the rule vector of the CA with a perturbation rate of "k" to shatter the deterministic notion of (hybrid) non-uniform CA system. It explores the dynamics of chaotic and locally chaotic non-uniform couples following the qualitative (space-time diagrams) and quantitative (density and Kolmogorov-Sinai entropy) experimental approach and establishes the fact that (a) Some chaotic non-uniform couples show strong resistance against the k perturbation; (b) The chaotic ECA 18 depicts dominating nature during interaction with the locally chaotic systems, however, ECA 126 is dominated due such perturbation; and (c) The class transition -that is, locally chaotic to chaotic and locally chaotic to fixed point exists during evolution of the k non-uniform system.

Keywords: Cellular Automata (CA) · Elementary Cellular Automata (ECA) · Non-uniformity · Chaos · Kolmogorov-Sinai Entropy

1 Introduction

Cellular automata (CA) are one of the oldest bio-inspired models of parallel and distributed computing [1]. The CA can be defined over a regular grid (here, we consider one-dimensional regular grid), where each cell of the system consists of a finite automaton, and the cells interact with its neighbors to reach its next state during evolution of the system. The beauty of CA is simplicity, where simple local interaction depicts complex global behavior. However, this decentralized model of computing also follows centralized idea in terms of uniform local rule (i.e. each cell follows same local rule); synchronous update (i.e. a global clock forces each cell to get updated synchronously); and uniformity in neighborhood dependency and lattice structure [1]. To break all this notion of centralized thinking, during the last three decades, the CA researchers have introduced non-classical CA models - non-uniform CA [3], asynchronous CA [6] and network automata [13].

S. Kamilya et al. (Eds.): ASCAT 2025, CCIS 2499, pp. 179–190, 2025.
https://doi.org/10.1007/978-3-031-94121-4_13

In this work, we explore non-uniform CA, where different cells can follow different local transition functions or local rules. Particularly, here, the target is to introduce *independence* during the local transition function selection process. To implement this notion of independence, the CA researchers have introduced non-classical models - non-uniform/hybrid CA [3], stochastic or probabilistic CA [7], temporally stochastic CA [12] etc. The hybrid CA are mostly studied in [1] to develop technological solutions. Due to hardware friendly structure of hybrid CA under *open* boundary condition, the CA community have used this system for VLSI design and test, cryptography and authentication, pattern recognition and classification [3]. Moreover, the CA researchers have also theoretically explored the global properties like reversibility, structural stability, decidability and equicontinuity of hybrid CA systems [1].

This scenario motivates us to explore the k non-uniform CA, a special case of hybrid CA, where the hybrid cellular system is associated with two ECA rules $\mathcal{R}_1$ and $\mathcal{R}_2$. The $\mathcal{R}_1$ acts as the default rule of the CA/system and $\mathcal{R}_2$ plays the role of noise or perturbation. Here, "k" depicts the noise rate -that is, k cells of the hybrid system follow the noise $\mathcal{R}_2$. In [5], the authors have identified single length l-attractor k non-uniform system ($k = 1$) after developing a tool, referred to as the NSRT diagram under periodic boundary ($l \in \{1, 2\}$). In a similar direction, in [4], the theoretical background for the same under open boundary condition is reported. However, the works of [4,5] mostly deal with the simple CA, considering the Wolfram's classification [14] and $k = 1$. On the other hand, the target of the current work is to explore the dynamics of chaotic ECA rules under k non-uniform model. Therefore, the following fundamental issues are addressed.

1. Whether two chaotic ECA together show simple behavior during evolution of the k non-uniform system.
2. Does the noise rate k has a significant effect?

Further, we also consider the locally chaotic ECA (following Li and Packard's classification [10]) to investigate the transition from simplicity to chaos in non-uniform system.

2 *k* Non-Uniform Cellular Automata

In this work, we consider the one-dimensional two-state three-neighborhood (left, self and right neighbors) cellular automata under periodic boundary condition-that is, the elementary cellular automata (ECA). Therefore, each cell can be represented by a set of indices $\mathcal{L} = \mathbb{Z}/n\mathbb{Z}$, where n is the number of cells or CA size, and a cell is associated with a state at each time step, where the set of states is $\mathcal{Q} = \{0, 1\}$. A *configuration* of CA represents the collection of the states of its cells at a given time. That is, $\varepsilon_n = \{0, 1\}^{\mathcal{L}}$ depicts the set of configurations. Here, the local transition function can be defined as $f : \{0, 1\}^3 \rightarrow \{0, 1\}$.

A CA cell updates its state following the self state and neighbors (left and right) states. Each argument (say, (x, y, z)) of local transition function is denoted by Rule Min Term (RMT) (say, r), where $r = 4 \times x + 2 \times y + z$. Moreover, the decimal equivalent of the eight outputs of the local transition function is called *rule* -that is, $f(1,1,1) \cdot 2^7 + f(1,1,0) \cdot 2^6 + \cdots + f(0,0,0) \cdot 2^0$. We have $2^8 = 256$ ECAs, out of which 88 are minimal representative ECAs and the remaining are their equivalent.

Traditionally, in hybrid CA, we follow the rule vector $\mathcal{R} = \langle \mathcal{R}_0, \mathcal{R}_1, \cdots, \mathcal{R}_i, \cdots, \mathcal{R}_{n-1} \rangle$ which configures the cells of an n-cell CA -that is, cell i follows rule $\mathcal{R}_i$. The $\mathcal{R}_0 = \mathcal{R}_1 = \cdots = \mathcal{R}_{n-1}$ holds for uniform (classical) CA. Otherwise, the cellular system is hybrid. Here, in $\mathtt{k}$ non-uniform CA, the hybrid cellular system is associated with two elementary CA rules $\mathcal{R}_1$ and $\mathcal{R}_2$. The $\mathcal{R}_1$ is the default rule of the system and $\mathcal{R}_2$ plays the role of noise or perturbation. The "$\mathtt{k}$" is the noise rate -that is, k cells of the hybrid system follow noise $\mathcal{R}_2$. To design the rule vector, we select k cells randomly and uniformly for noise rule $\mathcal{R}_2$. For example, $\mathcal{R} = \langle 18, 18, 22, 18, 22, 18 \rangle$ depicts an evidence rule vector for the 6-cell $\mathtt{k}$ non-uniform system, where $\mathcal{R}_1 = 18$, $\mathcal{R}_2 = 22$, and $\mathtt{k} = 2$.

We denote the $\mathtt{k}$ non-uniform system as $(\mathcal{R}_1, \mathcal{R}_2)[\mathtt{k}]$, where $\mathtt{k} \in [0, n]$. As a special case, $\mathcal{R}_1 = \mathcal{R}_2$ denotes the classical uniform system with rule $\mathcal{R}_1$ -that is, $(\mathcal{R}_1, \mathcal{R}_2)[0]$, and $(\mathcal{R}_1, \mathcal{R}_2)[n]$ is the uniform system with $\mathcal{R}_2$. The dynamics of $\mathtt{k}$ non-uniform system following qualitative and quantitative experimental approach follow next.

3 Qualitative Dynamics of K Non-Uniform System

The dynamics of chaotic (following Wolfram's classification [14]) and locally chaotic (following Li and Packard's classification [10]) ECA rules of the proposed $\mathtt{k}$ non-uniform system have been reported in this section. We consider only the minimal representative ECA [10]. As per Wolfram's classification [14], evolution of the following set of minimal ECA leads to chaotic structure (in other words, shows strange attractors).

$$\mathtt{Chaotic} = \{18, 22, 30, 45, 60, 90, 105, 122, 126, 146, 150\}.$$

Moreover, in a details analysis, Li and Packard's [10] has identified the following set of minimal locally chaotic rules, where the chaotic dynamics confined by the domain walls.

$$\mathtt{Locally\ chaotic} = \{26, 73, 154\}.$$

Following the combination of both of these classification [10, 14], the dynamics of classical ECA can be broadly grouped into - null rules (leads to homogeneous fixed points), fixed point rules (leads to non-homogeneous fixed points), periodic rules, locally chaotic, chaotic, and complex rules (with "glider" dynamics). These are denoted respectively as $\mathtt{N}$, $\mathtt{FP}$, $\mathtt{P}$, $\mathtt{LC}$, $\mathtt{C}$, $\mathtt{CO}$. To explore the dynamics of $\mathtt{k}$ non-uniform system, this broad classification is followed. Here, the class of ECA $\mathcal{R}$ is denoted as $\mathbb{C}(\mathcal{R})$. Therefore, the class of chaotic rules can be represented by $\mathbb{C}(\mathcal{R}) = \mathtt{C}$, where $\mathcal{R} \in \mathtt{chaotic}$. Similarly, $\mathbb{C}(\mathcal{R}) = \mathtt{LC}$, where $\mathcal{R} \in$

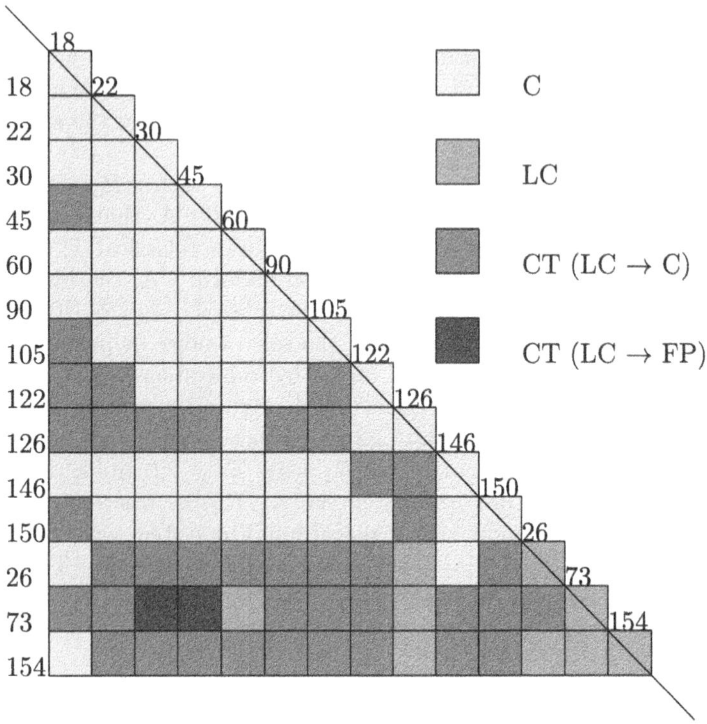

Fig. 1. Summarized dynamics of k non-uniform system after considering chaotic and locally chaotic ECA.

`Locally chaotic`. Following this, class of `k` non-uniform system $(\mathcal{R}_1, \mathcal{R}_2)[\mathtt{k}]$ is denoted by $\mathbb{C}((\mathcal{R}_1, \mathcal{R}_2)[\mathtt{k}])$. If the system shows static dynamics for changing `k` $\in [1, n-1]$, it can be denoted as $\mathbb{C}((\mathcal{R}_1, \mathcal{R}_2))$.

As both the qualitative and quantitative approach are to be followed to investigate the dynamics of `k` non-uniform system, in the first experiment, the space-time diagram of the system (during evolution to get an visual comparison) is analyzed (qualitative approach). The system is evolved for 2000 time steps starting from a random initial configuration (of density 0.5), considering $n \in [40, 50]$. The non-uniform $(\mathcal{R}_1, \mathcal{R}_2)[\mathtt{k}]$ system is constructed 2000 times to cross-verify the dynamics where the model may select different rule vectors for different run. Moreover, to validate the result for large CA size, the system is evolved for 2000 time steps after considering $n = 2500$. Figure 1 depicts the overall qualitative results of the `k` non-uniform systems after considering chaotic and locally chaotic ECA, where the chaotic and locally chaotic dynamics are respectively marked with yellow and red. In Fig. 1, `k` non-uniform system also shows the *class transition* dynamics (marked with green and blue). It reports the dynamics of $\frac{14\times 13}{2} = 91$ `k` non-uniform couples following the exchange symmetry after considering 11 chaotic and 3 locally chaotic ECA.

In Fig. 1, the box (i, j) shows the dynamics of k non-uniform couple $(\mathcal{R}_1, \mathcal{R}_2)$, where $\mathcal{R}_1$, $\mathcal{R}_2$ ECA rules are receptively represented by the ith and j^{th} box in the vertical and horizontal line. The box (i, j), where $i = j$, depicts the dynamics of uniform system -that is, $\mathcal{R}_1 = \mathcal{R}_2$. The following can be established from the results shown in Fig. 1.

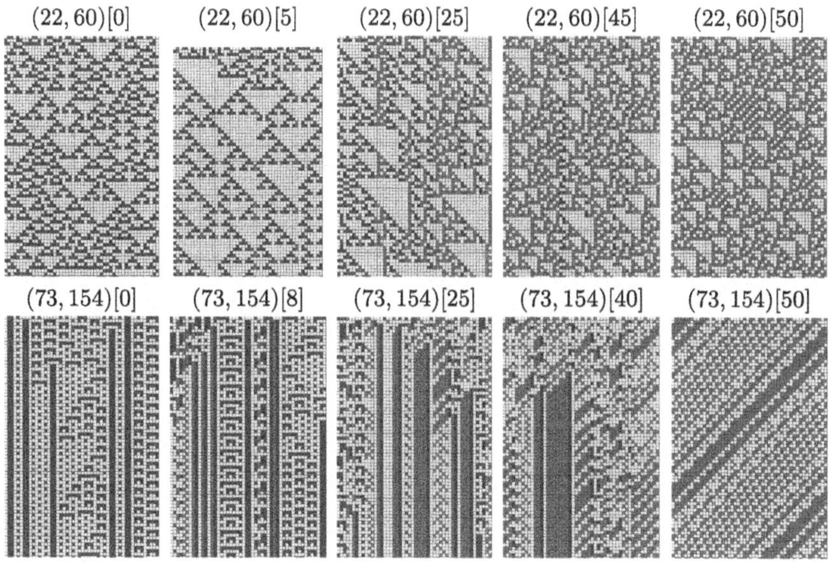

Fig. 2. Dynamics of k non-uniform system, where $\mathbb{C}(\mathcal{R}_1) = \mathbb{C}(\mathcal{R}_2) = \mathbb{C}(\mathcal{R}_1, \mathcal{R}_2)$. Here, n is 50 and the system runs for 75 time steps. Blue and red squares represent cells in states 1 for rule $\mathcal{R}_1$, $\mathcal{R}_2$ respectively whereas, white represents cells in state 0 for both the rules. This convention is kept in rest of the text.

1. When, $\mathbb{C}(\mathcal{R}_1) = \mathbb{C}(\mathcal{R}_2) = \mathbb{C}(\mathcal{R}_1, \mathcal{R}_2)$, the two participating ECA rules of k non-uniform system individually follow the same class dynamics -that is, $\mathbb{C}(\mathcal{R}_1) = \mathbb{C}(\mathcal{R}_2)$. Moreover, their k non-uniform couple also follows the same class. Therefore, this system shows strong resistance against this k non-uniform perturbation. Now, there can be two possibilities $\mathcal{R}_1, \mathcal{R}_2 \in$ C and $\mathcal{R}_1, \mathcal{R}_2 \in$ LC.
   - In Fig. 2, couple (22, 60) represents the system where $\mathbb{C}(22) = \mathbb{C}(60) = \mathbb{C}(22, 60) =$ C. This couple shows chaotic dynamics for k $\in [0, n]$ (see k $\in \{0, 5, 25, 45, 50\}$ in Fig. 2).
   - Similarly, the k non-uniform system (73, 154) is locally chaotic where $\mathbb{C}(73) = \mathbb{C}(154) =$ LC (see the space-time diagram for k $\in \{0, 8, 25, 40, 50\}$ in Fig. 2).

The $\mathbb{C}(\mathcal{R}_1) = \mathbb{C}(\mathcal{R}_2) \neq \mathbb{C}(\mathcal{R}_1, \mathcal{R}_2)$ can be referred as a contrast to $\mathbb{C}(\mathcal{R}_1) = \mathbb{C}(\mathcal{R}_2) = \mathbb{C}(\mathcal{R}_1, \mathcal{R}_2)$. However, there exists no chaotic/locally chaotic couple which follows this possibility.

2. For the case, where $\mathbb{C}(\mathcal{R}_1) \neq \mathbb{C}(\mathcal{R}_2)$, we first explore the possibility where $\mathbb{C}(\mathcal{R}_1, \mathcal{R}_2) = \mathbb{C}(\mathcal{R}_1)$ or $\mathbb{C}(\mathcal{R}_1, \mathcal{R}_2) = \mathbb{C}(\mathcal{R}_2)$. That is, two participating ECA rules of the non-uniform system follow different class dynamics. In the `k` non-uniform system, class of one of the participating ECA dominates. The study shows there can be two possibilities.

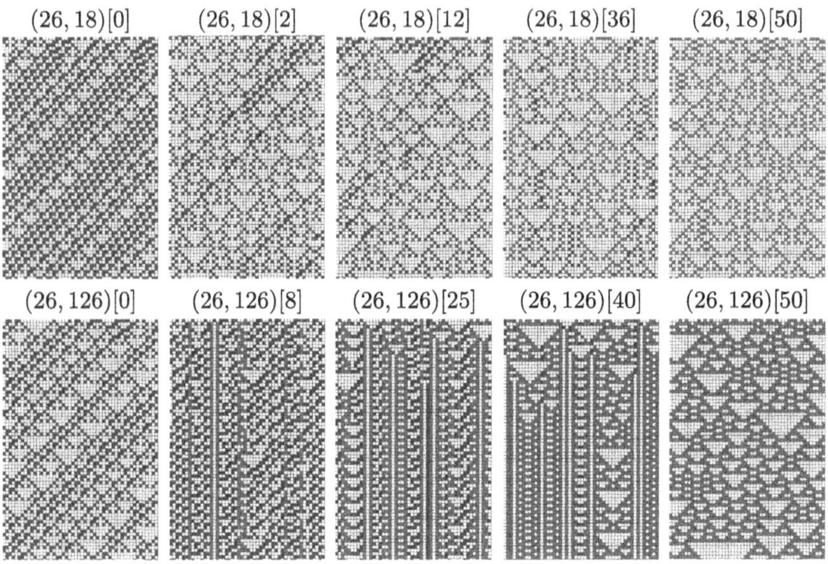

Fig. 3. Dynamics of k non-uniform system, where $\mathbb{C}(\mathcal{R}_1, \mathcal{R}_2) = \mathbb{C}(\mathcal{R}_1)$ or $\mathbb{C}(\mathcal{R}_1, \mathcal{R}_2) = \mathbb{C}(\mathcal{R}_2)$ after considering $\mathbb{C}(\mathcal{R}_1) \neq \mathbb{C}(\mathcal{R}_2)$.

- $\mathbb{C}(\mathcal{R}_1) = \texttt{LC}$, $\mathbb{C}(\mathcal{R}_2) = \texttt{C}$ and, in the `k` non-uniform system, class of chaotic ECA dominates -that is, $\mathbb{C}(\mathcal{R}_1, \mathcal{R}_2) = \texttt{C}$. Figure 3 shows the chaotic dynamics of couple (26, 18) for $\texttt{k} \in \{2, 12, 36\}$, where $\mathbb{C}(26) = \texttt{LC}$ and $\mathbb{C}(18) = \texttt{C}$. It can be observed that for small perturbation $\texttt{k} = 2$ (i.e. two cells of the system follow chaotic ECA 18), the system shows chaotic dynamics. The same is also true for high perturbation (see $\texttt{k} = 36$ in Fig. 3). As per Fig. 1, the couples (26, 146) and (154, 18) also depict the similar possibility. Therefore, the ECA 18 is the most dominating in nature while interacting with the locally chaotic system.
- The $\mathbb{C}(\mathcal{R}_1, \mathcal{R}_2) = \texttt{LC}$ points to the other possibility, where $\mathbb{C}(\mathcal{R}_1) = \texttt{LC}$, $\mathbb{C}(\mathcal{R}_2) = \texttt{C}$. In Fig. 3, the `k` non-uniform couple (26, 126) shows locally chaotic dynamics for $\texttt{k} \in \{8, 25, 40\}$, where $\mathbb{C}(26) = \texttt{LC}$ and $\mathbb{C}(126) = \texttt{C}$. In fact, chaotic ECA 126 is always dominated by the locally chaotic ECA -that is, for couples (73, 126) and (154, 126), the same is also true.

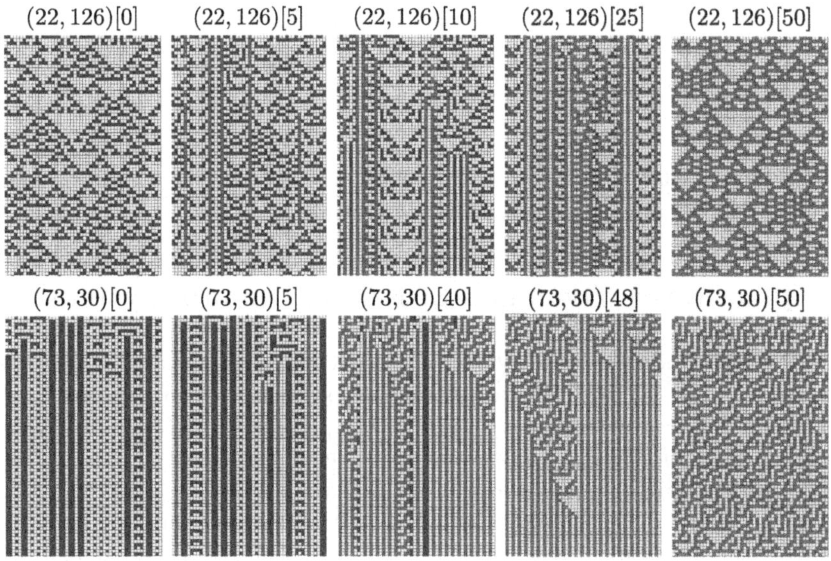

Fig. 4. Class transition dynamics of k non-uniform system where $\mathbb{C}((\mathcal{R}_1, \mathcal{R}_2)[k_1]) \neq \mathbb{C}((\mathcal{R}_1, \mathcal{R}_2)[k_2])$. Here, $k_1 \in [1, k_c]$ and $k_2 \in [k_c, n-1]$.

It is to be noted that for $\mathbb{C}(\mathcal{R}_1) \neq \mathbb{C}(\mathcal{R}_2)$, the other possibility can be $\mathbb{C}(\mathcal{R}_1, \mathcal{R}_2) \neq \mathbb{C}(\mathcal{R}_1)$ and $\mathbb{C}(\mathcal{R}_1, \mathcal{R}_2) \neq \mathbb{C}(\mathcal{R}_2)$. That is, none of the participating ECA rules dominates during evolution of the k non-uniform system. However, there exists no such (LC,C) couple.

The systems investigated in the earlier discussions follow static dynamics while changing the noise rate $\texttt{k} \in \{1, n-1\}$. That is, if k is changed progressively, the dynamics of the system remain unchanged. Such a system is not k (noise rate) sensitive.

3. For some k non-uniform rules, the system shows behavioral change in terms of class dynamics after a critical k value k_c -that is, for $\mathbb{C}((\mathcal{R}_1, \mathcal{R}_2)[k_1]) \neq \mathbb{C}((\mathcal{R}_1, \mathcal{R}_2)[k_2])$, where $k_1 \in [1, k_c]$ and $k_2 \in [k_c, n-1]$. This abrupt change in class dynamics is referred to as the *class transition* (denoted as CT). The following two types of class transition dynamics can be observed in the present case.
   - In Fig. 4, k non-uniform couple (22, 126) shows a class transition from *chaos* to *local chaos* for progressive change in the noise rate k. Here, $\mathbb{C}(22) = \mathbb{C}(126) = \texttt{C}$ (that is, $\mathbb{C}(\mathcal{R}_1) = \mathbb{C}(\mathcal{R}_2)$). The (22, 126) couple shows chaotic dynamics for $\texttt{k} = 5$, however, for high noise rate $\texttt{k} \in \{10, 25\}$, the non-uniform (22, 126) system depicts locally chaotic dynamics (see Fig. 4). Here, it is denoted as $\mathbb{C}(22, 126) = \{\texttt{C} \rightsquigarrow \texttt{LC}\}$.
   - On the other hand, if the change is for the non-uniform couple (73, 30), then the class transition is from *local chaos* to *fixed point*. In Fig. 4, the non-uniform (73, 30) system shows locally chaotic dynamics for k

$\in \{5, 40\}$. However, it shows fixed point dynamics for high noise rate $\mathtt{k} = 48$. The same is also true for the couple (73, 45). Here, we refer $\mathbb{C}(73, 30) = \mathbb{C}(73, 45) = \{\mathtt{LC} \rightsquigarrow \mathtt{FP}\}$.

The next section reports the quantitative dynamics of k non-uniform system following the notion of *density* and *entropy* profile.

4 Quantitative Dynamics of K Non-Uniform System

This section introduces the quantitative approach to verify the qualitative observations of Sect. 3. In the formal quantitative approach, we compute the *density* [7,12] and *Kolmogorov-Sinai entropy* [9] of the proposed k non-uniform system.

The density of a configuration c during the evolution of the system at time t can be defined as $d_{c_t} = c_1/n$, where c_1 denotes the number of 1s in the configuration c at time t and n denotes the CA size (here, for simplicity, we denote d_{c_t} as d). In this experiment, the system starts with random initial configuration of density 0.5 following *Bernoulli process* -that is, every cell is associated with 0.5 (resp. $(1 - 0.5)$) probability to have the state 1 (resp. 0). For the qualitative study, the CA size is considered as $n \in [40, 50]$ to capture the impact of divisibility of CA size by 2, 3, 4, following the notion of primary RMT sets [11]. However, for the quantitative approach, divisibility of CA size by $m \in \{2, 3, 4\}$ shows no impact on the system. Therefore, we only consider $n = 50$ following the initial results of space-time diagrams. The non-uniform system is run for 2000 time steps, and the densities of the configurations is calculated during evolution. These are then plotted (Fig. 5) as a function of time after considering the fixed noise rate k.

Figure 5 shows the profile of density as a function of time for the k non-uniform couples (18, 22), (26, 154), (26, 18), (105, 126, (73, 30), and (73, 45). The couples (18, 22) and (26, 154) depict the possibility of $\mathbb{C}(\mathcal{R}_1) = \mathbb{C}(\mathcal{R}_2) = \mathbb{C}(\mathcal{R}_1, \mathcal{R}_2)$. That is, $\mathbb{C}(18) = \mathbb{C}(22) = \mathbb{C}(18, 22) = \mathtt{C}$ and $\mathbb{C}(26) = \mathbb{C}(154) = \mathbb{C}(26, 154) = \mathtt{LC}$. Here, in Fig. 5 (26, 154), $\mathtt{k} = 0$ (resp. $\mathtt{k} = 50(n)$) shows the dynamics of individual participating ECA 26 (resp. 154) which shows periodicity in density profile. Moreover, for k non-uniform perturbation ($\mathtt{k} = \{5, 20, 40\}$), the density profiles also show a kind of periodicity with the progressive change of time (Fig. 5 (26, 154)). It can be noted that this periodicity in density profile indicates the periodic or locally chaotic dynamics. On the other hand, Fig. 5 (18, 22) shows fluctuating density profile during evolution which points to the chaotic dynamics for noise rate $\mathtt{k} \in \{0, 5, 20, 40, 50\}$. The couple (26, 18) depicts the case where $\mathbb{C}(\mathcal{R}_1) \neq \mathbb{C}(\mathcal{R}_2)$ and $\mathbb{C}(\mathcal{R}_1, \mathcal{R}_2) = \mathbb{C}(\mathcal{R}_2)$. In Fig. 5, for the couple (26, 18), the system only follow periodicity in density profile for $\mathtt{k} = 0$ -that is, the individual ECA 26. However, the system shows fluctuating density for k $\in \{5, 20, 40, 50\}$ which points to the chaotic nature ($\mathbb{C}(26, 18) = \mathbb{C}(18) = \mathtt{C}$).

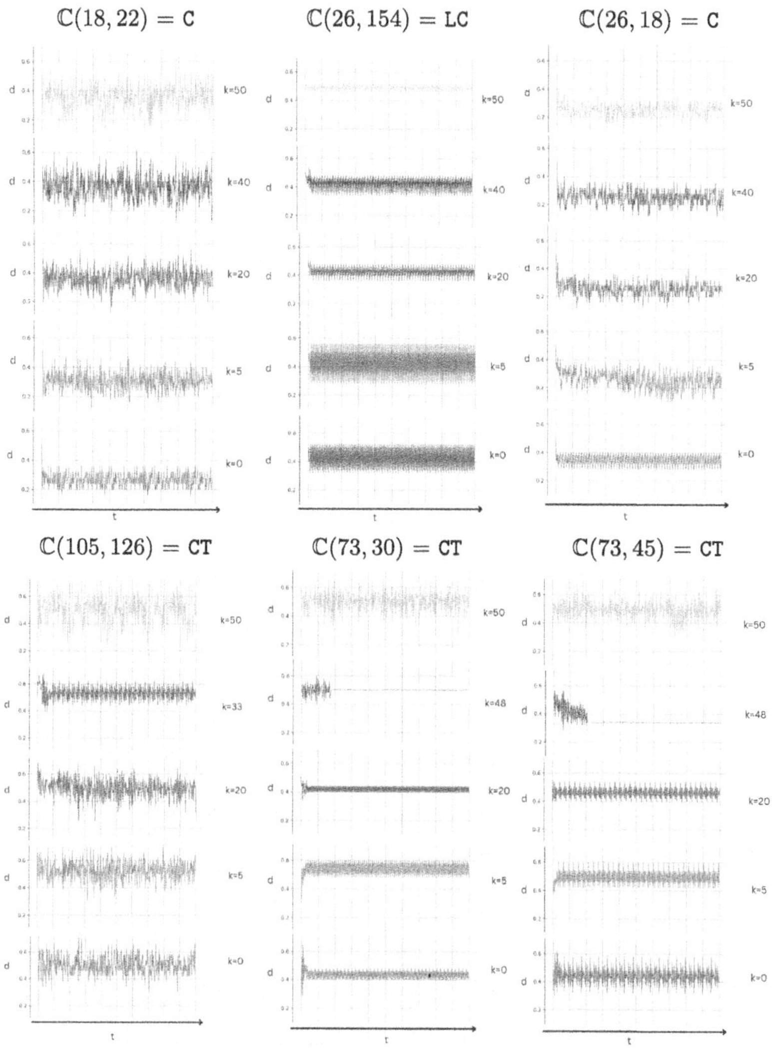

Fig. 5. Profile of density (d) as a function of time (t) for non-uniform couples $(18, 22)$, $(26, 154)$, $(26, 18)$, $(105, 126)$, $(73, 30)$ and $(73, 45)$.

The couple $(105, 126)$ has the class transition ($\{\texttt{C} \rightsquigarrow \texttt{LC}\}$) dynamics as $\mathbb{C}(\mathcal{R}_1) = \mathbb{C}(\mathcal{R}_2) = \texttt{C}$. In Fig. 5 $(105, 126)$, the $\mathbf{k}$ non-uniform system shows fluctuating density profile during evolution for $\mathbf{k} \in \{5, 20\}$, however, it shows periodicity in density profile for $\mathbf{k} = 33$. Similarly, the couples $(73, 30)$ and $(73, 45)$ follow the class transition ($\{\texttt{LC} \rightsquigarrow \texttt{FP}\}$) dynamics, where $\mathbb{C}(\mathcal{R}_1) \neq \mathbb{C}(\mathcal{R}_2)$. In this context, $\mathbb{C}(73) = \texttt{LC}$ and $\mathbb{C}(30) = \mathbb{C}(45) = \texttt{C}$. It is to be observed that the couple $(73, 30)$ shows constant non-zero density profile during evolution for $\mathbf{k} = 48$ and indicates the fixed point class dynamics. However, in Fig. 5, the system shows periodicity (that is, local chaos) in density profile for $\mathtt{k} \in \{5, 20\}$. The same is also true for the couple (73, 45). Therefore, the density profile of the system during evolution validates the observations of space-time diagrams (see Fig. 1). However, it is not possible to distinguish between the periodic and locally chaotic class dynamics following this experimental approach.

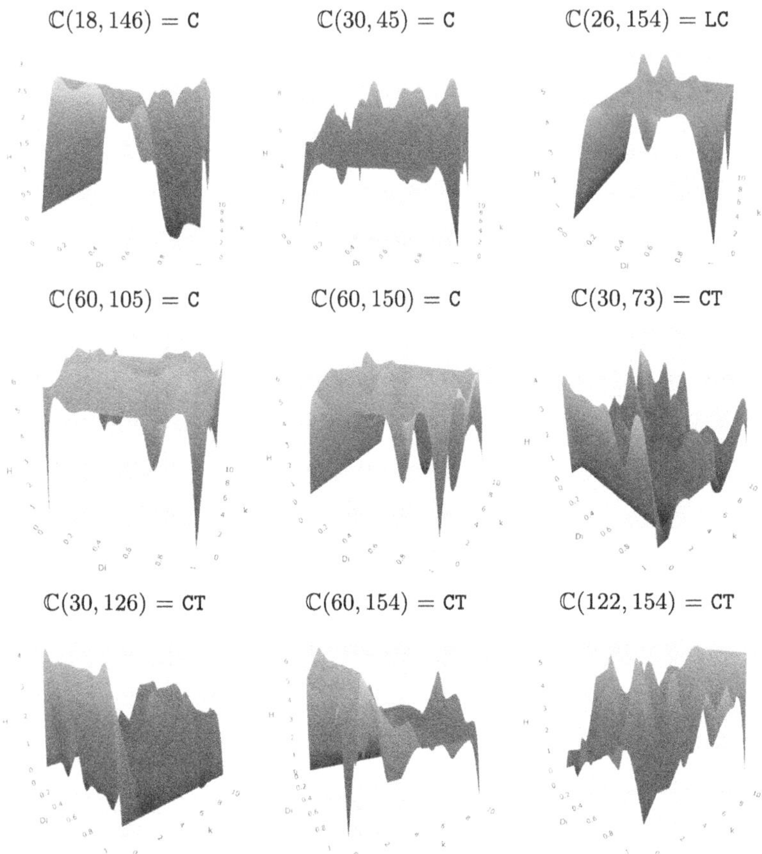

Fig. 6. Sampling surface according to entropy profile (initial density(D_i)-noise rate(k)-entropy (H)) of the non-uniform system.

The *Kolmogorov-Sinai entropy* [9] can also be a measure for exploring the dynamics of the proposed k non-uniform system. To compute Kolmogorov-Sinai entropy, the system is run for t time steps. The c_t denotes the configuration at time t. Therefore, $c_0, c_1, \cdots, c_t$ represent the sequence of configurations (denoted as Γ_t) during the evolution of the system. The number of occurrences of (any) configuration $c \in \mathcal{E}_n$ in the sequence of configurations Γ_t is denoted as Γ_{t_c}. Therefore, $\mu(c) = \frac{\Gamma_{t_c}}{t+1}$ gives the probability of occurrences of the configuration c

in the sequence Γ_t. Now, following the notion of Shannon's entropy, we compute the entropy $H = - \sum_{c \in \varepsilon_n} \mu(c) log(\mu(c))$ considering the probability space. To get a two-dimensional sampling surface of the entropy, the initial density (resp. noise rate k) is varied from 0 to 1 with 0.1 interval (resp. from 0 to n with 1 interval). In this first experiment, we consider $n = 10$ and $t = n \times 2^n$. It is to be noted that the larger entropy H reflects larger uncertainty of the cellular system, and vice versa. Figure 6 shows the two-dimensional sampling surface following entropy H profile for the non-uniform couples (18, 146), (30, 45), (26, 154), (60, 105), (60, 150), (30, 73), (30, 126), (60, 154) and (122, 154). It can be observed that the couple (122, 154) shows k-dependent entropy surface (that is, large entropy H for large k) which indicates the class transition ({C ⇝ LC}) dynamics of the system. The same is also true for the non-uniform couples (30, 73), (30, 126) and (60, 154). However, the couple (60.105) shows horizontal surface -that is, the system shows static chaotic dynamics for changing noise rate $k \in [0, n]$. Similarly, the couples (18, 146), (30, 45), (26, 154) and (60, 150) also depict horizontal surface. Here, the entropy surface directly captures the difference between k-sensitive (i.e. CT) and k-invariant dynamics.

All the above discussions point to the fact that the quantitative results (density and Kolmogorov-Sinai entropy) validate the qualitative findings reported in Fig. 1.

5 Conclusion

In this work, we explore the dynamics of k non-uniform system following qualitative dynamics (space-time diagram) and quantitative dynamics (density profile and Kolmogorov-Sinai entropy). This first experiment on `chaotic` and `locally chaotic` non-uniform couples explores the following interesting dynamics.

1. Some `chaotic` couples show solid resistance against this k non-uniform perturbation.
2. The `chaotic` ECA 18 is dominating in nature while interacting with the `locally chaotic` system. On the contrary, the `chaotic` ECA 126 is always dominated by the `locally chaotic` ECA rules.
3. The k non-uniform system shows rich class transition dynamics for changing k non-uniform noise rate -that is, `chaotic` ⇝ `locally chaotic` and `locally chaotic` ⇝ `fixed point`.

However, the work only follows experimental approach to investigate the dynamics of k non-uniform system. A proper theoretical understanding [2,8] behind these signature behaviors (resistance against non-uniform perturbation and class transition dynamics) is still open and points to the future direction of research.

References

1. Bhattacharjee, K., Naskar, N., Roy, S., Das, S.: A survey of cellular automata: types, dynamics, non-uniformity and applications. Nat. Comput. Int. J. **19**(2), 433–461 (2020)
2. Cattaneo, G., Formenti, E., Margara, L., Mauri, G.: On the dynamical behavior of chaotic cellular automata. Theor. Comput. Sci. **217**(1), 31–51 (1999)
3. Chaudhuri, P.P., Chowdhury, D.R., Nandi, S., Chattopadhyay, S.: Additive Cellular Automata – Theory and Applications, vol. 1. IEEE Computer Society Press, USA, 1997. ISBN 0-8186-7717-1
4. Dalui, M., Chakraborty, B., Das, N., Sikdar, B.K.: Nsrt diagram for identification of saca and taca rules in null-boundary. Int. J. Mod. Phys. C **33**(06), 2250071 (2022)
5. Das, B., Dalui, M., Saha, M., Ghosh, K., Das, N., Sikdar, B.K.: Exploring periodic boundary cellular automaton rules for one and two fixed points. Complex Systems **33**, 31–60 (2024)
6. Fatès, N.: Guided tour of asynchronous cellular automata. J. Cell. Autom. **9**(5–6), 387–416 (2014)
7. Fatès, N.: Diploid cellular automata: first experiments on the random mixtures of two elementary rules. In: Dennunzio, A., Formenti, E., Manzoni, L., Porreca, A.E. (eds.) AUTOMATA 2017. LNCS, vol. 10248, pp. 97–108. Springer, Cham (2017). https://doi.org/10.1007/978-3-319-58631-1_8
8. Kamilya, S., Das, S.: A study of chaos in non-uniform cellular automata. Commun. Nonlinear Sci. Numer. Simul. **76**, 116–131 (2019)
9. Lei, Q., Lee, J., Huang, X., Kawasaki, S.: Entropy-based classification of elementary cellular automata under asynchronous updating: an experimental study. Entropy **23**(2) (2021)
10. Li, W., Packard, N.: The structure of the elementary cellular automata rule space. Complex Systems **4**, 281–297 (1990)
11. Roy, S., Fatès, N., Das, S.: Reversibility of elementary cellular automata with fully asynchronous updating: an analysis of the rules with partial recurrence. Theor. Comput. Sci. **1011**, 114721 (2024)
12. Roy, S., Paul, S., Das, S.: Temporally stochastic cellular automata: classes and dynamics. Int. J. Bifurc. Chaos **32**(12), 2230029 (2022)
13. Watts, D.J., Strogatz, S.H.: Collective dynamics of 'small-world' networks. Nature **393**(6684), 440–442 (1998)
14. Wolfram, S.: Cellular Automata and Complexity: Collected Papers (1st ed.). CRC Press, Boca Raton (1994)

Despeckling Images Using Elementary Cellular Automata

Swarna Aishwarya Twinkle(✉), Supreeti Kamilya, and Jit Mukherjee

Department of Computer Science and Engineering, Birla Institute Of Technology, Mesra, Ranchi, India
{phdcs10056.22,supreeti.kamilya17,jit.mukherjee}@bitmesra.ac.in

Abstract. Speckle noise significantly degrades the quality of images in domain especially in Synthetic Aperture Radar (SAR) and medical imaging, complicating the detection of critical features such as terrain details or tumors. The task of suppressing the speckle noise from images, known as *despeckling*, has significant importance in the field of image processing. In this study, we propose an innovative algorithm using Elementary Cellular Automata (ECAs) to reduce speckle noise. The methodology involves dividing an image into patches and converting pixel intensities within each patch into binary states. Using ECA rules, pixel values are iteratively adjusted to reduce noise while preserving details. To assess the effectiveness of our approach, we compare its performance with that of traditional filters using parameters like Peak-Signal-to-Noise Ratio (PSNR), Structural Similarity Index (SSIM), Edge Preservation Index (EPI), and Perceptual Index (PI). The results of our ECA-based model achieved a PSNR of 28.532, surpassing traditional methods by significant margins: 8.1% higher than the Lee Filter, 12.8% above the Median Filter, 22.9% better than Anisotropic Diffusion, and over 57% higher than the Wiener Filter. The performance highlights the capability of our model in noise reduction compared to conventional approaches, offering a promising solution for applications in SAR and medical imaging.

Keywords: Elementary Cellular Automata · Speckle Noise · SAR Images · Ultrasound Images · PSNR

1 Introduction

Speckle noise remains a major challenge especially in satellite and medical images, often hiding critical details and compromising image clarity essential for accurate analysis. In satellite imaging, particularly with Synthetic Aperture Radar (SAR), speckle noise is a result of the interference from multiple scatterers within the observed area, making it challenging to detect fine details such as terrain features [13]. Similarly, in medical images such as in ultrasound, speckle appears due to the scattering of sound waves by tissue structures, which can hide key diagnostic features like tumors or blood vessels [14]. The removal of speckle noise from these images is crucial for improving the clarity of the data

S. Kamilya et al. (Eds.): ASCAT 2025, CCIS 2499, pp. 191–202, 2025.
https://doi.org/10.1007/978-3-031-94121-4_14

and enhancing the accuracy of analysis of scientific research as well as medical diagnosis. Various speckle suppression techniques such as Lee filter, Frost filter, Kuan filter [1], and Wiener filter [7] are commonly used to reduce speckle noise while preserving image edges and fine details. These filters adapt to the local variance of the image, smoothing the noise without significantly affecting important features. Deep learning technique, for example, DnCNN model [3], with its residual learning and batch normalization, is effective in handling Gaussian noise. Similarly, Feed Forward Network (FFDNET) [4] adapts to different noise levels by employing a tunable noise map, which improves denoising efficiency, especially down-sampled image segments. ID-CNN [5] uses convolutional layers, batch normalization, and residual connections to manage speckle noise effectively. Wavelet-based thresholding [6] works to separate noise from essential image structures, thereby retaining critical details that are crucial for diagnostic purposes. Anisotropic diffusion [8] reduces noise by selectively smoothing areas based on gradient information, which helps maintain edges and important tissue boundaries.

The discrete dynamical system, cellular automata (CAs) is used as effective tools in image processing due to their ability to model local interactions between neighboring cells. A recent research has explored the use of Cellular Automaton (CA) for image denoising, particularly through 2D-CA models [9]. These methods often work by treating image pixels as cells, where the state of each cell is influenced by its neighbors, allowing for noise suppression via local interactions [9]. While these approaches have shown good results in reducing noise, they primarily focus on other types of noise, such as Gaussian or Salt-and-Pepper noise. To the best of the authors' knowledge, there is no work on denoising speckle noise using 1-D cellular automata that leaves a gap in the existing literature for further exploration. Denoising such noise is challenging in image processing, especially in applications like remote sensing and medical imaging.

The present work uses the simplest kind of CAs, i.e., Elementary Cellular Automata (ECAs) for despeckling images. In this study, we propose an algorithm that uses ECA to reduce speckle noise, especially in SAR and medical images. We use SAR images and ultrasound breast cancer images from [15,16] and [17], that are prone to speckle noise. A controlled noise is introduced initially to the images. The images are then divided into small patches and average pixel intensity is calculated for each patch. The new pixel value is determined based on the pixel mean. Based on the new pixel value, a sum of neighborhood value is calculated. The next state value of the sum of an ECA rule decides if the pixel value of the image is to be incremented or decremented. This change in pixel values in the image reduces the speckle noise from the image. To evaluate its performance, we compare our proposed approach with traditional filters, assessing it on key parameters such as Peak-Signal-to-Noise Ratio (PSNR), Structural Similarity Index Measure (SSIM), Edge Preservation Index (EPI), and Perceptual Index (PI). These comparisons allow us to measure the effectiveness of our denoising algorithm quantitatively while maintaining image quality, especially in the context of SAR and medical imaging.

2 Preliminaries

2.1 Elementary Cellular Automata

A Cellular Automaton (CA) is an autonomous finite state machine (FSM) which functions in discrete time and space. Each cell in a CA updates its state at every time step, based on a set of rules. The current state of a cell at time t (known as the present state, PS) is used to compute its state at the next timestep or the next state (NS).

In a CA with 3-neighborhood structure (cell i, its left neighbor, and its right neighbor), the next state S_i^{t+1} of the ith cell is determined by:

$$S_i^{t+1} = f_i(S_{i-1}^t, S_i^t, S_{i+1}^t)$$

where S_{i-1}^t, S_i^t, and S_{i+1}^t are the present states of the left neighbor, the cell itself, and the right neighbor at time t. The configuration of states at time t, denoted by, $S^t = (S_1^t, S_2^t, \ldots, S_n^t)$, represents the present state of the entire CA with n cells. Each cell's next state function can also be represented as a truth table (Table 1), with the decimal equivalent of the outputs.

Elementary cellular automata are the simplest type of CA. They comprise of an array of cells, each of which can be in one of two possible states: 0 (white) or 1 (black) [11]. The state of each cell in the next generation/iteration is determined by its current state and the states of its immediate neighbors to the right and left. The rules governing this evolution can be represented by a lookup table (Table 1), specifying the next state of the cell based on the current configuration of itself and its two neighbors [11]. Since each neighboring cell configuration has three binary positions, there are $2^3 = 8$ possible combinations for these three cells. Each rule can be defined by assigning a 0 or 1 to each of these configurations, resulting in $2^8 = 256$ unique rules [10]. These rules are typically numbered from 0 to 255 in binary.

For example, "Rule 30" (Table 1) specifies that a cell will become 1 only in certain configurations, creating characteristic patterns over time. In our study we have worked with finite elementary cellular automata and used null boundary condition where left (right) neighbor of the leftmost (rightmost resp.) is null.

Table 1. Truth Table for Rule 30

PS	111	110	101	100	011	010	001	000	**Rule**
NS	0	1	1	1	1	0	1	0	30

2.2 Speckle Noise

Speckle noise is a prevalent form of interference that can reduce the image quality in both SAR and medical imaging. In SAR imagery, speckle occurs due to the coherent nature of radar waves, resulting in random intensity fluctuations throughout the image [13]. This type of noise emerges from the interference of scattered waves from numerous small scatterers within the scanned region, producing a textured pattern. Since speckle noise is multiplicative [20], it often hides finer details and complicates image analysis. In medical imaging, particularly ultrasound images, speckle acts as random pixel variations that blur the image, caused by sound wave scattering across different tissue structures. This can hinder essential features, such as tumors, blood vessels, or other pathologies [14], making it more difficult to diagnose and analyze conditions accurately. Figure 1 and 2 shows the speckled SAR and ultrasound image respectively.

Fig. 1. Speckled SAR Image.

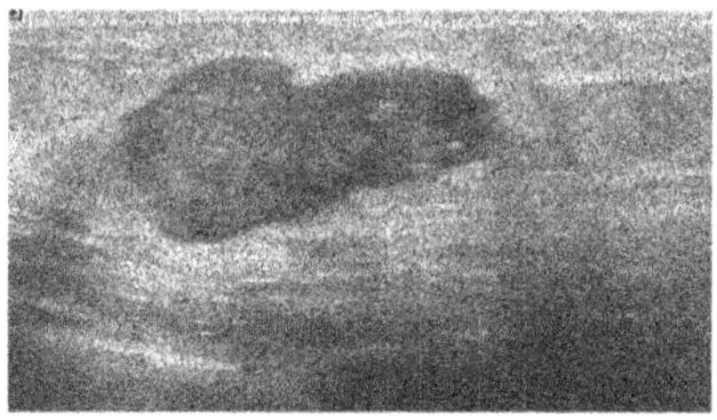

Fig. 2. Speckled Ultrasound Image.

3 Materials and Methods

In this study, we present a novel approach to denoise speckled images, specifically targeting SAR images and ultrasound breast cancer images. Speckle noise is a well-known challenge in satellite and medical imaging systems. By reducing this noise, we aim to enhance image quality making it easier to interpret and analyze crucial details in both remote sensing and medical applications. The following sections describe the data sources, preprocessing steps, and methodology used in this study.

3.1 Data

The ground truth dataset includes SAR images provided in papers [15] & [16] and ultrasound breast cancer images in [17], both of which are prone to speckle noise [18,19]. Speckle noise, which is multiplicative in nature [20], is a granular disturbance common in these types of images. It significantly impacts image clarity, making it difficult to interpret fine details. In the preprocessing step, we take clean images as ground truth images and add speckle noise to it. The introduction of controlled noise enables a direct assessment of denoising performance. The noisy image of size 256×256 is then divided into smaller non-overlapping patches of size 32×32 pixel. Dividing the image into smaller patches allows to address localized noise pattern within each region of the image, which results in effective denoising performance.

3.2 Methodology

As discussed, the image is first divided into smaller patches, and for each patch, the average pixel intensity is computed. This average intensity serves as a dynamic threshold for determining the new pixel value of each pixel within that patch. The new pixel value of each pixel is determined as follows:

$$\text{New Pixel Value} = \begin{cases} 1 & \text{if } I_{i,j} > \text{Patch Mean}, \\ -1 & \text{if } I_{i,j} < \text{Patch Mean}, \\ 0 & \text{if } I_{i,j} = \text{Patch Mean} \end{cases} \quad (1)$$

This conversion to new pixel values helps distinguish areas of higher intensity from areas of lower intensity, making it easier to apply the ECA rules effectively.

As an example, lets consider a 3×3 neighborhood with a central pixel (value 127) in a patch where the mean pixel intensity is calculated as 98. With the threshold of 98, the new pixel value for each pixel in the neighborhood are calculated following Eq. 1 as illustrated in Fig. 3.

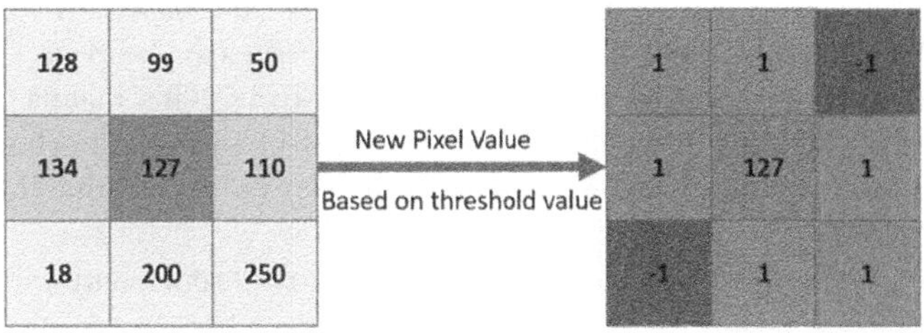

Fig. 3. Conversion to new pixel value based on threshold.

Here,

- Pixels with values greater than 98 will be assigned a new value of 1.
- Pixels with values less than 98 will be assigned a new value of –1.
- Pixels with values equal to 98 will be assigned a new value of 0.

The new pixel value of the neighboring pixels (excluding the central pixel) are summed to reflect the local contrast around the central pixel. The sum indicates whether the surrounding area has higher or lower intensities compared to the mean value. The sum of the neighborhood is computed as shown in Eq. 2

$$\text{Sum} = \sum_{k=-1}^{1} \sum_{l=-1}^{1} B_{i+k,j+l} - B_{i,j} \quad (2)$$

Here $B_{i,j}$ is the central pixel. Substituting the new pixel values of the example (shown in Fig. 3) in Eq. 2, we get,

$$Sum = (1) + (1) + (-1) + (1) + (1) + (-1) + (1) + (1) = 4 \tag{3}$$

This sum will then be used to determine the transformation of the central pixel. Once the sum is calculated, the next step is to apply the local transition of an ECA rule. The central pixel value is updated based on the sum, according to the following rules.

$$I_{i,j} = \begin{cases} I_{i,j} + 1 & \text{if Sum is positive\&Next State (NS) of the Sum is 1,} \\ I_{i,j} - 1 & \text{if Sum is negative\&Next State (NS) of the Sum is 1,} \\ I_{i,j} & \text{Next State (NS) of the Sum is 0} \end{cases} \tag{4}$$

These ECA-based rules are applied iteratively, allowing for the reduction of speckle noise while preserving the important features of the image. The more we get the difference while calculating the sum, the more change in color is there in a region. When the color is being denoised, it is checked whether any color is dominating the others. If the domination (i.e., the difference) is more, we smoothen it by incrementing or decrementing, depending on the sum value. Higher difference means a higher value of the ECA rule, based on whose NS the change is being made. Note that, the value of *Sum* can be 8 or –8 if all the neighbors of the central pixel are 1 or –1 respectively. This means, the central pixel is having different intensity where all values in the neighborhood are uniform. Therefore, we increment (decrement) the central pixel value if the *Sum* is 8 (–8 respectively).

As an example, let us consider ECA rule 245. It is represented in Table 2.

Table 2. Binary Input and Corresponding Output

111	110	101	100	011	010	001	000
1	1	1	1	0	1	0	1

In Eq. 3, we get the value of *Sum* as 4. We first convert 4 into binary representation of 3-bits i.e. 100. Now, according to Rule 245, next state of 100 is 1. Since the *Sum* in the equation is positive, so it indicates that the cental pixel should be incremented. Thus the central pixel 127 is changed to 128. This transformation process is applied to each patch across all patches, iteratively repeating over multiple iterations. Each pass applies ECA rules, allowing gradual noise reduction across the image. After processing, the denoised patches are reassembled into a single 256 × 256 image. Initially, all ECA rules i.e. 0 to 255 have been applied and tested. Out of which, few rules produced effective denoising results and are chosen for the method (provided in Sect. 4). The chosen rules show superior noise reduction while preserving image details, making them a good choice for this ECA-based denoising approach.

The steps of the methodology is summarized as follows.

1. Partition the image into non-overlapping $P \times P$ patches.
2. Calculate the average pixel intensity for each patch.
3. For each pixel in the patch, assign a new pixel value based on comparison with the patch mean, as given in Eq. 1.
4. Calculate Sum of the new pixel values of its 3×3 neighborhood, as shown in Eq. 2.
5. Based on the 3-bit binary representation of the sum, map it to the local transition of the ECA rule to determine if the central pixel should be incremented, decremented, or remain unchanged, as provided in Eq. 4.
6. Repeat Steps 3 to 5 for each patch, iterating over the entire image.
7. Combine all processed patches to reconstruct the final denoised image.

4 Results and Discussion

To thoroughly evaluate the performance of our method, we have tested it across all 256 possible rules. Out of these, specific rules such as 221, 223, 225, 227, 229, 231, 233, 235, 237, 239, 241, 245, 247, 249, and 253 gave notably high-quality denoising results. This is because, in our method, each patch's sum determines the transition applied to the central pixel by mapping it to a specific ECA rule. These rules assign a binary output, either 0 or 1, for each possible 3-bit pattern, which then governs whether the central pixel value is incremented, decremented, or remains unchanged.

Upon testing all 256 ECA rules, we observe that certain rules, particularly higher-value odd rules produced the most effective denoising results. This effectiveness arises because, in higher rules, the next state typically includes more 1 s, resulting in a higher incremental changes to pixel values across the image. This tendency towards the change contributes to improved noise reduction by actively adjusting pixel intensities in response to neighboring pixels. Conversely, in lower-value rules, the most common outcome of the next state is 0, leading to fewer pixel adjustments and a less pronounced denoising effect.

Rules with next state value as 0 for the 000 local transition (i.e., the even rules) generally prevent change in pixel values across the image. The unchanged pixel value indicates that the noise is not reduced. This explains the reason behind the effectiveness of odd rules in this study. Mostly odd rules greater than 221 are effective in this work. On the other hand, rule 255 (represented by 11111111) contains the maximum number of '1' as next state values. It proves less effective because it enforces changes even in areas that are already smooth, introducing excessive alterations that degrade image quality slightly. Thus, while Rule 255 can reduce noise, it also introduces unnecessary changes across the image, leading to over-processing. Therefore, our selected rules balance noise reduction with detail preservation, enabling our approach to achieve effective denoising without sacrificing important image features.

We compare our model with traditional filtering techniques, demonstrating its effectiveness in noise reduction. However, we did not extend the comparison to other deep learning (DL) models due to their higher time complexity. While typical DL models require 1 to 2 min per execution, the proposed ECA-based approach completes in just 5 to 10 s, achieving a speedup of approximately 85–95%. This emphasizes the potential of the proposed technique for real time processing. The significant reduction in execution time highlights the efficiency of our method. Further, training of complex networks, such as deep learning techniques, is time consuming and can have issues related to overfitting. The Fig. 4 illustrates the results achieved by applying various filters compared to our proposed method on SAR dataset. Figure 5 and 6 shows the performance of our model across multiple iterations (15, 25 and 40 iterations) on both datasets, highlighting the effectiveness of our method in reducing noise and enhancing image quality. At 15 iterations, noise is reduced but is still visible. By 25, features are clearer, and at 40, the image is smooth with well-defined edges. Beyond 40, PSNR stabilizes, indicating no further improvement in image quality.

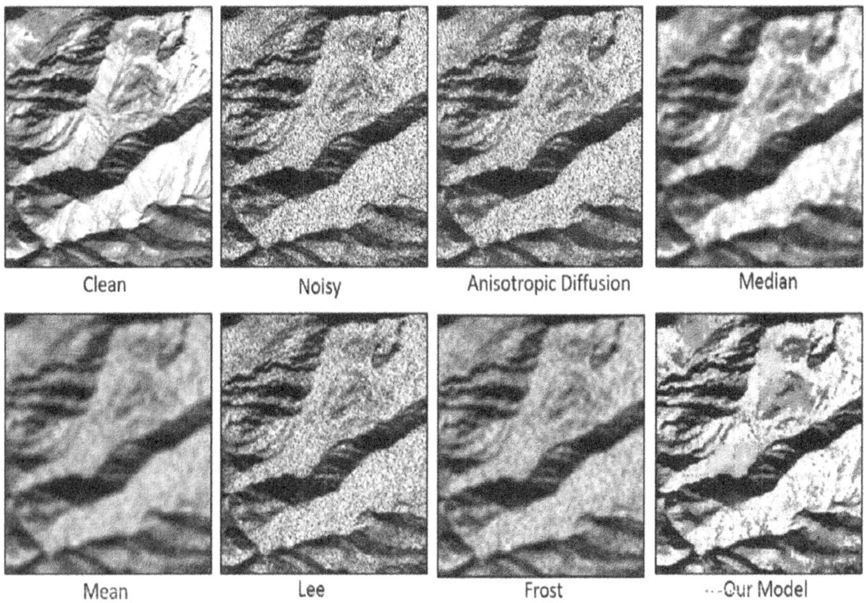

Fig. 4. Comparison with traditional filters.

To quantitatively assess the denoising effectiveness, we utilize the following image quality metrics.

1. *PSNR (Peak Signal-to-Noise Ratio)*: PSNR measures the ratio between the maximum possible signal power and the power of distorting noise. Higher PSNR values indicate better quality, indicating image denoising effectiveness.
2. *SSIM (Structural Similarity Index Measure)*: SSIM evaluates the similarity between the original and processed images by comparing luminance, contrast, and structure. It ranges from 0 to 1, where values closer to 1 indicate higher similarity and better image quality.

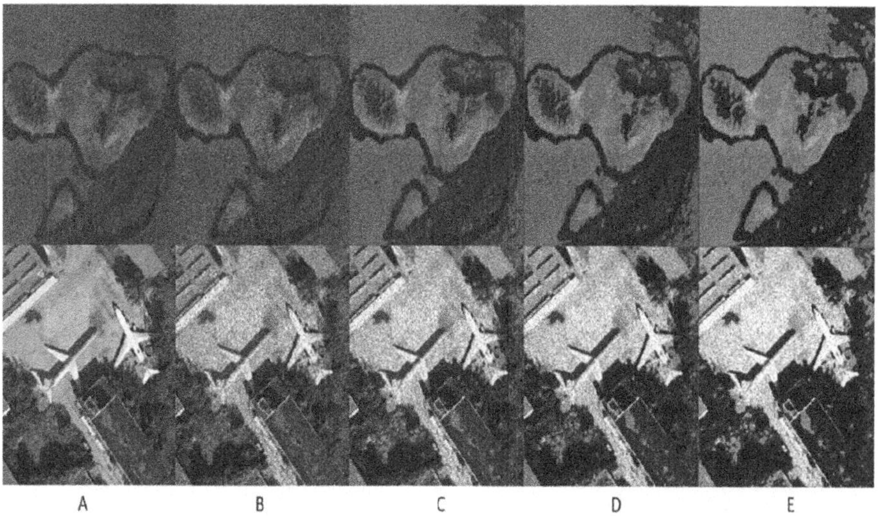

Fig. 5. Denoising Results on SAR Dataset; A: Clean Image; B: Noisy Image; C: After 15 iterations; D: After 25 iterations; E: After 40 iterations.

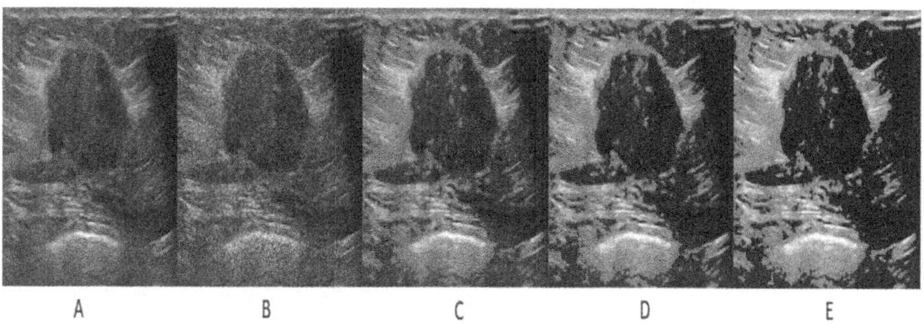

Fig. 6. Denoising Results on Ultrasound Dataset; A: Clean Image; B: Noisy Image; C: After 15 iterations; D: After 25 iterations; E: After 40 iterations

3. *EPI (Edge Preservation Index)*: EPI measures the effectiveness of the algorithm in preserving edges, an important factor in retaining image features during denoising. Higher EPI values means better edge preservation.
4. *PI (Perceptual Index)*: PI combines several perceptual quality measures to provide a more holistic view of image quality from a human perspective. Lower PI values indicate better perceived quality of the image.

By applying these metrics, we quantitatively validated that the selected rules performed well in enhancing image quality while effectively suppressing noise. Table 3 presents all the evaluation metrics used in this study. Graphical representations of these metrics are shown in Fig. 7.

Table 3. Performance Comparison of Speckle Suppression Filters

Method	PSNR	SSIM	EPI	PI
Our Model	**28.532**	**0.5872**	**0.5147**	**6.84**
Lee Filter	26.403	0.4733	0.3102	9.12
Median Filter	25.294	0.2058	0.1825	9.75
Anisotropic Diffusion	23.196	0.4303	0.2387	7.12
Mean Filter	22.804	0.4679	0.1457	7.93
Frost Filter	21.854	0.3493	0.2709	8.63
Wiener Filter	18.365	0.1368	0.2401	8.04
Noisy Image	12.117	0.3993	0.2473	7.52

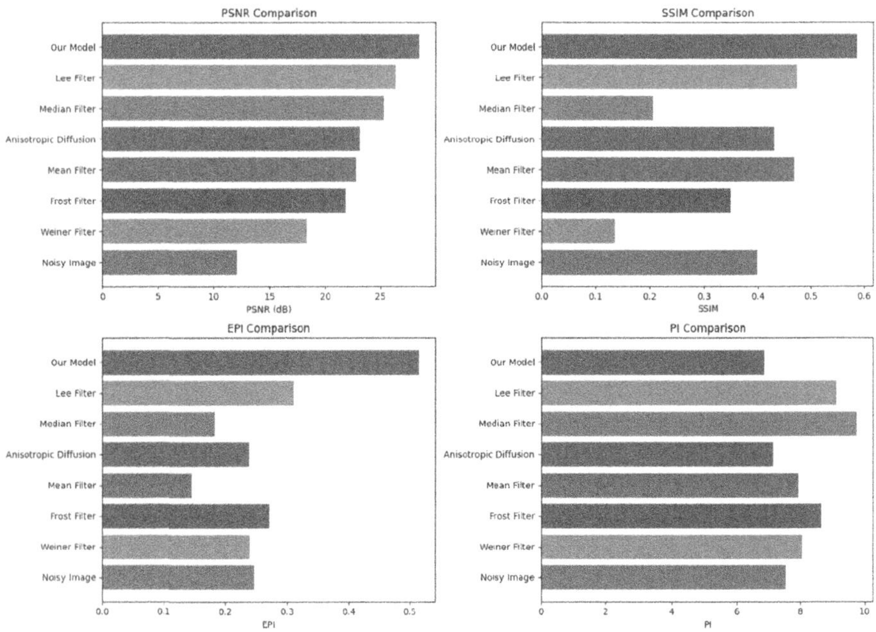

Fig. 7. Metrics (PSNR, SSIM, EPI and PI) Comparison of Our Model (Blue) with Lee Filter (Orange), Median Filter (Green), Anisotropic Diffusion (Red), Mean Filter (Purple), Frost filter (Brown), Wiener Filter (Pink), Noisy image (Grey). (Color figure online)

The potential limitation of this approach is using non-overlapping patches for denoising, as it can cause visible grid-like structure at the patch boundaries. To improve this, we aim to explore overlaps between patches to further reduce boundary artifacts and improve the overall denoising process. Additionally, we plan to implement non-uniform elementary cellular automata as future work, which has the potential to enhance both PSNR and image quality by adapting more effectively to different image structures.

5 Conclusion

This study presents a novel elementary cellular automata based approach for effective speckle noise reduction, especially important in SAR and medical imaging. By dividing the image into patches and applying ECA rules based on local transitions, our method achieves significant noise reduction while preserving essential image features. Through extensive testing, we identified specific ECA rules that enhance noise suppression, providing a balance between noise reduction and detail retention. Comparative results indicate that our model outperforms traditional filtering methods in terms of PSNR, SSIM, EPI and Perceptual Index. Our approach proves computationally efficient, requiring only 5 to 10 s per image. This quick processing time makes our approach practical for real-time applications, offering both quality and speed. Overall, our ECA-based approach provides a reliable, accessible alternative to more complex methods, making it a promising solution for applications that demand efficient, high-quality denoising.

Disclosure of Interests. The authors have no competing interests to declare that are relevant to the content of this article.

References

1. Singh, P., Diwakar, M., Shankar, A., Shree, R., Kumar, M.: A review on SAR image and its despeckling. Arch. Comput. Methods Eng. **28**(7), 4633–4653 (2021). https://doi.org/10.1007/s11831-021-09548-z
2. Baselice, F., Ferraioli, G., Johnsy, A.C., Pascazio, V., Schirinzi, G.: Speckle reduction based on wiener filter in ultrasound images. In: Annual International Conference of the IEEE Engineering in Medicine and Biology Society (EMBC), pp. 3065–3068 (2015). https://doi.org/10.1109/EMBC.2015.7319039. PMID: 26736939
3. Zhang, K., Zuo, W., Chen, Y., Meng, D., Zhang, L.: Beyond a Gaussian denoiser: residual learning of deep cnn for image denoising. IEEE Trans. Image Process. **26**(7), 3142–3155 (2017). https://doi.org/10.1109/TIP.2017.2662206
4. Zhang, K., Zuo, W., Zhang, L.: FFDNet: toward a fast and flexible solution for CNN-based image denoising. IEEE Trans. Image Process. **27**(9), 4608–4622 (2018). https://doi.org/10.1109/TIP.2018.2839891
5. Wang, P., Zhang, H., Patel, V.M.: SAR image despeckling using a convolutional neural network. IEEE Signal Process. Lett. **24**(12), 1763–1767 (2017). https://doi.org/10.1109/LSP.2017.2756566
6. Sudha, S., Suresh, G.R., Sukanesh, R.: Speckle noise reduction in ultrasound images by wavelet thresholding based on weighted variance. Int. J. Comput. Theory Eng. (2009)
7. Shruthi, B., Renukalatha, S., Siddappa, M.: Speckle noise reduction in ultrasound images - a review. Int. J. Eng. Res. Technol. (IJERT) **4**(2) (2015)

8. Flores, W.G., de Albuquerque Pereira, W.C., Infantosi, A.F.C.: Breast ultrasound despeckling using anisotropic diffusion guided by texture descriptors. Ultrasound Med. Biol. **40**(11), 2609–2621 (2014). https://doi.org/10.1016/j.ultrasmedbio.2014.06.005
9. Chandra, N., Teotia, S., Chaudhary, A., Jain, V., Vashishtha, T., Dwivedi, N.N.: Noise reduction using cellular automata. Nanosci. Quant. Technol. **20**(6), 4650–4672 (2022). https://doi.org/10.14704/nq.2022.20.6.NQ22456
10. Kamilya, S., Adak, S., Das, S., Sikdar, B.: SACAs: (Non-uniform) cellular automata that converge to a single fixed point. J. Cell. Autom. **14**, 27–49 (2019)
11. Wolfram, S.: A New Kind of Science. Wolfram Media, 2002. ISBN 1579550088, https://www.wolframscience.com
12. Lee, J., Jurkevich, L., Dewaele, P., Wambacq, P., Oosterlinck, A.: Speckle filtering of synthetic aperture radar images: a review. Remote Sens. Rev. **8** (1994). https://doi.org/10.1080/02757259409532206
13. Kumar, L.A., Jebarani, M.E.: A comprehensive review on speckle denoising techniques in satellite images. In: 2019 International Conference on Communication, Control, and Signal Processing (ICCSP), pp. 245–248, 2019. https://doi.org/10.1109/ICCSP.2019.8697926
14. Pradeep, S., Nirmaladevi, P.: A review on speckle noise reduction techniques in ultrasound medical images based on spatial domain, transform domain and CNN methods. In: IOP Conference Series: Materials Science and Engineering, vol. 1055, no. 1, p. 012116, 2021. IOP Publishing
15. Schmitt, M., Hughes, L.H., Zhu, X.X.: The SEN1-2 dataset for deep learning in SAR-optical data fusion. In: ISPRS Annals of the Photogrammetry, Remote Sensing and Spatial Information Sciences, vol. IV-1, pp. 141–146 (2018). https://doi.org/10.5194/isprs-annals-IV-1-141-2018
16. Dabhi, S., Soni, K., Patel, U., Sharma, P., Parmar, M.: Virtual SAR: A Synthetic Dataset for Deep Learning Based Speckle Noise Reduction Algorithms, 2020. arXiv preprint arXiv:2004.11021
17. Al-Dhabyani, W., Gomaa, M., Khaled, H., Fahmy, A.: Dataset of breast ultrasound images. Data Brief **28**, 104863. Elsevier (2020). https://doi.org/10.1016/j.dib.2019.104863
18. Singh, P., Shree, R.: Analysis and effects of speckle noise in SAR images. In: 2016 2nd International Conference on Advances in Computing, Communication, & Automation (ICACCA) (Fall), pp. 1–5 (2016). https://doi.org/10.1109/ICACCAF.2016.7748978
19. Li, X., Wang, Y., Zhao, Y., Wei, Y.: Fast speckle noise suppression algorithm in breast ultrasound image using three-dimensional deep learning. Front. Physiol. **13**, 880966 (2022). https://doi.org/10.3389/fphys.2022.880966
20. Lee, L., Jurkevich, P., Dewaele, P., et al.: Speckle filtering of synthetic aperture radar images: a review. Remote Sens. Rev. **8**(4), 313–340 (1994)

Information Propagation Based Data Classification with Reversible Non-Uniform Elementary Cellular Automata

C. J. Baby(✉) and Kamalika Bhattacharjee

Department of Computer Science and Engineering, National Institute of Technology, Tiruchirappalli620015, Tamil Nadu, India
babycj1120@gmail.com, kamalika@nitt.edu

Abstract. In this work, the cycles in reversible non-uniform elementary cellular automata (CAs) are used as a tool for data classification. For the non-uniform lattice, based on the individual information propagation of the rule for each cell, a graph theoretic network flow approach is applied to find the overall information propagation in the lattice. Considering this overall information propagation, a greedy selection strategy is devised that drastically reduces the possible number of effective rules suitable for data classification. To apply these CAs for classification, configurations within the cycles of the appropriate cellular automaton (CA) are assigned class labels to group the cycles into distinct classes. Finally, through experimentation and analysis, the performance of cellular automaton in classifying data based on accuracy and execution time is evaluated and shown to be comparable to the traditional classification algorithms.

Keywords: Non-Uniform Cellular Automata · Elementary Cellular Automaton (ECA) · Reversibility · Information Propagation · Cyclic Space · Classification

1 Introduction

Classification is a supervised machine learning approach where data is categorized into predefined classes based on patterns learned from labeled training datasets. Existing work on data classification using one-dimensional cellular automata (CAs) includes approaches based on Multiple Attractor Cellular Automata (MACA) [9], Asynchronous Cellular Automata and fixed point attractors [13,14] and reversible First Degree Cellular Automata (FDCAs) [1], among other techniques. Most of these approaches predict the class label based on a single configuration. In this work, we use information from multiple configurations within a cycle of reversible binary CAs to predict the class label.

This work introduces a data classification algorithm using the cyclic space of reversible non-uniform elementary cellular automata (ECAs). Data classification

S. Kamilya et al. (Eds.): ASCAT 2025, CCIS 2499, pp. 203–218, 2025.
https://doi.org/10.1007/978-3-031-94121-4_15

using reversible cellular automata was initially introduced in Ref. [1] using First Degree Cellular Automaton (FDCA). In that approach, CA configurations are represented in decimal form, which requires high computation time to process. Alternatively, non-uniform elementary CA (ECA) uses binary configurations, which allows faster bitwise operations. Therefore, the proposed data classification method in this work aims to achieve much lower execution time than the method that uses first-degree decimal CAs.

Here, class labels assigned to configurations of each cycle are used to group cycles into distinct classes. Assume a reversible cellular automaton has two cycles, $Cycle1$ and $Cycle2$. According to the training data, if the majority of configurations in $Cycle1$ are associated with $ClassA$, and the majority in $Cycle2$ are associated with $ClassB$, then we can label all configurations in $Cycle1$ as $ClassA$ and those in $Cycle2$ as $ClassB$. During prediction, if we have data with an unknown class label, we can determine its class based on its cycle. For example, if the testing data belongs to $Cycle1$, we assign it to $ClassA$.

As the number of possible reversible non-uniform Elementary Cellular Automata (ECA) grows exponentially over the cell size n, a strategy is required to reduce the rule space to the ones appropriate for data classification. Selecting an appropriate reversible cellular automaton for classification requires an understanding of how information propagates when a cell's state changes. To measure this, we calculate the information propagation to the neighboring cells [8].

In this work, we introduce a measure for assessing the overall information propagation across non-uniform cellular automata by considering the entire rule vector. This propagation is analogous to data movement in a graph, where cells act as nodes and the information transfer between them is represented by weighted edges. We use this measure of information propagation to identify effective rules for data classification using reversible non-uniform elementary cellular automata. Hereafter, by a "CA", we will refer exclusively to reversible non-uniform elementary CA.

Using the selected rules for data classification, we implement a CA-based classification algorithm utilizing the cyclic space. We then evaluate the performance of non-uniform elementary cellular automata in data classification in terms of accuracy and execution time, comparing it with traditional classification algorithms [11] such as Decision Tree [15], KNN [7], and Naive Bayes [12]. Our result shows that our algorithm gives a comparable result both in terms of accuracy and execution time.

2 Definitions and Terminologies

In this work, we use one-dimensional n-cell non-uniform elementary cellular automata arranged in a finite sequence under null boundary condition. Each cell $i \in n$ have a state belonging to the set $S = \{0, 1\}$. The next state of the cell i is determined by its immediate neighbors on the left and right and the current state. Specifically, the next state is computed by the local transition function $R_i : S^3 \rightarrow S$. The collection of local rules for all cells is represented by a rule vector $\mathbb{R} = \langle R_0, R_1, \ldots, R_{n-1} \rangle$ where the rule R_i is applied to the cell i.

Table 1 shows the tabular form representation of the rule R_i. In this table, if we take the string 00001001 of the next states (NS) in the first row corresponding to the present states (PS) from 111 to 000, and then convert it to decimal, we get the value 9. This indicates that the first row corresponds to the rule 9. As there are 8 possible combinations of present state, using an 8-length binary string, rules from 0 to 255 are possible. Here, Table 1 shows only 4 rules. Applying these rules, if the rule vector $\langle 9, 43, 90, 20 \rangle$ represents a non-uniform CA with a cell length of 4, then in this sequence, rule 9 is applied to cell 0, rule 43 to cell 1, rule 90 to cell 2, and rule 20 to cell 3.

Table 1. 2-state CA rules 9, 20, 43 and 90

PS	111	110	101	100	011	010	001	000	Rule
RMT	(7)	(6)	(5)	(4)	(3)	(2)	(1)	(0)	
NS	0	0	0	0	1	0	0	1	9
NS	0	0	0	1	0	1	0	0	20
NS	0	0	1	0	1	0	1	1	43
NS	0	1	0	1	1	0	1	0	90

Definition 1. *In a d-state CA, the combination of neighborhoods x, y, z as input to $R_i : S^3 \rightarrow S$, the local rule of the CA, with respect to the value $R_i(x, y, z)$, is called a Rule Min Term (RMT). Each RMT is associated with a number $r = x \times d^2 + y \times d + z$. We denote the value $R_i(x, y, z)$ by $R[r]$ [3].*

The possible combinations of the present states (PS) of three neighboring cells of a 2-state CA are listed in the Table 1. Each combination has its decimal equivalent, which represents the rule's RMT.

Definition 2. *(Equivalent RMTs)* [3]. *A set of d RMTs, $r_1, r_2, \ldots, r_d$, for a d-state CA rule are said to be equivalent if $r_1 d \equiv r_2 d \equiv \ldots \equiv r_d d \pmod{d^3}$.*

Definition 3. *(Sibling RMTs)* [3]. *A set of d RMTs $s_1, s_2, \cdots, s_d$ of a d-state CA rule are called sibling to each other if $\lfloor \frac{s_1}{d} \rfloor = \lfloor \frac{s_2}{d} \rfloor = \cdots = \lfloor \frac{s_d}{d} \rfloor$.*

Definition 4. *(Self-Equivalent RMTs). A set of d RMTs $q_1, q_2, \cdots, q_d$ of a d-state CA are called self-equivalent to each other if $q_1 d \equiv q_2 d \equiv \cdots q_d d \equiv \pmod{d^2}$.*

Table 2. The categorization of the RMTs for Rule R_i

Sibling RMTs		Self-Equivalent RMTs		Equivalent RMTs	
$\text{Sibl}_{i.0}$	{000, 001}	$\text{Self}_{i.0}$	{000, 010}	$\text{Equi}_{i.0}$	{000, 100}
$\text{Sibl}_{i.1}$	{010, 011}	$\text{Self}_{i.1}$	{001, 011}	$\text{Equi}_{i.1}$	{001, 101}
$\text{Sibl}_{i.2}$	{100, 101}	$\text{Self}_{i.2}$	{100, 110}	$\text{Equi}_{i.2}$	{010, 110}
$\text{Sibl}_{i.3}$	{110, 111}	$\text{Self}_{i.3}$	{101, 111}	$\text{Equi}_{i.3}$	{011, 111}

For a rule R_i, the sets of equivalent and sibling RMTs are represented as $\text{Equi}_{i.k}$ and $\text{Sibl}_{i.j}$ respectively, where $\text{Equi}_{i.k} = \{k, d^2+k, 2d^2+k, \ldots, (d-1)d^2+k\}$, $(0 \le k \le d^2-1)$ and $\text{Sibl}_{i.j} = \{d \cdot j, d \cdot j+1, \ldots, d \cdot j+(d-1)\}$ $(0 \le j \le d^2-1)$. The set of Self-Equivalent RMTs of rule R_i is represented as $\text{Self}_{i.l}$, where $Self_{i.l} = \{s \mid s = \frac{l}{d} \times d^2 + l \pmod d + d \times t, 0 \le t \le d-1\}$ $(0 \le l \le d^2-1)$. For any rule R_i, the sibling RMTs, self-equivalent RMTs, and equivalent RMTs for a 2-state CA are presented in the Table 2.

The local rules $\{R_i\}_{i \in n}$ defines the global transition function $G_n : S^n \to S^n$ for the CA expressed as $G_n(x)_i = R_i(x_{i-1}, x_i, x_{i+1})$, $0 \le i \le n-1, \forall x \in S^n$. In this expression, each $x \in S^n$ is the configuration of the cellular automata (CA), with $x = (x_i)_{i \in n}$.

Definition 5. *A finite CA of size n with global transition function $G_n : C_n \to C_n$ is reversible, if for each $y \in C_n$, where C_n is the set of all configurations of length n, there exists exactly one $x \in C_n$, such that, $G_n(x) = y$. Otherwise, the CA is irreversible for that n [2].*

In reversible cellular automata (CA), each configuration has a unique predecessor due to the bijective nature of the global transition function. As a result, configurations are organized into distinct cycles. Consider the transition diagram in Fig. 1. Here, every configuration has a unique predecessor, meaning there are no non-reachable configurations and all configurations are organized into two different cycles. This implies that the non-uniform CA $\langle 9, 43, 90, 20 \rangle$ is reversible.

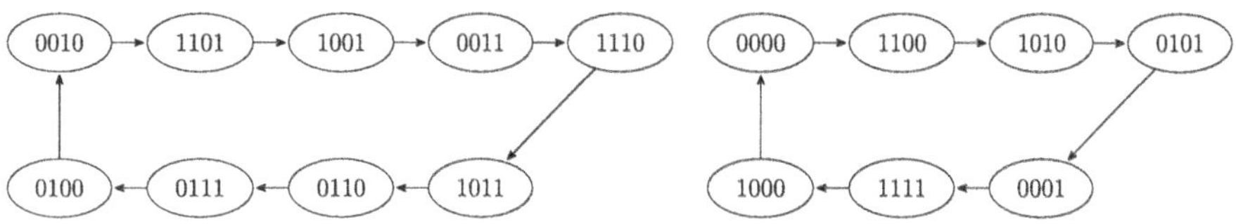

Fig. 1. Transition diagram of 4-cell reversible CA $\langle 9, 43, 90, 20 \rangle$.

A synthesis method introduced in Ref. [4] can be used to generate the rule vector for an n-cell non-uniform CA, ensuring that the rules generated from these tables (see Table 3) are always reversible. Example 1 illustrates the generation of a reversible rule vector using Table 3.

Example 1. Suppose we want to generate a reversible rule vector $\langle R_0, R_1, R_2, R_3 \rangle$ for a four-cell CA using Table 3. To select a rule for R_0, refer to Table 3b, the first rule table, and choose one rule from it. Let us select Rule 9 as R_0. After selecting Rule 9, the class of R_1 corresponds to the row of Rule 9 in Table 3b, which is Class III. So, we need to select a rule for R_1 from Class III of Table 3a, the class-relationship table. Let us select Rule 43 as R_1. The row of Rule 43 in Table 3a indicates that the class of R_2 is Class IV. Therefore, the rule for R_2 has to be selected from Class IV in Table 3a. Let us select Rule 90 as R_2. The row of Rule 90 as R_2 in Class IV of Table 3a indicates that the class of R_3 is Class IV. Therefore, we need to select the last rule for R_3 from Class IV in Table 3c. Let us select Rule 20 as R_3. Finally, the reversible rule vector $\langle R_0, R_1, R_2, R_3 \rangle$ is $\langle 9, 43, 90, 20 \rangle$.

3 Non-Uniform Cellular Automata as a Classifier

In this work, we use reversible non-uniform elementary CA as a data classifier. As an initial pre-processing step, we need to represent the dataset in binary form. To do this, we use the frequency encoding technique [10] to convert the data set into a binary representation.

Table 3. Tables to generate rule vector for reversible non-uniform CA

Class of R_i	R_i Rules	Class of R_{i+1}
I	51, 204, 60, 195	I
	85, 90, 165, 170	II
	102, 105, 150, 153	III
	53, 58, 83, 92, 163, 172, 197, 202	IV
	54, 57, 99, 108, 147, 156, 198, 201	V
	86, 89, 101, 106, 149, 154, 166, 169	VI
II	15, 30, 45, 60, 75, 90, 105, 120	I
	135, 150, 165, 180, 195, 210, 225, 240	II
III	51, 204, 15, 240	I
	85, 105, 150, 170	II
	90, 102, 153, 165	III
	23, 43, 77, 113, 142, 178, 212, 232	IV
	27, 39, 78, 114, 141, 177, 216, 228	V
	86, 89, 101, 106, 149, 154, 166, 169	VI
IV	60, 195	I
	90, 165	IV
	105, 204	V
V	51, 204	I
	85, 170	II
	102, 153	III
	86, 89, 90, 101, 105, 106	VI
	149, 150, 154, 165, 166, 169	VI
VI	15, 240	I
	105, 150	IV
	165, 90	V

(a) Relationship between R_i and R_{i+1}

Rules for R_0	Class of R_1
3, 12	I
10, 5	II
9, 6	III

(b) Table for First Rule

Rule Class for R_{k-1}	Rule Set for R_k
I	20, 17, 68, 65
II	20, 5, 80, 65
III	17, 5, 80, 68
IV	65, 20
V	68, 17
VI	80, 5

(c) Table for Last Rule

Table 4. Example for Frequency based encoding

Data ID	Attribute 1		Attribute 2		Class	Target configuration
	Eyes	Encoding	Height	Encoding		
1	blue	00	172	11	Positive	0011
2	black	11	160	01	Positive	1101
3	brown	10	161	01	Positive	1001
4	blue	00	150	00	Negative	0000
5	black	11	154	00	Negative	1100
6	brown	10	149	00	Negative	1000
7	brown	10	166	11	Positive	1011
8	blue	00	156	01	Negative	0001

3.1 Frequency Encoding

In frequency encoding, the attribute values are converted to their binary form based on the frequency of data elements within the attributes. Consider an example of frequency encoding in Table 4. This simple dataset is constructed from a hypothetical example belonging to two different classes. This dataset has two attributes: eyes and height. The first attribute eyes is categorical, while the second attribute height is continuous. The categorical attribute has three distinct values encoded using 2 bits as follows: 00 for blue, 10 for brown, and 11 for black. The continuous attribute values for height are divided into three sub-intervals: 00 for $height < 155$, 01 for $155 \leq height \leq 165$, and 11 for $height > 165$.

Using this frequency encoding, we can convert any attribute value into its binary representation. After encoding the attributes in binary form, concatenating these attributes gives us the target configuration of the CA and corresponding class label. The length of this target configuration determines the cell size n of the CA used for classification. In the dataset given in Table 4, the target configuration length is 4, so the cell size of CA required for classification is 4.

3.2 Data Classification Using Cyclic Space

In data classification using CA, the class information related to each configuration in the cycles allows us to determine the class label of each cycle. Figure 2 shows some hypothetical cycles of the cyclic space of a reversible cellular automaton. The number of configurations belonging to a cycle will vary depending on the cellular automaton used. In Fig. 2, red dots represent configurations belonging to Class A, blue dots represent configurations belonging to Class B which we get from the training data; however, black dots represent configurations that do not have class information from the training data. In the cycles shown in Fig. 2, Cycle 1 $\in$ Class A since the majority of its configurations are in Class A, Cycle 2 $\in$ Class A for the same reason, and similarly, Cycle 3 $\in$ Class B.

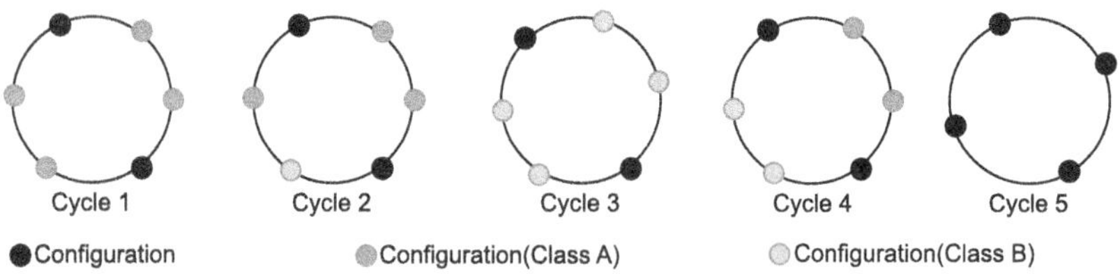

Fig. 2. Cycle 1 $\in$ Class A, Cycle 2 $\in$ Class A, Cycle 3 $\in$ Class B, Cycle 4&5 $\in$ Class A or B based on XOR of configurations.

In Cycle 4, both classes have the same majority, and in Cycle 5, there is no information related to the class. So the class of Cycle 4 and 5 is determined based on closest labeled cycle with respect to the average of all configurations in the cycles. As the configurations are in binary, the average of each cycle can be calculated as XOR of all the configurations in the cycle. Now, if two cycles' average values are closer, and one has an associated class level while the other does not have, we assign the same class level to it. To do this, the XOR values of all cycle configurations are first calculated and sorted in ascending order. Class labels are then assigned to unlabeled cycles based on the class labels of nearby cycles with similar XOR values. In this way, the cycles of reversible cellular automata can be used as a tool for classifying different classes within a dataset by following these steps:

1. **Step 1:** *Data Encoding-* Convert the dataset into binary strings using frequency encoding. Next, associate each data point with a target configuration and class label.
2. **Step 2:** *Cyclic Space Generation using rule vector*
3. **Step 3:** *Training Phase-* Assign the class label to each cycle using majority voting. When a clear majority cannot be determined, assign class labels based on the XOR of elements within each cycle, using the class labels of neighboring cycles to ensure accurate classification.
4. **Step 4:** *Prediction Phase-* Assign the predicted class label for each test instance using the labeled cycles.

Figure 3. outlines the training and prediction phases for classification using CA.

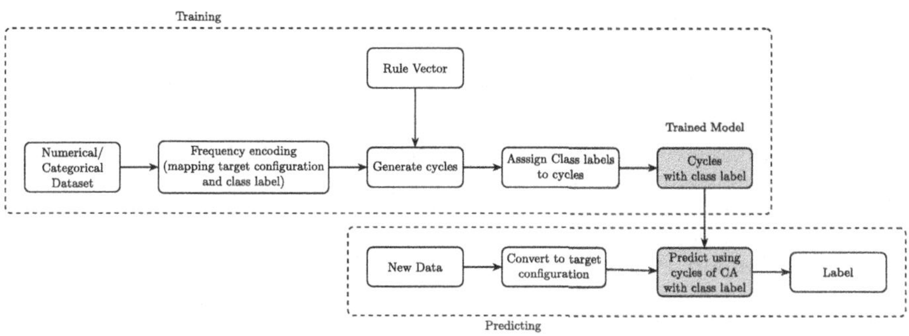

Fig. 3. Classification using Cellular Automata.

Consider the dataset in Table 4, which has a total of 8 data points. Let us use the first 6 data points as the training dataset and the remaining 2 as the testing dataset. Suppose we use the reversible CA $\langle 9, 43, 90, 20 \rangle$ for classification. Data IDs 1, 2, and 3, belong to the positive class in the training dataset, are part of the target configuration in Cycle 1 (left cycle of Fig. 1). This means we can label all configurations in Cycle 1 as positive. Similarly, Data IDs 4, 5, and 6, which belong to the negative class in the training dataset, are part of the configuration in Cycle 2 (right cycle of Fig. 1). This means we can label all configurations in Cycle 2 as negative. During testing, we identify the cycle to which the unlabeled data belongs and predict its label based on the cycle label. For example, the class label of testing data ID 7 can be predicted as positive since it belongs to Cycle 1.

However, the number of cells n in the cellular automata used for classification increases as the number of attributes in the dataset increases. As a result, when the cell size increases, the number of reversible rules that can be generated using Table 3 will increase exponentially with the size of n. So we need to select the appropriate rules for data classification as discussed in the following section.

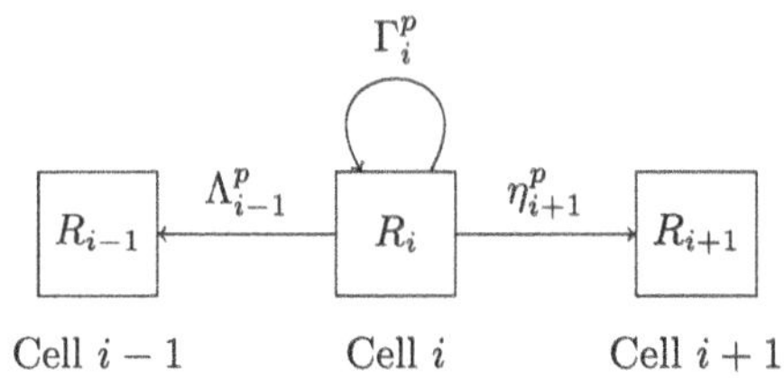

Fig. 4. Information propagation from Cell i to Cell $i-1$ and Cell $i+1$.

4 Role of Information Propagation in Data Classification

Any classification algorithm performs well when the distance between elements in the same class is small, while the distance between elements in different classes is larger. To classify data using the cycles of reversible cellular automata, we have to use rules that produce similar configurations within the same cycle. If a CA produces cycles with similar configurations, that means, in that CA, the information propagation from one cell to the other is low. To find such rules we now measure the rate of overall information propagation in the CA.

4.1 Information Propagation in Non-Uniform CA

In non-uniform cellular automata, when the state of a cell i is altered, the impact on its neighboring cells in the subsequent time can be calculated using the information Propagation to the left neighbor (Λ^p_{i-1}), information propagation to right neighbor (η^p_{i+1}), and self information propagation (Γ^p_i) (see Fig. 4).

Information Propagation to Left Neighbor: [8] Using the following parameter Λ^p_{i-1} we can calculate the overall probability of information propagating to the left of cell i.

$$\Lambda^p_{i-1} = \frac{1}{d^2} \sum_{j=0}^{d^2-1} \Lambda^{(-1)}_{i-1.j}$$

where, $\Lambda^{(-1)}_{i-1.j} = \frac{1}{d^2-d} \sum_{r,s \in \text{Sibl}_{i-1.j}, r \neq s} \delta^{(-1)}_{i-1.j}(r,s)$ measures the likelihood that the change in cell i affects cell $i-1$. Using the function $\delta^{(-1)}_{i-1.j}(r,s)$ we can determine how cell $i-1$ changes its state when cell i changes where

$$\delta^{(-1)}_{i-1.j}(r,s) = \begin{cases} 1, & \text{if } R_{i-1}(r) \neq R_{i-1}(s),\ r,s \in \text{Sibli}_{i-1.j} \\ 0, & \text{otherwise} \end{cases}$$

Now, let us consider an example to calculate Λ^p_{i-1}.

Example 2. Suppose cell $i-1$ has rule 43 (see Table 1). Here d = 2, to find $\delta^{(-1)}_{i-1.0}$, consider the sibling RMT set $\text{Sibli}_{i-1.0} = \{000, 001\}$ (see Table 2). In this case, $\delta^{(-1)}_{i-1.0}(000, 001) = 0$ because the next state values of 000 and 001 are same. Similarly calculate $\delta^{(-1)}_{i-1.j}$ for all sibling RMT sets. Then, we can find $A^{(-1)}_{i-1.0} = 0$, $A^{(-1)}_{i-1.1} = 1$, $A^{(-1)}_{i-1.2} = 1$, $A^{(-1)}_{i-1.3} = 0$. From all these values, we have $\Lambda^p_{i-1} = 0.5$.

Information Propagation to Right Neighbor: [8] Using the following parameter η^p_{i+1}, we can calculate the effect of cell i on cell $i+1$ when the state of cell i changes:

$$\eta^p_{i+1} = \frac{1}{d^2}\sum_{k=0}^{d^2-1} B^{(+1)}_{i+1.k}$$

where $B^{(+1)}_{i+1.k} = \frac{1}{d^2-d}\sum_{r,s\in\text{Equi}_{i+1.k},\, r\neq s}\delta^{(+1)}_{i+1.k}(r,s)$ measures how many times cell $i+1$ changes due to a change in cell i and using the function $\delta^{(+1)}_{i+1.k}(r,s)$ we can determine how cell $i+1$ changes its state when cell i changes where

$$\delta^{(+1)}_{i+1.k}(r,s) = \begin{cases} 1 & \text{if } R_{i+1}(r) \neq R_{i+1}(s),\ r,s \in \text{Equi}_{i+1.k} \\ 0 & \text{otherwise} \end{cases}$$

Now, let us take an example to calculate η^p_{i+1}.

Example 3. Suppose cell $i+1$ has rule 43 (see Table 1). Here d = 2, to find $\delta^{(+1)}_{i+1.0}$, we consider the equivalent RMT set $\text{Equi}_{i.0} = \{000, 100\}$ (see Table 2). In this case, $\delta^{(+1)}_{i+1.0}(000, 100) = 1$ because the next state values of 000 and 100 are different. We can similarly calculate the values of $\delta^{(+1)}_{i+1.k}$ for all equivalent RMT sets. Then, we can find $B^{(+1)}_{i+1.0} = 1$, $B^{(+1)}_{i+1.1} = 0$, $B^{(+1)}_{i+1.2} = 0$, $B^{(+1)}_{i+1.3} = 1$. From all these values, we have $\eta^p_{i+1} = 0.5$.

Self Information Propagation: To measure the self information propagation of the cell i, we can use the function $\delta^{(0)}_{i.l}$. Here the focus is on self-equivalent RMT sets related to cell i. The calculation for $\delta^0_{i.l}$ is as follows:

$$\delta^0_{i.l}(r,s) = \begin{cases} 1 & \text{if } R_{i+1}(r) \neq R_{i+1}(s),\ r,s \in \text{Self}_{i.l} \\ 0 & \text{otherwise} \end{cases}$$

How many times that cell i changes in the next transition due to a change in cell i is denoted as $C^0_{i.l}$, which is defined as:

$$C^0_{i.l} = \frac{1}{d^2 - d} \sum_{r,s \in \text{Self}_{i.l},\, r \neq s} \delta^0_{i.l}(r, s)$$

Finally, we can measure the self information propagation, denotes as Γ^p_i, of a cell i, using the following formula:

$$\Gamma^p_i = \frac{1}{d^2} \sum_{l=0}^{d^2-1} C^0_{i.l}$$

Consider an example for the calculation of self information propagation Γ^p_i as follows:

Example 4. Suppose ECA rule 43 (see Table 1) is applied to cell i. Here d=2, to measure the value of $\delta^0_{i.0}$, we consider the self-equivalent RMT set $\text{Self}_{i.0} = \{000, 010\}$ (see Table 2). In this case, $\delta^0_{i.0}(000, 010) = 1$ because the next state values of 000 and 010 are different. Then calculate $\delta^0_{i.l}$ for all self-equivalent RMT sets. Next we can find $C^0_{i.0} = 1$, $C^0_{i.1} = 0$, $C^0_{i.2} = 0$, $C^0_{i.3} = 1$. Finally, we have $\Gamma^p_i = 0.5$.

Overall Information Propagation Of Non-Uniform CA: In a non-uniform CA, the rules are represented as rule vector $\langle R_0, R_1, \cdots, R_i, \cdots, R_{n-1}\rangle$, where

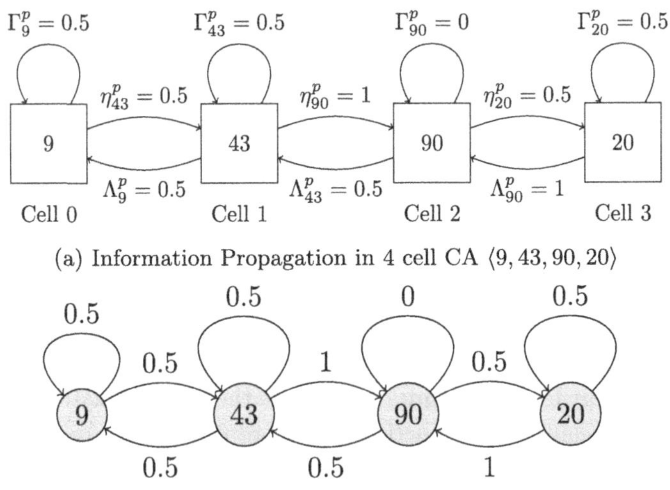

(a) Information Propagation in 4 cell CA $\langle 9, 43, 90, 20\rangle$

(b) Graph representation of $\langle 9, 43, 90, 20\rangle$

Fig. 5. Graphical representation of non-uniform CA $\langle 9, 43, 90, 20\rangle$.

rule R_i is applied to cell i. The overall information propagated across the non-uniform cellular automata can be measured by considering the entire rule vector. This propagation is analogous to the network flow in a graph, where cells act as nodes and the information transfer between them is represented as weighted edges. So, for the non-uniform lattice, based on the individual information propagation of the rule for each cell, we apply a graph theoretic network flow approach to find the overall information propagation in the lattice.

Example 5. Consider a non-uniform CA with rule vector $\langle 9, 43, 90, 20 \rangle$. The information propagation from each cell is represented in Fig. 5a. In this non-uniform CA, four cells represent the four nodes, and information propagation from each cell corresponds to a weighted directed edge in the graph. The graphical representation of non-uniform CA with rule vector $\langle 9, 43, 90, 20 \rangle$ is shown in Fig. 5b.

In graph theory, the maximum flow [6] in a flow network from source to destination is determined by the minimum-weight edge whose removal would disconnect the source from the destination. In a non-uniform CA, if we set cell 0 as the source and cell $n-1$ as the destination, calculating the maximum flow from source to destination will represent the overall information propagation from left to right. This means we need to take the minimum from the right information propagation (η^p_{i+1}) of cell 0 to cell $n-2$ (the last cell $n-1$ is not considered because of null boundary). Similarly, if we set cell $n-1$ as the source and cell 0 as the destination, calculating the maximum flow will give the overall information propagation from right to left. This means we need to take the minimum left information propagation (Λ^p_{i-1}) of cell $n-1$ to cell 1 (again, cell 0 is not considered because of null boundary).

Example 6. In Fig. 5a, to find the overall information propagation of the non-uniform CA from left to right, we take cell 0 as the source and cell 3 as the destination. The node sequence in the path from source to destination is (9, 43, 90, 20), and the minimum edge weight in this path is 0.5. This means the maximum overall information propagation from left to right is 0.5. Similarly, by considering cell $n-1$ as the source and cell 0 as the destination, we can determine that the maximum overall propagation from right to left of the non-uniform CA is 0.5.

4.2 Selection of Appropriate Rules for Classification

When using the reversible rule vectors for data classification, we need to ensure that the distance between configurations in the cycle is low. To achieve this, rules from Table 3 that produce rule vectors generating cycles with dissimilar configurations have to be removed. We achieve this by aggressively selecting rule vectors such that the overall information propagation on either side is minimum and self information propagation is maximum.

Table 5. Tables to generate rule vector for data classification

Class of R_i	R_i Rules	Class of R_{i+1}
I	51, 204	I
	85	II
	54, 57, 99, 108, 147, 156, 198, 201	V
II	15	I
III	204	I
	85	II
	27, 39, 78, 114, 141, 177, 216, 228	V
V	51, 204	I
	85	II

(a) Relationship between R_i and R_{i+1}

Rules for R_0	Class of R_1
3, 12	I
6, 9	III

(b) Table for First Rule

Rule Class for R_{k-1}	Rule Set for R_k
I	17, 20, 65, 68
II	20, 65
III	17, 68
V	17, 68

(c) Table for Last Rule

Filtering Based on Self Information Propagation: Self information propagation refers to the rate of information propagation due to changes within the cell itself. The maximum self information propagation rate is 100%, while the minimum is 0%. The classification algorithm performs well when the distance between elements in the same class is small, while the distance between elements in different classes is larger. If we select rules with high self information propagation, then the reversible CA has similar configurations in the cycles. Thus, our first criterion for filtering rules for data classification from the set of rules given in Table 3 is set as follows:

Criterion 1: *Select rules such that the self information propagation rate is greater than* 50% .

Based on Criterion 1, out of 73 rules in Table 3, we select 64 rules by removing rules 5, 10, 15, 80, 85, 90, 165, 170, and 240, as these 9 rules have a self information propagation rate less than 50%.

Filtering Based on Overall Information Propagation: As shown already, in a CA, the overall information propagation from left to right is determined by the rule in the rule vector with the lowest right information propagation. Similarly, the information propagation from right to left is determined by the rule with the lowest left information propagation. The classification algorithm performs well when the distance between elements in the same class is small. So if we select rules with low overall information propagation, then the reversible CA has more chance of having similar configurations in the cycles. Thus our second criteria for select rules is as follows:

Criterion 2: *Select rules such that the left information propagation rate is less than* 50% *and the right information propagation rate is also less than* 50%.

Based on Criterion 2, out of the remaining 64 rules, we select 42 rules by removing rule 102, 105, 150, 153, 86, 89, 101, 106, 149, 154, 166, 169, 60, 195, 30, 45, 75, 120, 135, 180, 210 and 225, as these 22 rules have either the left information propagation rate as greater than 50% or the right information propagation rate as greater than 50%.

Finally, we form Table 5 by removing the rules that do not satisfy Criteria 1 and 2 from Table 3. As a result, the rows corresponding to Class IV and

VI in Table 3a, the rows corresponding to Class II in Table 3b, and the rows corresponding to Class IV and VI in Table 3c are removed. The rules that can be used in data classification to generate the rule vector are shown in Table 5.

To perform classification using cycles, we ensure that the rule vector has high self information propagation and low information propagation to neighbors. To achieve this, we choose rule vectors based on the following greedy strategy:

Strategy I: *In the synthesized rule vector* $\langle R_0, R_1, \cdots, R_{i-1}, R_i, R_{i+1}, \cdots, R_{n-1}\rangle$, *at least one rule has self information propagation rate of* 100% , *one rule has left information propagation rate of* 0% , *and one rule has right information propagation rate of* 0% .

Considering Strategy I, we generate the rule vector using Table 5 for classification and implement the classification model using our non-uniform elementary cellular automata. Following section shows the results.

Table 6. Description about datasets used

Dataset	Instances	Attributes
Balloons	16	4
Haber-man	306	3
Iris	150	4
Bank Note Authentication	1372	4
Balance Scale	625	4
Predictive Maintenance	10000	8
Gender Classification	50001	7
PIMA Indians Diabetes	768	8

5 Implementation and Performance Analysis

We implement our proposed algorithm using the datasets shown in Table 6 which are available in the online data repository [5]. Since we are using binary non-uniform CA, we encode each dataset into binary form using frequency encoding, allocating 3 bits for continuous attributes and $\log_2(k)$ bits for categorical attributes with k number of categories. We randomly generate different rule vectors using Table 5 and Strategy I, then apply them to the dataset. The best result for each dataset and its corresponding non-uniform cellular automaton in our random trial is shown in Table 7.

Our implementation is converted into a Python module[1] for reusability. We evaluate our algorithm's performance using accuracy and execution time, with a focus on accuracy, which is defined as the ratio of correct predictions to total

[1] GitHub repository: https://github.com/kamalikaB/ClassificationECA.

Table 7. Performance of Reversible Non-Uniform CA for Classification

Dataset	Accuracy	Rule vector
Balloons	0.800	$\langle 9, 27, 204, 85, 15, 201, 51, 147, 85, 15, 201, 17\rangle$
Haber-man	0.761	$\langle 9, 204, 85, 15, 198, 51, 108, 85, 65\rangle$
Iris	0.962	$\langle 12, 57, 85, 15, 57, 51, 204, 108, 204, 204, 85, 65\rangle$
Bank Note Authentication	0.977	$\langle 12, 156, 51, 204, 156, 51, 85, 15, 198, 51, 85, 65\rangle$
Balance Scale	0.826	$\langle 3, 156, 51, 204, 156, 204, 201, 204, 108, 85, 15, 20\rangle$
Predictive Maintenance	0.957	$\langle 12, 85, 15, 204, 85, 15, 85, 15, 57, 204, 57, 51, 108, 85, 65\rangle$
Gender Classification	0.854	$\langle 6, 102, 23, 105, 169, 150, 195, 154, 240, 156, 68\rangle$
PIMA Indians Diabetes	0.603	$\langle 3, 204, 204, 57, 85, 15, 204, 204, 85, 15, 85, 15, 147, 85, 65\rangle$

predictions, as the main measure of classification ability. It indicates how well the algorithm makes predictions, while execution time assesses efficiency and overhead. It can be observed that, by both metrics, our algorithm achieves high accuracy while operating efficiently in real-world applications.

Table 8. Comparison of performance with other models.

DataSet	Non-uniform Binary CA		Decision Tree		KNN		Gaussian Naive Bayes	
	Accuracy	Execution time(sec)	Accuracy	Execution time(sec)	Accuracy	Execution time(sec)	Accuracy	Execution time(sec)
Balloons	0.800	0.098	0.800	0.008	0.800	0.012	0.800	0.008
Haber-man	0.761	0.015	0.706	0.007	0.717	0.004	0.717	0.017
Iris	0.962	0.075	0.955	0.003	0.977	0.015	0.977	0.007
Bank Note Authentication	0.977	0.056	0.980	0.007	1.000	0.028	0.830	0.005
Balance Scale	0.826	0.075	0.078	0.005	0.808	0.025	0.910	0.004
Predictive Maintenance	0.957	0.680	0.989	0.025	0.973	0.116	0.992	0.007
Gender Classification	0.854	0.036	0.962	0.014	0.961	0.066	0.962	0.012
PIMA Indians Diabetes	0.603	0.057	0.727	0.009	0.696	0.021	0.770	0.005

The performance of our classification model using non-uniform cellular automata, along with a comparison to other classical machine learning algorithms, is shown in Table 8. The results clearly show that our algorithm is both effective and comparable to other machine learning algorithms.

We also compare the proposed data classification method in this work with the existing CA-based method(s). For most of the works, reproducibility of the results is difficult due to unavailability of the codes. So, we compare with first-degree decimal cellular automata (CA) [1] based on execution time (measured in seconds) and accuracy. In the first-degree decimal CA model, the encoding method is just a concatenation of attributes, which restricts its applicability to small datasets. Therefore, we select only two datasets—Balloon and Haberman—for comparison, as they are compatible with the first-degree decimal CA approach and do a random trial. For the Haberman dataset, the proposed model requires only 0.015 s, whereas the first-degree decimal CA-based method takes 23.807 s under the same execution environment. Similarly, for the Balloon dataset, the first-degree decimal CA approach required 0.148 s, while the proposed model completed execution in 0.098 s. These results clearly demonstrate that the proposed algorithm significantly reduces execution time. In the case of accuracy, the first-degree decimal CA shows an accuracy of 1.0 on the Balloon dataset and 0.96 on the Haberman dataset which are slightly higher than our random trial result.

6 Conclusion

This work has reported a data classification method that uses the cyclic space of reversible non-uniform elementary cellular automata. The datasets are converted to binary using frequency encoding, allowing us to use binary cellular automata for classification. To select the effective rules for data classification, we have calculated the overall information propagation of a reversible cellular automaton using a graph network flow theoretic approach. This method for calculating overall information propagation is applicable for any n and can be used in any application that requires information propagation of non-uniform CA as a component. In this work, we evaluated our model with a maximum cell size n of 15. Our model performs well compared to other classical machine learning algorithms, based on accuracy and execution time.

All of the CA-based classification algorithms require some type of data encoding method. This chosen encoding method plays a significant role in the classification process. Due to differences in the encoding methods used in most of the existing CA-based approaches, we are limited in our ability to compare our model with methods other than FDCAs. Another limitation occurs when the number of attributes increases, as the cell length also increases, leading to higher complexity. In the future, we can extend this work to classify high-dimensional datasets that require larger cell-size cellular automata for effective classification.

Acknowledgment. The authors are grateful to Mr. Mayank Kumar and Mr. Prajal Singhal for implementing and verifying the initial version of our algorithm.

References

1. Baby, C., Bhattacharjee, K.: Reversible decimal first degree cellular automata for data classification. In: International Conference on Cellular Automata for Research and Industry, pp. 147–162. Springer (2024)
2. Bhattacharjee, K.: Cellular automata: reversibility, semi-reversibility and randomness. Ph.D Thesis, IIEST, Shibpur (2019). http://arxiv.org/abs/1911.03609
3. Bhattacharjee, K., Das, S.: Reversibility of d-state finite cellular automata. J. Cell. Automata **11** (2016)
4. Das, S.: Theory and applications of nonlinear cellular automata in VLSI design. Ph.D. thesis, PhD thesis, BE College (2006)
5. Dua, D., Graff, C., et al.: UCI machine learning repository (2017)

6. Ford, L.R., Fulkerson, D.R.: Maximal flow through a network. Can. J. Math. **8**, 399–404 (1956)
7. Guo, G., Wang, H., Bell, D., Bi, Y., Greer, K.: KNN model-based approach in classification. In: On The Move to Meaningful Internet Systems 2003: CoopIS, DOA, and ODBASE: OTM Confederated International Conferences, CoopIS, DOA, and ODBASE 2003, Catania, Sicily, Italy, 3–7 November 2003. Proceedings, pp. 986–996. Springer (2003)
8. Kamilya, S., Das, S.: A study of chaos in non-uniform cellular automata. Commun. Nonlinear Sci. Numer. Simul. **76**, 116–131 (2019)
9. Maji, P., Shaw, C., Ganguly, N., Sikdar, B.K., Chaudhuri, P.P.: Theory and application of cellular automata for pattern classification. Fund. Inform. **58**(3–4), 321–354 (2003)
10. Mukherjee, S., Bhattacharjee, K., Das, S.: Reversible cellular automata: a natural clustering technique. J. Cell. Automata **16** (2021)
11. Ray, S.: A quick review of machine learning algorithms. In: 2019 International Conference on Machine Learning, Big Data, Cloud and Parallel Computing (COMIT-Con), pp. 35–39. IEEE (2019)
12. Rish, I., et al.: An empirical study of the naive bayes classifier. In: IJCAI 2001 Workshop on Empirical Methods in Artificial Intelligence, vol. 3, pp. 41–46. Seattle, WA, USA; (2001)
13. Sethi, B., Das, S.: Modeling of asynchronous cellular automata with fixed-point attractors for pattern classification. In: 2013 International Conference on High Performance Computing & Simulation (HPCS), pp. 311–317. IEEE (2013)
14. Sethi, B., Roy, S., Das, S.: Asynchronous cellular automata and pattern classification. Complexity **21**(S1), 370–386 (2016)
15. Song, Y.Y., Ying, L.: Decision tree methods: applications for classification and prediction. Shanghai Arch. Psychiatry **27**(2), 130 (2015)

Synthesis and Analysis of Regular Clocking-Based Full Subtractor in QCA

Amit Kumar Pramanik[1(✉)], Jayanta Pal[2], and Bibhash Sen[3]

[1] Department of CSE, Dumka Engineering College, Dumka, Jharkhand, India
amit.pramanik21@gmail.com
[2] Department of Information Technology, Tripura University, Agartala, India
jayantapal@tripurauniv.ac.in
[3] Department of CSE, National Institute of Technology, Durgapur, WB, India
bsen.cse@nitdgp.ac.in

Abstract. Due to CMOS technology's higher power consumption and lower device density, it faces serious challenges in digital logic design. Overcoming the restrictions of conventional CMOS technology, Quantum-dot Cellular Automata (QCA) presents a viable way to construct ultra-small circuits. The fundamental building block for building QCA logic circuits is the QCA cell, which is made up of four two-electron quantum dots. The majority gate and inverters are the essential parts of QCA. In order to ensure correct synchronization and information transmission inside a circuit, clocking serves a key part. Regular clocking also encourages scalability and tackles manufacturing issues in the nanoscale area. This article aims to realize the Full Subtractor circuit using various regular clocking schemes. This work also provides a detailed comparison-based analysis between the clocking-based circuits with respect to the basic parameters in QCA. According to this study, the circuits' performance is remarkable when realized with modified 1-D clocking.

Keywords: Full Subtractor · Regular Clocking · QCA Cost · QCADesigner

1 Introduction

Conventional devices are being altered to meet the size, speed, and accuracy requirements of the present generation. A few disadvantages of CMOS technology include issues with heat dissipation, channel shortening, and physical scaling. A methodology based on QCA may be the most effective way to address CMOS technology's drawbacks [1]. An investigation into QCA technology is crucial to creating the required strategies. A QCA cell consists of four quantum dots and two electrons arranged diagonally. The electrons are positioned this way due to electrostatic repulsion between them. The two fundamental QCA logics that can be employed to realize any QCA circuit are the majority gate and the inverter.

A. K. Pramanik, J. Pal and B. Sen—Contributed equally to this work.

S. Kamilya et al. (Eds.): ASCAT 2025, CCIS 2499, pp. 219–228, 2025.
https://doi.org/10.1007/978-3-031-94121-4_16

By assigning fixed values to any one of the three inputs, the majority gate is able to realize traditional AND and OR gates.

The implementation of a QCA logic circuit relies significantly on clocking [2] because it permits electrons to flow between quantum dots. A QCA clocking cycle commonly contains four clock zones: switch, hold, release, and relax. The concept of clocking was first introduced in [3]. In order to successfully build QCA logic circuits and guarantee their scalability and stability, regular clocking is essential. Regarding regular clocking, the 1-D clocking idea was introduced by [4], and a revised form of this clocking was presented by [5]. Many regular clocking approaches have been introduced, including USE [6], RES [7], Optimized 2-D [8], Zig-Zag [9] clocking, and others. The RES [7] and Zig-Zag [9] clocking schemes facilitate three-way data flow, whereas the USE [6] clocking method permits bidirectional data flow, and 1-D [5] clocking allows one-directional information flow.

The subtractor is an essential circuit in a digital circuit design. Numerous logical processes, including ALUs, CPUs, GPUs, etc., can be designed utilizing a subtractor. Several approaches to designing subtractors have been presented in the literature [10–15]. Most of the previous research is only concerned with the realization of subtractor circuits by considering a few important parameters in QCA. However, none of them have given any idea about which regular clocking scheme is suitable for a full subtractor circuit with respect to the basic parameters (like number of cells, circuit area, latency, and cost) in QCA.

Inspired by the aforementioned factors, well-known regular clocking-based full subtractors are realized, and a suitable clocking scheme is determined. In this direction, the key aspects of this research are summarized below:

- Regular clocking-based full subtractor circuits are realized.
- An analysis is carried out to identify a suitable regular clocking scheme to realize a full subtractor.
- Fundamental parameters of QCA design and cost are considered for result analysis.
- This work reportedly exhibits one of its kind to realize a full subtractor using a regular clocking scheme.

The article is organized as follows: Sect. 2 provides a brief summary of the basics of QCA, clocking, and full subtractor. The realization of full subtractor using well-known clocking is provided in Sect. 3. Section 3.1 demonstrates the comparison analysis between the various clocking-based circuits. Lastly, a conclusion is drawn in Sect. 4.

2 QCA Fundamentals

The basic details of the several components used in this study are presented in this section.

2.1 Basic Elements

As seen in Fig. 1a, a cell is the fundamental unit in QCA. A cell consists of a square lattice with four quantum dots, each holding two electrons capable of moving within the cell to represent information. In a cell, the direction of the electrons is dictated by the intra-cell coulombic force. Electrons preferentially go along the largest diagonal span in a QCA cell. In a cell, quantum dots can be arranged in two different ways: symmetrically and regularly. The QCA cell appears to have two polarization states (depicted by P): +1 denotes a logic 1, and −1 denotes a logic value of 0, as seen in 1b. The majority gate and inverter are the fundamental gates of QCA. Figure 1c shows that the majority gate has three inputs and one output.

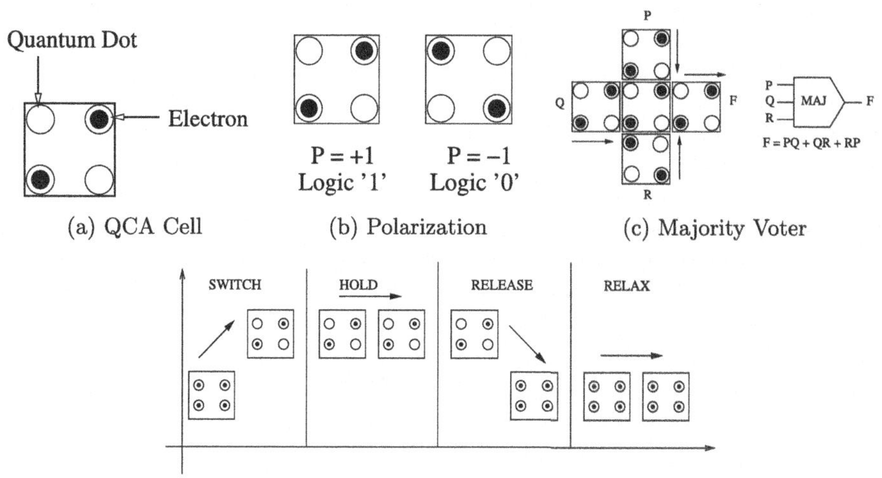

(a) QCA Cell (b) Polarization (c) Majority Voter

(d) QCA Clocking

Fig. 1. Fundamental Elements of QCA.

2.2 Clocking

A crucial component of QCA technology is clocking. Information flow is precisely determined by clocking systems. Therefore, it is impossible to overlook the function that clocking plays in QCA information flow [16]. Additionally, the KINK state must be taken into account. When several cells are switched together, this is known as the KINK state. A QCA circuit is divided into several clock domains to address the KINK state problem. In QCA circuits, the clock signal is essential for both synchronization and data manipulation. It efficiently regulates the information flow by dictating the velocity of the electrons inside the quantum dots. The four separate phases of the QCA clock cycle are Hold,

Release, Switch, and Relax, as depicted in Fig. 1d. Each phase affects electron tunneling and interaction by regulating the energy barriers separating nearby quantum dots. To guarantee correct operation, the four steps take place in a particular order. The data stored in the electron positions can be manipulated due to its sequential character. It is reasonable to divide the QCA circuit into zones with different clock cycles. Clock zoning is a technique that can help preserve energy consumption and signal integrity. By giving the cells distinct colors, each zone can be easily identified, which facilitates analysis and design.

2.3 Regular Clocking

Regular clocking plays a significant role in the realization of scalable and efficient circuits. Lent et al. [3] introduced the concept of clocking, which has subsequently developed into a crucial component of QCA circuit design. Regarding regular clocking, the $1-D$ clocking concept was introduced by Rani et al. in [4], and the updated form of this clocking was presented by Ebrahimy et al. in [5]. There are several popular clocking methods available today for implementing QCA circuits, and USE [6] clocking Scheme (shown in Fig. 2(a)) is one of them. USE utilizes multilayer wire crossing to provide two-directional data flow in the QCA circuit. Due to multilayer wire crossing, fabrication of USE clocking circuits is very difficult. Conversely, Fig. 2(b) shows the RES Clocking Scheme [7], which can permit three-directional information flow at certain points. But attaining this benefit comes at the expense of the clock zone's underlying structure becoming less uniform. As illustrated in Fig. 2(c), the Optimized $2-D$ Clocking Scheme [8] features a 2:1 aspect ratio clocking region in a tile-like pattern. Three-directional input and output are made possible by the Zig-Zag Clocking Scheme [9], which uses a bidirectional feedback path, presented in Fig. 2(d).

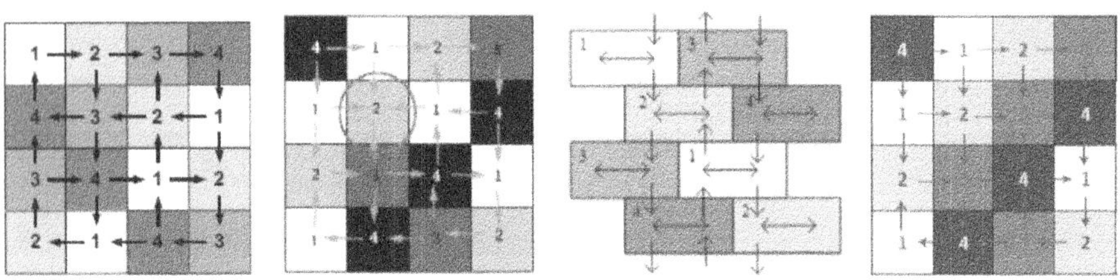

(a) USE [6] (b) RES [7] (c) Optimized 2-D [8] (d) Zig-Zag [9]

Fig. 2. Regular Clocking Schemes.

3 Realization of Full Subtractor

This article demonstrates a realization of various designs of QCA based full subtractor circuits. The Half Subtractor is utilized to subtract only two values. In order to solve this issue, a full subtractor had been developed. A full subtractor is a combinational circuit that carries out binary subtraction on three input bits: the minuend (A), subtrahend (B), and a borrow-in (Bin) bit, as depicted in Fig. 3(a) and the schematic diagram is presented in Fig. 3(b). The full subtractor provides a difference (D) bit and a borrow-out (Bo) bit, which indicates if a borrow is needed for the subsequent higher-bit subtraction.

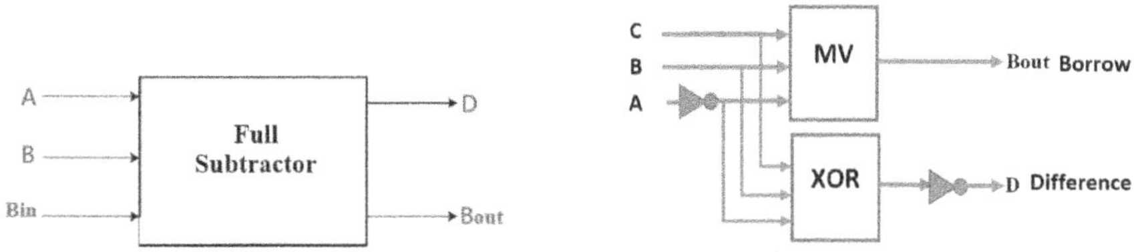

(a) Block Diagram (b) Schematic Diagram

Fig. 3. Representation of Full Subtractor.

In this direction, a few articles have reported the full subtractor circuit realization [10–15]. It is to be noted that the full subtractor logic has almost all similarities with a full adder, except the inverted input (A) and output ($Difference$). Subsequently, the logic outcome for each operation (Sum, $Carry$, $Difference$ & $Borrow$) is represented in the form of a Truth Table (Table 1) and the outputs of full subtractor can be expressed as presented in Eq. 1.

Table 1. Truth Table

Inputs			Adder		Subtractor	
A	B	C	Sum	Carry	Difference	Borrow
0	0	0	0	0	0	0
0	0	1	1	0	1	1
0	1	0	1	0	1	1
0	1	1	0	1	0	1
1	0	0	1	0	1	0
1	0	1	0	0	0	0
1	1	0	0	0	0	0
1	1	1	1	1	1	1

$$\begin{aligned} f_{(Sum/Difference)} &= A \oplus B \oplus C \\ f_{(Carry)} &= A.B + A.C + B.C \\ f_{(Borrow)} &= \overline{A}.B + \overline{A}.C + B.C \end{aligned} \quad (1)$$

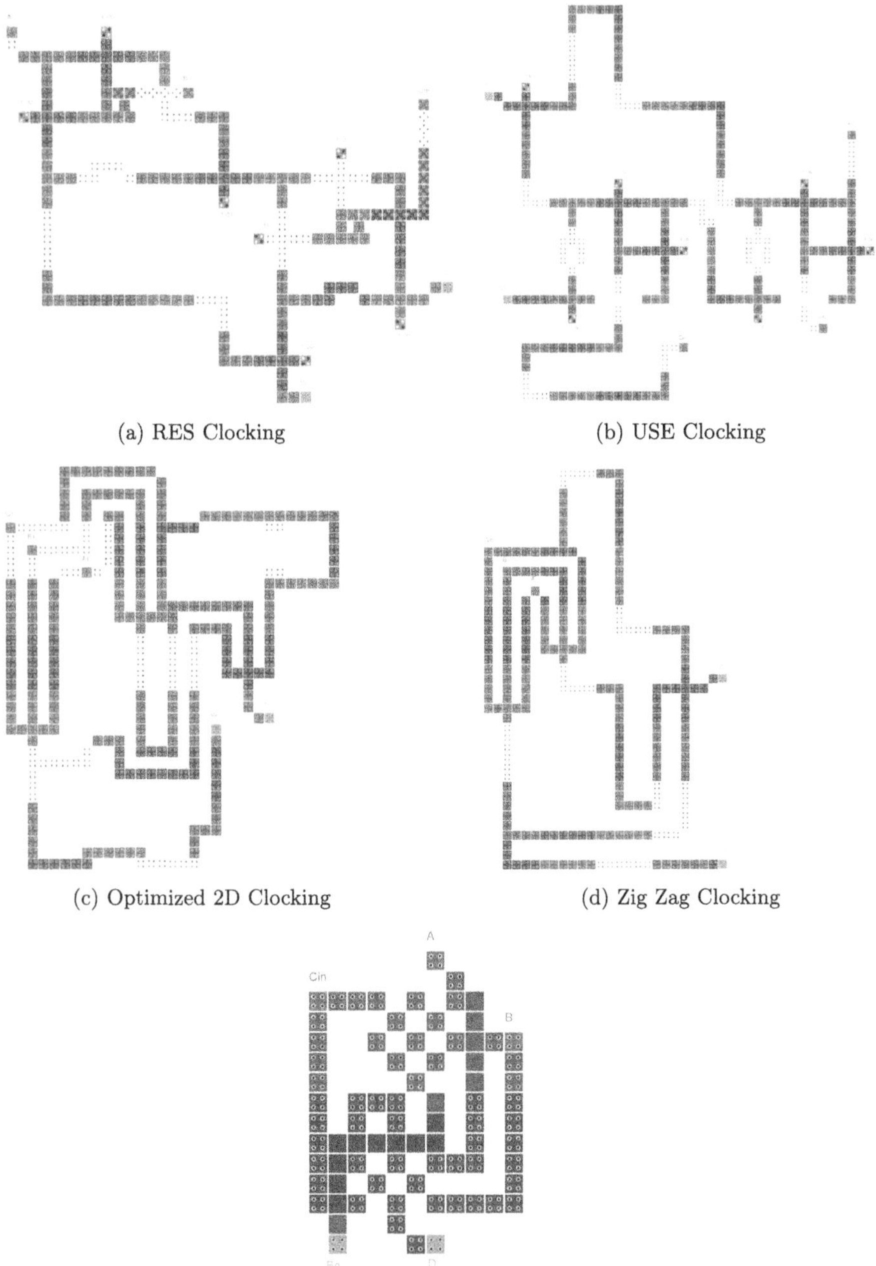

(a) RES Clocking

(b) USE Clocking

(c) Optimized 2D Clocking

(d) Zig Zag Clocking

(e) Modified 1D Clocking

Fig. 4. QCA Layout of Full Subtractor using Regular Clocking.

Observing the truth table as presented in Table 1 and Eq. 1, it is understood that the outputs of a full adder or full subtractor can easily be realized in QCA. Whereas, the same logic of full subtractor can also be represented in terms of majority gate and XOR gate with respect to the full adder logic with slight modifications (Eq. 2).

$$\begin{aligned} f_{(Difference)} &= \overline{\overline{A} \oplus B \oplus C} \\ f_{(Borrow)} &= M(\overline{A}, B, C) \end{aligned} \tag{2}$$

It is observed that an underlying clocking scheme is necessary for standard cell placement, routing algorithms and fabrication. In a regular clocking scheme, continuous zone numbering can identify the neighbouring clock zones. As an example, clock 0 is followed by 1 and so on, up to the last zone (Relax phase). Information flows sequentially through the sequence of clock zones, starting from clock zone 0.

Accordingly, the logic circuit for full subtractor is realized in different well known clocking schemes, as depicted in Fig. 4. The simulated output of the full subtractor is found in Fig. 5. In Fig. 5, the input C_{in} represents the B_{in} input as depicted in Fig. 3(a). In the case of synthesis, the universal, scalable and efficient (USE) clocking scheme [6], the robust, efficient and scalable (RES)

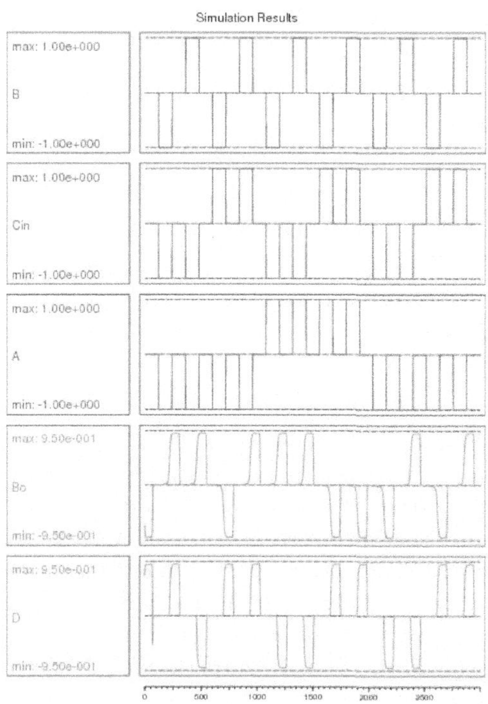

Fig. 5. Simulation Result of Full Subtractor.

clocking scheme [7], Optimized 2D clocking scheme [8], Zig-Zag clocking scheme [9] and the modified clocking scheme [4] are utilized.

3.1 Analysis

The analysis of the logic synthesis of the underlying clocking-based full subtractor is presented in this section. The structural parameters like cell count, and basic gates (majority gate, inverter, crossover, latency) are presented in Table 2. The area consumed by each of the designs is also represented. Apart from that, the cost analysis of the circuits is also investigated. An Area-Delay product to find the cost was helpful in optimum VLSI design, which was reconsidered in [17] and proposed CMOS technology cost function as Eq. 3. Here, A indicates area and T indicates latency.

$$Cost_1 = A \times T^2 \tag{3}$$

Another cost function was proposed in [18] for the QCA circuit considering cell, gate, area, latency and crossover, as in Eq. 4. Here M indicates Majority voters gate, I indicates inverter, C indicates crossover and T indicates Latency.

$$Cost_2 = (M^2 + I + C^2) \times T^2 \tag{4}$$

Table 2. Comparative Analysis

Clocking	No. of Cell	M	I	C	A (μm^2)	T	$Cost_1$ (Eq. 3)	$Cost_2$ (Eq. 4)
USE [6]	290	7	4	2	0.71	2.75	5.37	400.81
RES [7]	207	7	4	2	0.5	3	4.50	477.00
Optimized 2-D [8]	300	3	5	0	0.44	3	3.96	126.00
Zig-Zag [9]	246	3	4	0	0.46	2.5	2.88	81.25
Modified 1-D [4]	86	2	3	2	0.08	0.5	0.02	1.75

As per the result analysis depicted in Table 2, it is noticed that the modified $1D$ clocking scheme has a significant reduction in the design parameters. However, in terms of fabrication, it is noteworthy to design the logic in a coplanar manner. In this regard, the Optimized $2D$ clocking and Zig-Zag clocking schemes propose the circuit in a single layer. At the same time, a multi-layer approach for QCA synthesis may be observed in a Modified $1D$ clock based design. On the other hand, it supersede its counterpart when compared in terms of QCA cost. Therefore, the modified $1D$ clock outperformed in structural design as well as cost analysis.

4 Conclusion

Due to its important applications in digital circuits and the necessity for multi-bit subtraction, a full subtractor is an integral part of digital logic design. This paper demonstrates a clocking-based mechanism for the synthesis of the full subtractor in QCA. The circuit realization comprises a majority voter gate, inverter, and xor gate. The result of the presented full subtractor in various underlying clocking schemes is analyzed with respect to the structural design and QCA cost. In this regard, QCADesigner is used for the synthesis of the circuits. According to the analysis, modified $1D$ clocking exhibits a significant impact on the design of the full subtractor as compared to the counterparts. In the best case, it achieves a noticeable gain in cell count, area, design parameters, and cost as well. However, this work represents the basic design for a regular clock-based 1-bit full subtractor, which can further be investigated for n-bit subtractor design.

References

1. Karim, F., Walus, K.: Efficient simulation of correlated dynamics in quantum-dot cellular automata (qca). IEEE Trans. Nanotechnol. **13**(2), 294–307 (2014)
2. Deng, F., Xie, G., Cheng, X., Zhang, Y.: A general and efficient clocking scheme for majority logic in quantum-dot cellular automata. Microelectron. J. **128**, 105544 (2022)
3. Lent, C.S., Tougaw, P.D.: A device architecture for computing with quantum dots. Proc. IEEE **85**(4), 541–557 (1997)
4. Rani, S., Sasamal, T.N.: Design of QCA circuits using new 1d clocking scheme. In: 2017 2nd International Conference on Telecommunication and Networks (TEL-NET), pp. 1–6 (2017). https://doi.org/10.1109/TEL-NET.2017.8343540
5. Ebrahimy, M., Gholami, M., Adarang, H., Yousefi, R.: A novel low-latency alu in the one-dimensional clock scheme in qca nanotechnology. Eur. Phys. J. Plus **139** (2024). https://doi.org/10.1140/epjp/s13360-024-04901-0
6. Campos, C., Marciano, A.L., Neto, O., Torres, F.S.: Use: a universal, scalable, and efficient clocking scheme for qca. IEEE Trans. Comput. Aided Des. Integr. Circuits Syst. **35**(3), 513–517 (2015)
7. Goswami, M., Mondal, A., Mahalat, M.H., Sen, B., Sikdar, B.K.: An efficient clocking scheme for quantum-dot cellular automata. Int. J. Electron. Lett. 1–14 (2019)
8. Wang, L., Xie, G., Zhu, R., Yu, C.: An optimized clocking scheme for nanoscale quantum-dot cellular automata circuit. In: 2019 IEEE 14th International Conference on Nano/Micro Engineered and Molecular Systems (NEMS), pp. 336–341. IEEE (2019)
9. Pal, J., Pramanik, A.K., Sharma, J.S., Saha, A.K., Sen, B.: An efficient, scalable, regular clocking scheme based on quantum dot cellular automata. Analog Integr. Circ. Sig. Process **107**(3), 659–670 (2021). https://doi.org/10.1007/s10470-020-01760-4
10. Karthigai Lakshmi, S., Athisha, G., Karthikeyan, M., Ganesh, C.: Design of subtractor using nanotechnology based QCA. In: 2010 International Conference on Communication Control and Computing Technologies, pp. 384–388 (2010). https://doi.org/10.1109/ICCCCT.2010.5670582

11. Radhouane Laajimi, L.T., Bahar, A.N.: A novel efficient coplanar qca full adder and full subtractor design. Int. J. Electron. **110**(8), 1431–1446 (2023). https://doi.org/10.1080/00207217.2022.2098386
12. Safoev, N., Jeon, J.-C.: Design of high-performance qca incrementer/decrementer circuit based on adder/subtractor methodology. Microprocess. Microsyst. **72**, 102927 (2020). https://doi.org/10.1016/j.micpro.2019.102927
13. Mosleh, M.: A novel full adder/subtractor in quantum-Dot cellular automata. Int. J. Theor. Phys. **58**(1), 221–246 (2018). https://doi.org/10.1007/s10773-018-3925-x
14. Marshal Raj, L.G., Ko, S.-B.: Design and analysis of novel qca full addersubtractor. Int. J. Electron. Lett. **9**(3), 287–300 (2021). https://doi.org/10.1080/21681724.2020.1726479
15. Sharma, V.: Single layer adder/subtractor using qca nanotechnology for nanocomputing operations. Eng. Res. Express **4**, 045002 (2022). https://doi.org/10.1088/2631-8695/ac9957
16. Pramanik, A.K., Pal, J., Sen, B.: Impact of genetic algorithm on low power qca logic circuit with regular clocking (2022)
17. Thompson, C.D.: A complexity theory for vlsi. PhD thesis, Carnegie-Mellon University Pittsburg, PA (1980)
18. Liu, W., Lu, L., O'Neill, M., Swartzlander, E.E.: A first step toward cost functions for quantum-dot cellular automata designs. IEEE Trans. Nanotechnol. **13**(3), 476–487 (2014). 10

Author Index

S. Kamilya et al. (Eds.): ASCAT 2025, CCIS 2499, p. 229, 2025.
https://doi.org/10.1007/978-3-031-94121-4

The manufacturer's authorised representative in the EU is Springer Nature Customer Service Centre GmbH, Europaplatz 3, 69115 Heidelberg, Germany. If you have any concerns regarding our products, please contact ProductSafety@springernature.com

Printed and bound by CPI Group (UK) Ltd, Croydon, CR0 4YY
30/09/2026
02225566-0010